The International Library of Ethics, Law and Technology

Volume 42

Technologies are developing faster and their impact is bigger than ever before. Synergies emerge between formerly independent technologies that trigger accelerated and unpredicted effects. Alongside these technological advances new ethical ideas and powerful moral ideologies have appeared which force us to consider the application of these emerging technologies. In attempting to navigate utopian and dystopian visions of the future, it becomes clear that technological progress and its moral quandaries call for new policies and legislative responses. Against this backdrop, this book series from Springer provides a forum for interdisciplinary discussion and normative analysis of emerging technologies that are likely to have a significant impact on the environment, society and/or humanity. These will include, but be no means limited to nanotechnology, neurotechnology, information technology, biotechnology, weapons and security technology, energy technology, and space-based technologies.

Elisabeth Hildt • Kelly Laas
Eric M. Brey • Christine Z. Miller
Editors

Building Inclusive Ethical Cultures in STEM

Springer

Editors
Elisabeth Hildt
Center for the Study of Ethics
in the Professions
Illinois Institute of Technology
Chicago, IL, USA

Kelly Laas
Center for the Study of Ethics
in the Professions
Illinois Institute of Technology
Chicago, IL, USA

Eric M. Brey
Department of Biomedical
Engineering, AET
The University of Texas at San Antonio
San Antonio, TX, USA

Christine Z. Miller
Savannah College of Art and Design
Savannah, GA, USA

ISSN 1875-0044 ISSN 1875-0036 (electronic)
The International Library of Ethics, Law and Technology
ISBN 978-3-031-51559-0 ISBN 978-3-031-51560-6 (eBook)
https://doi.org/10.1007/978-3-031-51560-6

This Springer imprint is published by the registered company Springer Nature Switzerland AG
The registered company address is: Gewerbestrasse 11, 6330 Cham, Switzerland

Acknowledgments

This collected volume has been made possible through a generous grant from the National Science Foundation, Award # 1635661.

Contents

Editors, Authors and Contributors

About the Editors

Elisabeth Hildt is Professor of Philosophy and Director at the Center for the Study of Ethics in the Professions, Illinois Institute of Technology, USA; ehildt@iit.edu. Her research focus is on bioethics, ethics of technology, research ethics, and science and technology studies. Her research interests include research ethics, philosophical and ethical aspects of neuroscience, and artificial intelligence.

Kelly Laas is Librarian and Ethics Instructor at the Center for the Study of Ethics in the Professions, Illinois Institute of Technology, USA; laas@iit.edu. Her research interests include the history and use of codes of ethics in professional fields, ethics education in STEM, research ethics, and integrating ethics into technical curricula.

Christine Z. Miller is Professor of Design Management at Savannah College of Art and Design. cmiller@scad.edu. Dr. Miller is a design educator, researcher, and practitioner working at the intersection of anthropology, design, and business. Her research interests include the ways in which society and culture influence socio-technical systems and technology-mediated communication.

Eric M. Brey is David and Jennifer Spencer Distinguished Chair, Edward E. Whitacre, Jr. Endowed Chair, at the Department of Biomedical Engineering, University of Texas, San Antonio. His research interests focus on the fields of tissue engineering, regenerative medicine, and biomaterials. He also has a significant interest in engineering education, especially in the area of undergraduate research and its influence on education and career trajectories.

About the Authors

Felix K. Abagale is Professor of Soil and Water Engineering at West African Centre for Water, Irrigation and Sustainable Agriculture (WACWISA), University for Development Studies, Ghana; fabagale@uds.edu.gh. His research is focused on the practice of irrigation science and engineering, water quality, ethical practice, and smart systems.

Mamudu Abunga Akudugu is Associate Professor of Agricultural Economics and Director at Directorate of Research, Innovation and Partnerships Services, University for Development Studies, Tamale, Ghana; makudugu@uds.edu.gh. His research focuses on how to promote inclusive and sustainable development in rural communities grounded on ethical principles. His research interests include rural livelihoods, climate change adaptation, and technology and agrarian change for inclusive and ethical development.

Josep-Maria Balaguer is a PhD candidate at the Department of Bioengineering and Rehab Neural Engineering Labs, University of Pittsburgh, USA; jbalaguer@pitt.edu. His research focuses on motor control, neuroprosthetics, paralysis rehabilitation, and computational modelling.

Rebecca A. Bates is a professor and Chair in the Department of Integrated Engineering at Minnesota State University, Mankato, USA; bates@mnsu.edu. Her research focuses on automatic speech recognition and engineering and computer science education. She also directs of the department's project-based learning programs, Iron Range Engineering and Twin Cities Engineering.

Louise Bezuidenhout is a senior data specialist at DANS (Data Archiving and Networked Services, Royal Netherlands Academy of Arts and Sciences). Her research focus is on data ethics and access to Open resources, specialising in issues of digital justice. As the co-chair of the CODATA-RDA Schools for Research Data Science, Louise is an active educator in the fields of Open and responsible research, digital justice, and reproducible research.

Jason Borenstein is the Director of Graduate Research Ethics Programs at Georgia Tech, USA; borenstein@gatech.edu. He is currently Program Director for Ethical and Responsible Research (ER2) at the National Science Foundation. His research interests include robot and artificial intelligence ethics, engineering ethics, research ethics, and bioethics.

Marietjie Botes is Postdoctoral Researcher at SnT Interdisciplinary Centre for Security, Reliability and Trust, University of Luxembourg, EU, marietjiebotes1@gmail.com. Her research focus is on the ethical, legal, and social implications (ELSI) of biotechnologies, neurotechnologies, and digital ethics.

Melinda Box is Laboratory Manager and Chemical Hygiene Officer at Elon University, Elon, NC, USA; mbox@elon.edu. Her research focus is on chemical education and safety culture development. Her work experience includes safety management for chemistry labs at an R1 university.

Alexandra Bradner is Visiting Assistant Professor of Philosophy at Kenyon College, USA, and Executive Director of the American Association of Philosophy Teachers; alexandrabradner@gmail.com. Her research concerns explanation and understanding broadly construed–within the philosophy of science, between incommensurable cultural paradigms, and within teaching and learning.

Reena Cheruvalath is Professor of Philosophy at the Department of Humanities and Social Sciences, Birla Institute of Technology and Science Pilani K.K. Birla Goa Campus, India. Her research focus is meta-moral cognition and research ethics. Her research interests include critical thinking, moral education, and artificial intelligence.

Sarah Dawod is a student in the Bioethics Masters Program at the University of Pittsburgh, USA. She was a founding member of the Rehab Neural Engineering Labs Bioethics Discussion Sessions. Her research interest surrounds Neuroethics, Personalism, and Phenomenology.

Raquel Diaz-Sprague is former Adjunct Instructor at Ohio State University, College of Medicine and former Visiting Scholar at the Computer Science Department and Electrical and Computer Engineering Department, University of Alabama at Birmingham, USA; diazspra@uab.edu. Her research interests include health and social justice issues for minorities, developing prosocial technologies to promote moral reasoning and intercultural understanding.

Juhi Farooqui is a PhD candidate in Neural Computation at Neuroscience Institute, Carnegie Mellon University, USA; farooqui@cmu.edu. Her research focuses on the development of inclusive and accessible neurotechnology. Other research interests include computational modeling, neuroprostheses, and ethical considerations in neuroscience and technology development.

Trijsje Franssen is Lecturer and Postdoc Researcher in the Ethics and Philosophy of Technology section of Delft University of Technology; T.M.Franssen@tudelft.nl. Her research concentrates on the role of narratives, art and creativity in ethics and philosophy of technology, and their role in the experience of and (societal) deliberation on emerging technologies.

Maria Gallardo Williams was a Teaching Professor and Director of the Organic Chemistry Teaching Labs in the Chemistry Department at North Carolina State University and is now a Senior Faculty Development Specialist in the NC State Office of Faculty Development. She received BS and MS degrees from Universidad

Simon Bolivar in her native Venezuela and a PhD from the University of South Florida. She is an educational researcher, with an interest in technology and student and faculty-generated teaching and learning materials.

Andrea Gammon is Assistant Professor of Ethics and Philosophy of Technology at TU Delft, the Netherlands; a.r.gammon@tudelft.nl. Andrea works in the areas of environmental philosophy and engineering ethics education, and develops curriculum and approaches for ethics teaching in STEM fields. She is involved in collaborative research projects on experiential engineering ethics education, funded through the 4TU Centre for Engineering Education (NL), and engineering ethics education in cross-cultural settings, funded through the National Science Foundation (USA).

Erinn M. Grigsby is Postdoctoral Associate at Rehab Neural Engineering Labs and Department of Physical Medicine and Rehabilitation at the University of Pittsburgh, USA; emg90@pitt.edu. Her research focuses on motor control, neuroprosthetics, stroke rehabilitation, and education reform in STEM.

Quintin Kreth is a PhD student at the School of Public Policy, Georgia Institute of Technology, USA; quintin@gatech.edu. His research examines scientific careers, training, and workplaces, with a focus on academic research productivity in lower-resourced and emergent research institutions in the United States.

Laetus O. K. Lategan is Research Professor of Research Education and Postgraduate Development at Central University of Technology, Bloemfontein, South Africa; llategan@cut.ac.za. His research focus is on next-generation researcher development and applied ethics. His research interests include research education, postgraduate supervision, postgraduate studies development, research ethics and integrity, and public health ethics.

Jeonghyun Lee is the director of research education at the Center for 21 Century Universities (C21U) at Georgia Institute of Technology, USA; jonnalee@gatech.edu. Her expertise is students' learning motivation and engagement from behavioral, cognitive, and socio-cognitive perspectives.

Lavinia Marin is Assistant Professor at the Ethics and Philosophy of Technology Section, TU Delft, the Netherlands (l.marin@tudelft.nl). Her current research investigates the conditions of possibility for epistemic and moral agency (both at the individual and group level) for users of social networking platforms using approaches from ethics, social epistemology, and situated cognition.

Clair Morrissey is Associate Professor of Philosophy and Director of the Undergraduate Research Center, Occidental College, USA; cmorrissey@oxy.edu. Her research interests are on practical ethics and political philosophy, with special focus on bioethics, environmental ethics, values in science, and the relationship between ethics and aesthetics.

Jennifer F. Nyland is Associate Professor of Biology at the Department of Biological Science, and Director of the Henson Honors Program in Science and Mathematics, Clarke Honors College, Salisbury University, Maryland, USA; jfnyland@salisbury.edu. She is an immunotoxicologist with research focused on environmental exposures and impacts to health and is interested in establishing best practices for the integration of ethics into STEM curricula.

Aive Pevkur is Senior Lecturer at the Department of Business Administration, Tallinn University of Technology, Estonia; aive.pevkur@taltech.ee. Her research focus is on professional, engineering, technology, and digital transformation ethics. Her research interests include bioethics and human research, research integrity, public ethics, and teaching ethics.

Emanuele Ratti is Lecturer in the Department of Philosophy, University of Bristol, UK; mnl.ratti@gmail.com. His area of research and teaching is ethics and philosophy of science and technology (with a focus on the life sciences and data science). He is particularly interested in the aspects of the natural sciences and data science that stand at the intersection of ethical and epistemic questions.

Arianna Rossi is a research scientist at the Interdisciplinary Center for Security, Reliability and Trust (SnT) of the University of Luxembourg; arianna.rossi@uni.lu. Arianna carries out research on online manipulation (dark patterns), usable privacy, and legal design. Arianna has a mixed background, with a joint international Doctoral Degree in Law, Science and Technology (University of Bologna) and a PhD degree in Computer Science (University of Luxembourg).

Hannah Sargeant is Research Fellow at Space Park Leicester, University of Leicester, UK; hs564@le.ac.uk. Her research interests include the characterisation and utilisation of resources on the Moon, and the development of policies and practices that ensure sustainable use of resources in space. She is also a science educator with an interest in the philosophy of space science.

Devapratim Sarma is Postdoctoral Research Fellow at NeuroMechatronics Lab, Carnegie Mellon University, USA; devsarma@cmu.edu. His research interests include understanding the complex neural mechanisms of somatosensory integration as well as the ethical development of human-centred technology and therapies for neurorehabilitation and recovery.

Daniel S. Schiff is Assistant Professor of Technology Policy and Co-Director of Governance and Responsible AI Lab, Purdue University, USA; dschiff@purdue.edu. Daniel studies the formal and informal governance of AI through policy and industry, as well as AI's social and ethical implications in domains like education, labour, finance, and criminal justice.

Michele M. Schlehofer is Professor of Psychology at the Department of Psychology, Salisbury University, Maryland, USA; mmschlehofer@salisbury.edu. A community psychologist, her training is in participatory action-research approaches, and her research and community practice work centre on pursuit of LGBTQ+ equality.

Alan P. Sprague was Professor of Computer Science (retired) at University of Alabama at Birmingham, USA; sprague@uab.edu. His PhD training is in Mathematics and also in Computer Science (PhD). His math research interests include multiple threshold graphs and combinatorics. His computer science interests include frequent set algorithms, data mining, cyberbullying and prosocial apps, and integrating ethics in computer science.

Timothy Stock is Professor and Chair at the Department of Philosophy, Salisbury University, Maryland, USA; testock@salisbury.edu. His background is in the philosophy of religion and literature, with an emphasis on maintaining communities of inquiry across ideological difference. His current research interests are in public philosophy, building curricular pathways between humanities and other fields, and community research around ethical and religious values.

Michael T. Stuart is Associate Professor of Philosophy at the Institute of Philosophy of Mind and Cognition at National Yang Ming Chiao Tung University in Taiwan, Research Associate at the Centre for Philosophy of Natural and Social Science at the London School of Economics, and Lecturer in Philosophy at the Department of Philosophy at the University of York. His research focuses on the tools that scientists use to assist them in their uses of imagination, including artificial intelligence, models, diagrams, computer simulations, and visualisations.

Janna van Grunsven is Assistant Professor at the Ethics and Philosophy of Technology Section, TU Delft. She conducts research at the intersection of embodied cognition, philosophy of technology, engineering ethics (education), and disability studies. In a project funded by the Dutch Research Council (NWO), entitled Mattering Minds: Understanding the Ethical Lives of Technologically Embedded Beings with 4E Cognition, she examines how different theoretical accounts of the mind and different technological developments can have decisive ethical implications for how disabled people are brought in view in a moral sense. Additionally, she is involved in COMET; a project on experiential engineering ethics education funded by the 4TU Centre for Engineering Education. Her work has appeared in journals such as Advances in Engineering Education, Social Epistemology, Ethics and Information Technology, Techné: Research in Philosophy and Technology, Phenomenology and the Cognitive Sciences, and the Journal of Consciousness Studies.

Karina Vielma is an assistant professor in the Department of Biomedical Engineering at the University of Texas, San Antonio, USA; karina.vielma@usta.edu. Her research focus is on engineering education, improving equity and retention in engineering fields, student engagement in meaningful research experience, and teaching and learning methodologies for engineering and STEM education.

Lina Wei is Assistant Professor of China Jiliang University, China; weilina12@163.com. Her research focus is on engineering ethics and engineering ethics education. Her research interests include professionalism among Chinese engineers, characteristics of engineering ethics in China, comparative studies of engineering ethics, and engineering ethics education.

Jian Yuan is Associate Professor of School of Management, Zhejiang Shuren University, China; yuanjian0204@163.com. His research focus is on engineering education and higher education. His research interests include engineering ideas of freshmen in engineering colleges and the cultivation of engineering students.

Ellen Zegura is Professor of Computer Science at Georgia Tech, USA; ewz@cc.gatech.edu. Her research focus is on computer networking and computing for social good. She is interested in the teaching and learning of ethics and other responsible computing concepts within undergraduate and graduate computer science curriculum.

Contributors

F. K. Abagale University for Development Studies, Tamale, Ghana

M. A. Akudugu University for Development Studies, Tamale, Ghana

Josep-Maria Balaguer Rehab Neural Engineering Labs, University of Pittsburgh, Pittsburgh, PA, USA
Center for the Neural Basis of Cognition, Pittsburgh, PA, USA
Department of Bioengineering, University of Pittsburgh, Pittsburgh, PA, USA

Rebecca A. Bates Department of Integrated Engineering, Minnesota State University, Mankato, MN, USA

Louise Bezuidenhout Centre for Science and Technology Studies (CWTS), Leiden University, Leiden, Netherlands

Jason Borenstein School of Public Policy, Georgia Institute of Technology, Atlanta, GA, USA
Office of Graduate and Postdoctoral Education, Georgia Institute of Technology, Atlanta, GA, USA

Marietjie Botes SnT Interdisciplinary Center for Security, Reliability and Trust, University of Luxembourg, Esch-sur-Alzette, Luxembourg
College of Law, University of KwaZulu Natal, Durban, South Africa

Melinda Box Department of Chemistry, Elon University, Elon, NC, USA

Alexandra Bradner Department of Philosophy, Kenyon College, Gambier, OH, USA

Eric M. Brey Department of Biomedical Engineering, AET, The University of Texas at San Antonio, San Antonio, TX, USA

Reena Cheruvalath Birla Institute of Technology and Science Pilani Goa, Zuari Nagar, Goa, India

Sarah Dawod Center for Bioethics and Health Law, University of Pittsburgh, Pittsburgh, PA, USA

Raquel Diaz-Sprague University of Alabama at Birmingham, Birmingham, AL, USA

Juhi Farooqui Rehab Neural Engineering Labs, University of Pittsburgh, Pittsburgh, PA, USA
Neuroscience Institute, Carnegie Mellon University, Pittsburgh, PA, USA
Center for the Neural Basis of Cognition, Pittsburgh, PA, USA

Trijsje Franssen Delft University of Technology, Delft, Netherlands

Maria Gallardo Williams Office for Faculty Excellence, North Carolina State University, Raleigh, NC, USA

Andrea Gammon Delft University of Technology, Delft, Netherlands

Erinn M. Grigsby Rehab Neural Engineering Labs, University of Pittsburgh, Pittsburgh, PA, USA
Center for the Neural Basis of Cognition, Pittsburgh, PA, USA
Department of Physical Medicine and Rehabilitation, University of Pittsburgh, Pittsburgh, PA, USA

Elisabeth Hildt Center for the Study of Ethics in the Professions, Illinois Institute of Technology, Chicago, IL, USA

Quintin Kreth School of Public Policy, Georgia Institute of Technology, Atlanta, GA, USA

Kelly Laas Center for the Study of Ethics in the Professions, Illinois Institute of Technology, Chicago, IL, USA

Laetus O. K. Lategan Central University of Technology, Bloemfontein, South Africa

Jeonghyun Lee Center for 21st Century Universities, Georgia Institute of Technology, Atlanta, GA, USA

Lavinia Marin Delft University of Technology, Delft, Netherlands

Christine Z. Miller Savannah College of Art and Design, Savannah, GA, USA

Clair Morrissey Occidental College, Los Angeles, CA, USA

Jennifer F. Nyland Department of Biological Sciences, Salisbury University, Salisbury, MD, USA

Aive Pevkur Department of Business Administration, Tallinn University of Technology, Tallinn, Estonia

Emanuele Ratti Department of Philosophy, University of Bristol, Bristol, UK

Arianna Rossi SnT Interdisciplinary Center for Security, Reliability and Trust, University of Luxembourg, Esch-sur-Alzette, Luxembourg

Hannah Sargeant Department of Philosophy, University of York, York, UK
Aerospace Engineering, University of Leicester, Leicester, UK

Devapratim Sarma Rehab Neural Engineering Labs, University of Pittsburgh, Pittsburgh, PA, USA
Center for the Neural Basis of Cognition, Pittsburgh, PA, USA
Department of Mechanical Engineering, Carnegie Mellon University, Pittsburgh, PA, USA

Daniel S. Schiff Department of Political Science, Purdue University, West Lafayette, IN, USA

Michele M. Schlehofer Department of Psychology, Salisbury University, Salisbury, MD, USA

Alan P. Sprague University of Alabama at Birmingham, Birmingham, AL, USA

Timothy Stock Department of Philosophy, Salisbury University, Salisbury, MD, USA

Michael T. Stuart Department of Philosophy, University of York, York, UK

Janna van Grunsven Delft University of Technology, Delft, Netherlands

Karina Vielma Department of Biomedical Engineering and Chemical Engineering, University of Texas at San Antonio, San Antonio, TX, USA

Lina Wei China Jiliang University, Hangzhou, China

Jian Yuan Zhejiang Shuren University, Hangzhou, China

Ellen Zegura School of Computer Science, Georgia Institute of Technology, Atlanta, GA, USA

Chapter 1
Building Inclusive Ethical Cultures in STEM

Elisabeth Hildt, Kelly Laas, Christine Z. Miller, and Eric M. Brey

1.1 Introduction

1.1.1 Importance of STEM and STEM Ethics Education

Science, technology, engineering, and mathematics (STEM) fields are central to any educational system. The term started with the National Science Foundation as "SMET" and was changed to STEM at a later date due to phonetic reasons. The term was not widely used until Virginia Tech University began offering a "STEM education" degree in 2005 (Friedman 2005). The term STEM covers a broad spectrum of different disciplines. While, in general, STEM is used as an umbrella term for the natural sciences, engineering, mathematics, and technology-related fields, there are some variations in its usage. For example, the US National Science Foundation (NSF) characterizes STEM as including chemistry, computers and information technology, engineering, geoscience, life sciences, mathematical sciences, physics and astronomy, psychology, social sciences, educational research, and STEM education (Gonzalez and Kuezi 2012). Whereas this approach considers the social sciences to be of central importance to STEM and the STEM workforce

E. Hildt · K. Laas (✉)
Center for the Study of Ethics in the Professions, Illinois Institute of Technology, Chicago, IL, USA
e-mail: ehildt@iit.edu; laas@iit.edu

C. Z. Miller
Savannah College of Art and Design, Savannah, GA, USA

E. M. Brey
Department of Biomedical Engineering, AET, The University of Texas at San Antonio, San Antonio, TX, USA
e-mail: eric.brey@utsa.edu

E. Hildt et al. (eds.), *Building Inclusive Ethical Cultures in STEM*, The International Library of Ethics, Law and Technology 42,
https://doi.org/10.1007/978-3-031-51560-6_1

(Spalter-Roth 2004), in the United Kingdom, STEM is seen to include fields such as medicine, dentistry, and architecture but exclude the social sciences and education (United Kingdom, House of Lords, Science, and Technology, Committee 2012).

STEM-related industries are essential for societal and technological innovation, with STEM professionals playing crucial roles in research and development. Along with helping prepare students to enter a knowledge-based economy, STEM students are uniquely positioned to help find solutions to the real-world problems our societies face in health, energy, and the environment (Struyf et al. 2019). In educational and academic contexts, the study of STEM fields involves acquiring theoretical knowledge, most often complemented by practical training. Through education and students' individual experiences, future researchers and practitioners absorb and begin to practice the knowledge, conventions, and norms of their disciplines.

Students learn general standards of conduct and research integrity in research environments in academic and training contexts. The ethical culture of research environments, especially in educational and training environments, plays an outsized role in the current and future success, responsibility, and validity of scientific research (Weil and Arzbaecher 1996; Haven et al. 2020; Schraudner et al. 2019; Woolston 2019; Hofmann and Holm 2019).

Ethical STEM, which is the ethically responsible undertaking of science, technology, engineering, and mathematics, has many dimensions. These include academic integrity, research ethics and responsible conduct of research (RCR), professional ethics, adequate communication with the public, and reflection on ethical and societal implications of science and technology and the value of the products resulting from STEM research.

Ideally, university ethics education should cover all these fields in a way relevant to student's current and future practice. Educating STEM students, future researchers, and STEM professionals in standards, norms, and ethical aspects and implications of their field is a vital component of STEM education.

Alongside ethics education, it is also essential for universities and other research institutions to foster an ethical research culture. This includes responsible conduct of research (RCR), professional ethics, and a reflection of technology's ethical and social implications. It also extends to how supervisors, graduate students, and undergraduate students interact with one another (Bird 2001).

1.1.2 STEM Ethics Involves a Broad Spectrum of Topics

STEM ethics is a field at the intersection of STEM disciplines and ethics. The interdisciplinarity of STEM ethics cannot be stressed enough. Ethics and STEM are interconnected in various ways, implying there are multiple perspectives on STEM ethics. STEM ethics and its subtopics have been characterized and defined in various ways around the globe, as seen in the chapters of this book.

One way to categorize STEM ethics is to distinguish between (1) ethical aspects of research and conducting research responsibly, i.e., a field that can be called research ethics, on the one hand, and (2) ethical and societal implications of science and technology on the other.

In the United States, Responsible Conduct of Research (RCR) is at the center of the first field (Steneck and Bulger 2007; Kalichman 2013). RCR comprises the standards and rules of research and a responsible research process. In his 2013 article "A Brief History of RCR Education," Michael Kalichman describes it this way. "In other words, the focus of RCR education should be the specialized practices that constitute the responsible planning, conduct, and reporting of research. In short, the focus is on specifics of research that are handled in ways that are not self-evident." (p. 381).

Responsible Conduct of Research (RCR) is the basis of everyday laboratory practice. In the laboratory environment, students primarily establish their understanding of ethics guiding STEM research culture. Issues of relevance include data collection and management, collaboration with fellow researchers and other stakeholders, data interpretation, plagiarism, conflict of interest, working with human or animal subjects, mentoring relationships, pressure and pressure coping strategies, safety, publication of results, and environmental effects.

In 2022, the National Institutes of Health updated their 2010 requirements and added new topics to include in RCR education. These new topics include the issues of peer review, secure and ethical data use, and safe research environments, which they define as environments "that promote inclusion and are free of sexual, racial, ethnic, disability and other forms of discriminatory harassment" (NIH 2022).

Realizing the importance of research ethics and responsible conduct of research, countries around the world have taken steps to strengthen education in these fields. In the United States, research funded by the National Science Foundation and the National Institutes of Health requires all undergraduate, graduate students and postdoctoral researchers to receive responsible conduct of research training (NIH 2022; NSF 2022).

Other countries rely less on the concept of Responsible Conduct of Research but see and discuss the norms and standards of the research process within the broader context of research ethics. For example, the European Commission states in a 2018 brief, "Research integrity is generally understood to mean the performance of research according to the highest standards of professionalism and rigour, in an ethically robust manner." (EC 2018). The All European Academies (ALLEA) recently adopted the European Code of Conduct for Research Integrity in 2017 to help foster a shared understanding of research integrity (ALLEA 2017). A recent study of 16 European countries and the research integrity training and education programs found an extreme diversity between countries. Some countries, like Austria, offer non-mandatory training except for some doctoral programs for those who want to qualify for academic positions. In many other countries such as Denmark, France, Ireland, and Moldova, training is mandatory for doctoral and postdoctoral students and all publicly-funded research organizations (Perković Paloš et al. 2023).

The state of research ethics education in Africa is changing quickly. Though there currently needs to be more institutional structures, systems, and guidelines on research integrity in many African countries, some encouraging developments have happened in the past few years. In 2003, South Africa established their National Health Research Ethics Council, which focuses on ensuring good research practices involving human and animal subjects, and Nigeria launched its Ethics Council in 2005. (Simiyu et al. 2021). In 2022, the African Research Integrity Network (ARIN) was launched during the seventh World Conference on Research Integrity hosted by the University of Cape Town in South Africa. Though there is a current lack of institutional structure, systems, and guidelines on research integrity in many African countries, the ARIN holds as its goal to bring together academic and non-academic partners to reflect on the current research integrity status of the African continent and to drive Africa-centric agenda to advance research integrity in the region. (Bain et al. 2021; ARIN 2022).

The second field, ethical and societal implications of science and technology, has a different focus. Ethics reflection and ethics education in the second area of ethical and societal implications of science and technology can loosely be divided into engineering ethics, and science and technology ethics, which overlap in manifold ways with bioethics, biomedical ethics, and Science and Technology Studies (STS). Whereas science and technology ethics more broadly addresses philosophical, ethical, and societal aspects of research, science, and technology, engineering ethics are more directly connected to engineers and ethical decision-making in engineering. According to scholar Joe Herkert, one way that more traditional STS scholars differ from engineering ethics scholars and educators are that the latter are often more engaged in studying ethical issues within the professions and participate in more activist and social change movements. Engineering ethics educators also tend to focus on the education of young professionals and engage in the contextual analysis of science and technology in society- often with the collaboration of engineers in the field (Herkert 2006).

Engineering ethics education has become a part of accreditation standards for engineering schools in many parts of the world. The Japan Accreditation Board for Engineering Education (JABEE) requires all undergraduates to take a mandatory course that includes ethics (Balakrishnan et al. 2019). Similarly, in the United States, ABET, the accreditation board for college and university programs in applied and natural science, computing, engineering, and engineering technology, requires that all students graduate with the "ability to recognize ethical and professional responsibilities in engineering situations and make informed judgments, which must consider the impact of engineering solutions in global, economic, environmental, and societal contexts," (ABET 2022). In 2013, when China joined the Washington Agreement that allows for mutual recognition of engineering degrees in both countries, the Chinese Engineering Education Accredited Association used the ABET requirements as a "startup template" and adopted similar ethics education standards (Zhu et al. 2014).

In research ethics and engineering ethics, a distinction has been established between micro and macro ethics. Microethics considers individuals and internal relationships in research or the profession and the standards and norms guiding individual behavior, and macroethics applies to the collective social responsibility of the profession, to societal decisions about technology, and societal implications of technology (Herkert 2005; Bird 2014).

In this context, the European concept of "Responsible Research and Innovation" can clearly be characterized as a macro ethics-oriented approach. The concept has been defined as follows (von Schomberg 2013, p. 63):

> Responsible Research and Innovation is a transparent, interactive process by which societal actors and innovators become mutually responsive to each other with a view to the (ethical) acceptability, sustainability, and societal desirability of the innovation process and its marketable products (to allow a proper embedding of scientific and technological advances in our society).

Accordingly, Responsible Research and Innovation is a process focused not primarily on researchers, scientists, or engineers but involves a broad spectrum of stakeholder groups, if not the entire society. In this process, the ethical, legal, and societal implications of STEM, especially the societal implications of new and emerging technologies, are crucial (von Schomberg 2013; Zwart et al. 2014).

Finally, Artificial Intelligence (AI) ethics has exploded in the past few years, with over 120 ethical codes and other normative documents being authored by professional associations, businesses, and non-profit groups being published (Schiff et al. 2021). With the pervasiveness of AI in all parts of society, scholars and leaders in AI ethics point to the need for students - the future developers and adopters of AI – to develop AI literacy, understand the underlying ethical questions, and be able to successfully engage with these ethical issues in a way that helps this technology benefit society (Borenstein and Howard 2021). In response to this, universities and other educational institutions across the Americas, Asia, and Europe are pioneering new ways of integrating a focus on AI ethics across their curriculums - from exploring issues around bias and data colonialism to experiential learning approaches where students evaluate AI-generated code for potential ethical uses (Zembylas 2023; Becker et al. 2023). The emergence of ChatGPT in November of 2022 and the expanding use of generative AI by students of all ages has only exacerbated the need to address these issues, and universities and educators are again responding to this call, some by adopting guidelines and others by actively using this technology as a chance for ethical reflection (Fuchs 2023; Stahl and Eke 2024).

1.1.3 Traditional Approaches to STEM Ethics Education

In the United States, before RCR education was mandated by federal agencies such as the National Institutes of Health and the National Science Foundation, the apprentice model of scientists teaching students good practices was the norm (Kalichman 2013). In this model, a novice researcher gets hands-on knowledge from an experienced researcher- usually a faculty member- in the practice and norms of research. The downside of this approach is the possibility of experienced research teaching the student how to circumvent good research practice. In the

1980s, several institutions established research ethics and scientific integrity courses. After a series of highly publicized research misconduct cases, NIH established its ethics education mandate in 1989 and 1992 and NSF in 2009 (NIH 1989, 1992; NSF 2009).

Since these mandates began, ethics education related to STEM research and RCR in the United States has been done mainly through classroom teaching, usually involving lectures.

In the US, the focus is often on RCR, as RCR is of central relevance regarding the quality and reliability of research, study design, and data management.

Some approaches have met with more success than others. Whereas there are many responsible conduct of research (RCR) education courses, seminars, workshops, and online tutorials in the United States, reports indicate that these traditional education measures lack effectiveness (Steneck and Bulger 2007; Holsapple et al. 2012; Kalichman 2014). In a seminal review of RCR education, authors Tyler J. Mulhearn and colleagues outlined eight different instructional approaches often used in teaching RCR. These include field-specific compliance, online, professional decision-making, general discussion, targeted, experimental interventions, and norm adherence. In this and follow-up studies, the authors found the common approach of general discussion training, where discussions are held in small and large groups with insufficient attention to content and processes, and norm adherence training, where the minimum amount of content is delivered through lectures (Steele et al. 2016). Limitations also arise when the content of these courses focuses on issues of rules and compliance and weeding out the few "bad apples" who engage in research misconduct rather than developing students' ethical sensitivity and moral decision-making skills (Mulhearn et al. 2017) (Table 1.1).

Course-based teaching that focuses on conveying knowledge about laws, rules, and guidelines resembles legal training, and many students need to find this more relevant and attractive. When ethics and ethics education in STEM is considered roughly equivalent to purely formal requirements, such as those related to IRBs or human subjects training, students may gain knowledge of the existence of these rules and institutions but set out in their career of researchers having only a sparse knowledge of the legal requirements framing research. They need to gain the required cognitive skills, such as ethical sensitivity and problem-solving, to address these issues when they arise effectively. They need to have the opportunity to practice the social and communication skills required to solve these issues in a collaborative environment.

Despite these drawbacks, many universities have adopted courses to comply with federal requirements. Other schools rely on online tutorials that save time or money but often focus on relaying knowledge and lack any form of interpersonal exchange and meaningful discussion (Kalichman and Plemmons 2007). For many aspiring researchers, ethics remains an abstract concept unrelated to everyday laboratory practice.

Generally speaking, there are four key goals of research ethics education: (1) increases in knowledge of issues and practices, (2) increases in skills related to ethical decision-making and conflict management, (3) improved attitudes toward open

Table 1.1 STEM ethics instructional approaches and characteristics

Instructional approaches	Characteristics
Exemplar based	Lectures, examples, personal responsibility, human rights, stakeholders, codes of conduct, privacy, and confidentiality, field-specific content.
Field-specific compliance training	Lecture, discussion, focus on field-specific examples. Content includes compliance, guidelines, field-specific content, protection of human participants, legality, and human rights.
General discussion training	Discussion in large and small groups, small to moderate amounts of moral philosophy. Extremely common type of ethics education approach.
Norm adherence training	Small to moderate lecturing, small to moderate moral philosophy content
Online training	Online delivery, sometimes a mixture of interactive and self-directed, web-based discussions. Content focuses on guidelines. Examples of this include CITI training.
Philosophical self-reflection	Moral philosophy, generality of guidelines, values, historical development, contemporary ethical issues, coursework often includes self-reflection, essays
Professional decision processes training	Emphasis on professionalism, focus on stakeholders in decision-making, lecture, problem-based learning, team-based learning, case-based instruction, and discussion
Targeted experimental interventions	Case-based instruction, active participation, self-directed learning. Often happens within labs, workplaces, or outside of the traditional classroom.

communication and respect for issues, and (4) improvements in behavior and choices (Kalichman 2012). These goals are relevant not only for RCR and research ethics education but also for educational approaches considering the ethical and societal implications of scientific and technological innovation. Newberry articulated similar goals in his 2004 article, namely that students should (a) be emotionally engaged and *want* to be ethical, (b) that they should have intellectual engagement and *know* how to be ethical, and (c) have particular knowledge or *discipline-specific* knowledge of codes or practices that allow students to make ethical decisions (Newberry 2004).

With these goals in mind, how can new approaches fulfill these learning objectives and help build more inclusive STEM education and research cultures?

1.1.4 Need for New Approaches

Ethics Education in STEM has changed drastically over the 30 or so years it has been taught (Kalichman 2013; Anderson et al. 2013). It has evolved from relying on scientists to self-regulate and learn solely from the apprentice model to addressing emerging and systemic issues such as sustainability, social justice, diversity, and inclusion, as well as integrating new approaches as modes of delivery.

Along with this development, there was increasing awareness that ethics education in STEM is a complex endeavor. It requires an interdisciplinary approach, an understanding of the STEM field, communication skills, and interpersonal interaction.

Over the past decades, more promising approaches to ethics education have been developed. They include target experimental interventions, where students are asked to deliberate about a case and apply moral reasoning skills, as does exemplar-based training (Steele et al. 2016). Professional decision processes training has also been shown effective, both because of its emphasis on teaching students strategies for ethical decision-making, and its ability to provide students with the tools necessary to manage the ambiguous, ill-defined nature of ethical decisions (Mumford et al. 2007; Steele et al. 2016).

The use of multiple delivery methods for ethics education is also essential, especially its ability to reach diverse learners. (Steele et al. 2016; Watts et al. 2017). Universities, research institutes, and other organizations have developed several methods for providing this education to future researchers and practitioners. These include developing role-play and theatrical presentations of scenarios where research ethics questions arise, immersing students in developing ethics guidelines for their labs, to community-based learning projects where students are asked to help solve multifaceted issues faced by a community and take into account different stakeholder views (Chen and Hester 2023; Hildt et al. 2019; Bielfeldt and Lima 2020).

There are several ways in which educational approaches to STEM ethics can be improved.

One way is to avoid having ethics in STEM taught by a teacher who lectures to students in a classroom but aims at getting students actively involved. This can take various forms, such as active classroom discussions, case study-based approaches, or role-play. Addressing diversity aspects and increasing DEI awareness helps build an inclusive, ethical culture. In addition, group projects or hands-on learning exercises provide chances for direct and practical learning experiences. This is particularly true for interdisciplinary projects and approaches. STEM courses may benefit from ethics by design, i.e., the strategy to directly integrate ethics aspects or ethical concepts in technology design.

Another way is to take STEM ethics education out of the classroom and move ethics education into real-life contexts where ethical issues around STEM play out and matter, for example, in research laboratories. Out-of-classroom experiences like direct interaction with stakeholders or community involvement allow for innovative projects and approaches.

Designing and developing new ethics education programs and approaches needs creativity, interdisciplinary competency, time, and energy. What unites the various authors of this book is that even though they work in different disciplines and stem from other parts of the planet, they all have pioneered innovative ethics education methods or been involved in restructuring STEM ethics education in their respective contexts.

The following chapters of this book introduce their innovative approaches to STEM ethics education.

1.1.5 About This Collected Volume

This collected volume is based on the two-day workshop "Building Inclusive Ethical Cultures in STEM" (April 23–24, 2021). The workshop was originally meant to disseminate the results of a National Science Foundation-funded research project[1] at the Illinois Institute of Technology, which sought to bring research ethics education out of the classroom and into technical labs. During this project, we encountered numerous novel approaches being developed around the world and thought that an online conference could help connect these disparate groups. Presenters shared many approaches, highlighting successes, challenges, and new questions around this topic. After the conference, it seemed important to share and expand upon the findings of this conference, so the research team asked presenters if they were willing to contribute a chapter and also sought out new contributors who could provide a more global view of ethics education in their field and community.

The book seeks to share innovative approaches to STEM ethics education and effectively engage students and faculty working in research labs, lab-based classrooms, and courses to build inclusive, ethical cultures. The frameworks and approaches presented move beyond traditional research ethics training or STEM ethics education. The book's chapters showcase best practices and approaches to embedding educational interventions in courses, research labs, departments, and workplace environments.

Moving beyond the two-day conference that inspired this collected volume, the various chapters address questions like: What are approaches and tools to integrate ethics education in STEM effectively? How can STEM ethics education be improved? What can researchers do to build more inclusive research environments? How can meaningful discussions about ethics be effectively integrated into STEM courses, research labs, and workplace environments? While each chapter takes a different perspective and is located in its respective context, the contributions are united by the goal of effectively including ethical reflection in STEM education.

In selecting authors to contribute, we sought authors from a broad spectrum of disciplines and countries who share their approaches toward innovative ethics education.. The approaches towards ethics education and building ethical cultures in STEM suggested by the various chapters and authors may inspire not only STEM, STEM ethics, and ethics educators but also STEM researchers, principal investigators, faculty, or administrators interested in programmatic approaches to improving

[1] Award # 1635661

mentoring, research ethics education, and the research culture of their research environment.

The collection has been organized into the following four sections. The first part, "Restructuring Ethics Education in STEM," discusses the current shortcomings of STEM ethics education as it currently exists in the United States and India and suggests some curricular approaches to address these shortcomings. The second section, "How Socio-political Context Influences STEM Ethics Education," explores how the challenges and goals of a society frame the topics, values emphasized, and what it means to successfully teach STEM ethics to students. By examining the context of ethics education in South Africa, Ghana, Eastern Europe, the U.S., and China, readers will gain a nuanced understanding of how delivery methods and areas of focus can be better adapted to meet the needs of a community.

The third section, "Embedding Ethics Education in Practice Contexts and Labs," explores novel approaches focusing on ethics in practices: out of the theoretical contexts of classrooms and directly into the laboratory or workplace. The final part, "New Approaches in Framing Ethical Issues," looks at new approaches to framing ethical issues and how these approaches help students explore ethical questions from a new perspective. These approaches seek to break students out of their routine by engaging them in active problem-solving, utilizing multidisciplinary approaches to learning, and helping them appreciate and develop essential but often underappreciated skills like imagination and storytelling in solving complex ethical questions.

References

ABET. 2022. *Criterion MI3 student outcomes, Outcome 4. Criteria for Accrediting Engineering Programs 2022–2023.* Accessed 24 Mar 2023. https://www.abet.org/accreditation/accreditation-criteria/criteria-for-accrediting-engineering-programs-2022-2023/

African Research Integrity Network. 2022. Accessed 24 Mar 2023. https://africarinetwork.wixsite.com/website/about

All European Academies (ALLEA). 2017. *The European code of conduct for research integrity.* Berlin: All European Academies. https://www.allea.org/wp-content/uploads/2017/05/ALLEA-European-Code-of-Conduct-for-Research-Integrity-2017.pdf32A.

Anderson, M.S., M.A. Shaw, N.H. Steneck, E. Konkle, and T. Kamata. 2013. Research integrity and misconduct in the academic profession. In *Higher education: Handbook of theory and research*, 217–261. Dordrecht: Springer. https://doi.org/10.1007/978-94-007-5836-0_5.

Bain, L.E., L.A. Tchuisseu-Kwangoua, O. Adeagbo, N.C. Nkfusai, H. Amu, F.I. Saah, and F. Kombe. 2021. Fostering research integrity in sub-Saharan Africa: Challenges, opportunities, and recommendations. *The Pan African Medical Journal* 43: https://doi.org/10.11604/pamj.2022.43.182.37804.

Balakrishnan, B., F. Tochinai, and H. Kanemitsu. 2019. Engineering ethics education: A comparative study of Japan and Malaysia. *Science and Engineering Ethics* 25: 1069–1083. https://doi.org/10.1007/s11948-018-0051-3.

Becker, B.A., P. Denny, J. Finnie-Ansley, Al Luxton-Reilly, J. Prather, and E.A. Santos. 2023. Programming is hard-or at least it used to be: Educational opportunities and challenges of AI code generation. In SIGCSE 2023: Proceedings of the 54th ACM Technical symposium on computer science education V. March 2023, 500–506. https://doi.org/10.1145/3545945.3569759.

Bielefeldt, and Marybeth Lima. 2020. Service-learning and civic engagement as the basis for engineering design education. In *New innovations in engineering education and navel engineering*, ed. Nur Md Sayeed Hassan and Sergio Antonio Neves Lousada. London: InTech Open.

Bird, S.J. 2001. Mentors, advisors and supervisors: Their role in teaching responsible research conduct. *Science and Engineering Ethics* 7: 455–468.

———. 2014. Social responsibility and research ethics: Not either/or but both. *Professional Ethics Report* 27 (2): 1–4.

Borenstein, J., and A. Howard. 2021. Emerging challenges in AI and the need for AI ethics education. *AI and Ethics* 1: 61–65. https://doi.org/10.1007/s43681-020-00002-7.

Chen, K.C., and L.L. Hester. 2023. A dramatized method for teaching undergraduate students responsible research conduct. *Accountability in Research* 30 (3): 176–198. https://doi.org/10.1080/08989621.2021.1981871.

European Commission. *Science with and for society policy brief*, No. 4, November 2018. https://ec.europa.eu/research/participants/documents/downloadPublic?documentIds=080166e5bf5c08aa&appId=PPGMS#:~:text=Research%20integrity%20can%20be%20understood,of%20research%20on%20the%20other

Friedman, T. L. 2005. *The world is flat: A brief history of the twenty-first century*. New York: Farra, Straus, and Giroux.

Fuchs, K. 2023. Exploring the opportunities and challenges of NLP models in higher education: Is Chat GPT a blessing or a curse? *Frontiers in Education* 8: 1166682. https://doi.org/10.3389/feduc.2023.1166682.

Gonzalez, H.B., and J. Kuenzi. 2012. *Science, technology, engineering, and mathematics (STEM): A primer*. Washington: Congressional Research Service. https://www.ccc.edu/departments/Documents/STEM_labor.pdf.

Haven, T., H.R. Pasman, G. Widdershoven, L. Bouter, and J. Tijdink. 2020. Researchers' perceptions of a responsible research climate: A multi focus group study. *Science and Engineering Ethics* 26 (6): 3017–3036. https://doi.org/10.1007/s11948-020-00256-8.

Herkert, J.R. 2005. Ways of thinking about and teaching ethical problem solving: Microethics and macroethics in engineering. *Science and Engineering Ethics*. 11: 373–385. https://doi.org/10.1007/s11948-005-0006-3.

———. 2006. Confessions of a Shoveler: STS subcultures and engineering ethics. *Bulletin of Science, Technology, and Society* 26 (5): 410–418.

Hildt, Elisabeth, K. Laas, C.Z. Miller, S. Taylor, and E.M. Brey. 2019. Empowering graduate students to address ethics in research environment. *Cambridge Quarterly of HealthCare Ethics* 28 (3): 542–550.

Hofmann, B., and Holm, S. 2019. Research integrity: environment, experience, or ethos? *Research Ethics, 15* (3–4): 1–13.

Holsapple, M.A., D.D. Carpenter, J.A. Sutkus, C.J. Finelli, and T.S. Harding. 2012. Framing faculty and student discrepancies in engineering ethics education delivery . *Journal of Engineering Education, 101* (2): 169–186.

Kalichman, M.W. 2012. Why teach research ethics? In *Practical guidance on science and engineering ethics education for instructors and administrators*, 5–16. Washington, DC: The National Academies Press.

Kalichman, M. 2013. A brief history of RCR education. *Accountability in Research* 20 (5–6): 380–394. https://doi.org/10.1080/08989621.2013.822260.

———. 2014. Rescuing responsible conduct of research (RCR) education. *Accountability in Research* 21: 68–83.

Kalichman, M.W., and D.K. Plemmons. 2007. Reported goals for responsible conduct of research courses. *Academic Medicine* 82 (9): 846–852.

Mulhearn, T.J., L.M. Steele, L.L. Watts, K.E. Mederios, M.D. Mumford, and S. Connelly. 2017. Review of instructional approaches in ethics education. *Science and Engineering Ethics* 23: 883. https://doi.org/10.1007/s11948-016-9803-0.

Mumford, M.D., S.T. Murphy, S. Connelly, J.H. Hill, A.L. Antes, R.P. Brown, and L.D. Devenport. 2007. Environmental influences on ethical decision making: Climate and environmental predictors of research integrity. *Ethics & Behavior* 17 (4): 337–366.

Newberry, B. 2004. The dilemma of ethics in engineering education. *Science and Engineering Ethics* 10 (2): 343–351.

Perković Paloš, A., R. Roje, V. Tomić, and A. Marušić. 2023. Creating research ethics and integrity country report cards: Case study from Europe. *Accountability in Research*. https://doi.org/10.1080/08989621.2022.2163632.

Schiff, D., J. Borenstein, J. Biddle, and K. Laas. 2021. AI ethics in the public, private, and NGO sectors: A review of a global document collection. *IEEE Transactions on Technology and Society* 2 (1): 31–42. https://doi.org/10.1109/TTS.2021.3052127.

Schraudner, M., K. Hochfeld, and C. Striebing. 2019. *Arbeitskultur und Arbeitsatmosphaere in der Max-Planck-Gesellschaft*. Kurzbericht, Fraunhofer Center for Responsible Research and Innovation. https://www.mpg.de/13631088/Kurzbericht_MPG-Arbeitskultur.pdf

Simiyn, P.R., E. Buraimoh, and I.E. Davidson. 2021. Fostering research integrity in African higher education. *African Journal of Inter/Multidisciplinary Studies* 3 (Special Issue): 97–109. https://doi.org/10.51415/ajims.v3i1.980.

Spalter-Roth, R. 2004. *Social sciences are key to developing the STEM workforce*. https://www.asanet.org/wp-content/uploads/savvy/footnotes/feb04/fn3.html

Stahl, B.C., and D. Eke. 2024. The ethics of ChatGPT: Exploring the ethical issues of an emerging technology. *International Journal of Information Management* 74: 102700.

Steele, L.M.T.J., K.E. Mulhearn, L.L. Medeiros, S. Connelly Watts, and M.D. Mumford. 2016. How do we know what works? A review and critique of current practices in ethics training evaluation. *Accountability in Research* 23 (6): 319–350. https://doi.org/10.1080/08989621.2016.1186547.

Steneck, N.H., and R.E. Bulger. 2007. The history, purpose, and future of instruction in the responsible conduct of research. *Academic Medicine* 82 (9): 829–834.

Struyf, A., H. De Loof, J. Boeve-de Pauw, and P. Van Petegem. 2019. Students' engagement in different STEM learning environments: Integrated STEM education as promising practice? *International Journal of Science Education* 41 (10): 1387–1407. https://doi.org/10.1080/09500693.2019.1607983.

United Kingdom, House of Lords, Science and Technology Committee. 2012. Chapter 2: Definition of STEM. In *Higher education in science, technology engineering and mathematics (STEM) subjects*. 17 Jul 2012. Accessed 23 Mar 2023. https://publications.parliament.uk/pa/ld201213/ldselect/ldsctech/37/3705.htm

United States, National Institutes of Health. 1989. Requirement for programs on the responsible conduct of research in national research service award institutional training programs. *NIH Guide for Grants and Contracts* 18 (45): 1. http://grants.nih.gov/grants/guide/historical/1989_12_22_Vol_18_No_45.pdf.

———. 1992. Reminder and update: Requirement for instruction in the responsible conduct of research in national research service award institutional training grants. *NIH Guide* 21 (43) http://grants.nih.gov/grants/guide/noticefiles/not92-236.html.

———. 2022. *FY 2022 updated guidance: Requirement for instruction in the responsible conduct of research*. Notice Number NOT-OD-22-055. February 17. https://grants.nih.gov/grants/guide/notice-files/NOT-OD-22-055.html

United States, National Science Foundation. 2009. *Responsible conduct of research. Proposal and award policies and procedures guide*. Part II – Award and administration guidelines, p. IV3. Last accessed 30 Jul 2013. http://www.nsf.gov/pubs/policydocs/pappguide/nsf10_1/nsf10_1.pdf

———. 2022. PAPPG requirements on RECR certification Chapter II.D. 1.d https://beta.nsf.gov/policies/pappg/23-1

Von Schomberg, R. 2013. A vision of responsible research and innovation. In *Responsible innovation. Managing the responsible emergence of science and innovation in society*, ed. R. Owen, J. Bessant, and M. Heintz, 51–74. Wiley.

Watts, L.L., K.E. Medeiros, T.J. Mulhearn, L.M. Steele, S. Connelly, and M.D. Mumford. 2017. Are ethics training programs improving? A meta-analytic comparison spanning 35 years of ethics instruction in the sciences. *Ethics and Behavior* 5 (27): 351–384.

Weil, V., and R. Arzbaecher. 1996. Ethics and relationships in laboratories and research communities. *Professional Ethics* 4 (3–4): 83–125. https://doi.org/10.5840/profethics199543/414.

Woolston, C. 2019. Ph.D. Poll reveals fear and joy, contentment and anguish. *Nature* 575: 403–406. https://doi.org/10.1038/d41586-019-03459-7.

Zembylas, M. 2023. A decolonial approach to AI in higher education teaching and learning: Strategies for undoing the ethics of digital neocolonialism. *Learning, Media and Technology* 48 (1): 25–37. https://doi.org/10.1080/17439884.2021.2010094.

Zhu, Q., B.K. Jesiek, and J. Yuan. 2014. *Engineering education policymaking in cross-national context: A critical analysis of engineering education in China*. Paper presented at the 121st *ASEE Annual Conference and Exposition*. Indianapolis, IN June 15–18. Paper 8896.

Zwart, H., L. Landeweerd, A. van Rooij, and A. 2014. Adapt or perish? Assessing the recent shift in the European research funding arena from 'ELSA' to 'RRI'. *Life Sciences, Society and Policy* 10: 11.

Part I
Introduction: Restructuring Ethics Education in STEM

Over the past decades, there has been an increase in awareness of STEM ethics education and the relevance of Responsible Conduct of Research (RCR) education, as outlined in the introduction to this volume. This has led to an impressive evolution of ethics education in STEM. Courses that address ethics in STEM and RCR were developed in the 1980s and 90s and have been increasingly taught at many universities. (Herkert 2000). While this is significant progress, there is a need to develop innovative approaches that go beyond conveying the dos and don'ts of responsible research in STEM but take a broader picture of the ethical aspects and implications of STEM, the ethical culture of STEM institutions, social responsibility, and justice. Furthermore, there is a need to increase the availability of ethics education in STEM. Overall, ethics education plays only a minor role in STEM contexts, with considerable differences between schools, universities, and countries.

As the conditions for STEM ethics education vary between institutions, a first step towards bringing about effective change in this context is an analysis of the status quo. Against the background of a characterization of the existing practice and its downsides, potential ways of overcoming the downsides can be devised and pursued.

A broad spectrum of factors may contribute to hindering a better qualitative or quantitative consideration of ethical aspects in STEM education. For example, STEM educators may feel uncomfortable teaching ethical aspects of their discipline as these may seem beyond their purview. Time or resource limitations or institutional factors may hamper bringing ethics reflection into courses or curricula.

Based on a thorough analysis, existing weaknesses can be identified, and approaches developed on how existing weaknesses can be overcome. The latter may include removing (purported) barriers, facilitating interdisciplinary collaboration, or explicitly encouraging STEM educators to have ethics topics in their teaching.

This section of the volume discusses some of the shortcomings of STEM ethics education as it currently exists and presents approaches that aim at improving and restructuring the existing situation of ethics education in STEM fields.

Authors from the United States and India discuss curricular strategies to address these shortcomings. They depict their work to innovate ethics in STEM education in their respective contexts. namely, efforts to integrate education about social responsibility, the development of ethics credentials to help students more smoothly integrate and gain better recognition of the importance of ethics STEM education, and efforts to enhance STEM education overall.

The authors explain how a thorough analysis helps identify weaknesses and overcome existing disadvantages. While specific to their respective contexts, their approaches can be implemented in a relatively broad spectrum of educational environments.

The chapter by Quintin Kreth, Daniel Schiff, Jeonghyun Lee, Jason Borenstein, and Ellen Zegura, "Social Responsibility and Ethics in STEM Education: The State of the Field," discusses some of the shortcomings traditional U.S. engineering education has in addressing issues of social responsibility. After providing an overview of some of the intellectual foundations of social responsibility and the critical distinctions between personal and professional social responsibility, the authors discuss their own work to identify influences and inhibitors within STEM undergraduate education, as well as similarities and differences between the development of personal and professional social responsibility in students. The authors discuss how efforts are needed to actively encourage faculty and departments to include discipline-specific education and activities directly tied to the cultivation of professional responsibility and the need for longitudinal studies of how social responsibility develops over time.

Alexandra Bradner and Rebecca A. Bates' chapter, "Developing an Ethics Credential for Undergraduate STEM Majors," discuss their development of a one to two-day ethics module that STEM faculty can easily incorporate humanities content into their courses. The modules are meant to empower faculty with limited expertise in the humanities to deliver this new content and help STEM students reflect, in a humanistic way, on the long-term ethical consequences of their work and career choices. The authors discuss the challenges of developing their modules and the learning goals met and describe a sample teaching module that focuses on the just distribution of limited medical resources.

Reena Cheruvalath's chapter, "Ethics Education in Engineering and Technological Institutes in India: Challenges and looking forward," provides an extensive overview of STEM ethics education in India and discusses the need to give students a stronger foundation in ethics and ethical decision-making. After thoroughly discussing current teaching methods, the author provides illustrative examples of how educators can better help students develop their meta-moral cognitive skills and see ethical questions from multiple perspectives.

Finally, the chapter, "Embedding Moral Reasoning and Teamwork Training in Computer Science and Electrical Engineering" by Allen and Raquel Diaz-Sprague describes an interactive ethics module that engaged students in interactive lectures and asked the students to lead the class in playing a game that exemplified the ethics of teamwork. The authors also describe a competition that asked students to create artwork or an app that promotes every day ethics and cooperative behaviors. The authors describe both of these activities and discuss some of the benefits of using these kinds of active learning sessions and competitions where student design "prosocial" apps.

Reference

Herkert, J.R. 2000. Engineering ethics education in the USA: Content, pedagogy and curriculum, *European Journal of Engineering Education*, 25(4): 303–313. https://doi.org/10.1080/03043790050200340.

Chapter 2
Social Responsibility and Ethics in STEM Education: The State of the Field

Quintin Kreth, Daniel S. Schiff, Jeonghyun Lee, Jason Borenstein, and Ellen Zegura

Abstract The relationship between ethics education and recent scholarship on social responsibility is crucial to explore. At times, ethics education has been designed to focus narrowly on compliance with rules and regulations. In contrast, other forms of ethics education emphasize direct attention to social responsibility and the types of obligations that future professionals have to society. In this chapter, we provide an overview of social responsibility, including some of its intellectual foundations, and discuss an important distinction to the realm of ethics education between *personal* and *professional* social responsibility. We also review some of the current social responsibility literature and describe connections between social responsibility and traditional approaches to ethics education. We conclude by highlighting opportunities for future research in the realm of social responsibility, including lessons learned from our own research. In particular, there is a need for further analysis of the long-term development of student social responsibility attitudes and their impact on professional practice.

Q. Kreth
School of Public Policy, Georgia Institute of Technology, Atlanta, GA, USA
e-mail: quintin@gatech.edu

D. S. Schiff (✉)
Department of Political Science, Purdue University, West Lafayette, IN, USA
e-mail: dschiff@purdue.edu

J. Lee
Center for 21st Century Universities, Georgia Institute of Technology, Atlanta, GA, USA
e-mail: jonnalee@gatech.edu

J. Borenstein
School of Public Policy, Georgia Institute of Technology, Atlanta, GA, USA

Office of Graduate and Postdoctoral Education, Georgia Institute of Technology, Atlanta, GA, USA
e-mail: borenstein@gatech.edu

E. Zegura
School of Computer Science, Georgia Institute of Technology, Atlanta, GA, USA
e-mail: ewz@cc.gatech.edu

E. Hildt et al. (eds.), *Building Inclusive Ethical Cultures in STEM*, The International Library of Ethics, Law and Technology 42,
https://doi.org/10.1007/978-3-031-51560-6_2

Keywords Social responsibility · Community engagement · Ethics education · Macroethics · Professional ethics · Engineering ethics

2.1 Introduction

In the twenty-first century, public perceptions of scientists, engineers, and technology professionals have, to some degree, grown more negative (Kennedy et al. 2022). While many factors can contribute to this, high-profile ethical lapses are an ongoing problem. Tragedies involving flawed implementation of sophisticated technology, such as the Boeing 737 MAX crashes (Herkert et al. 2020), and even intentional wrongdoing by scientists or engineers, including the Volkswagen emissions scandal (Patel 2015) and practices within the blood testing company Theranos (Allyn 2022), have shaken public confidence in STEM professionals. Given these and other troubling incidents, combined with the power that science and technology can exert over our lives, it is an important time to re-examine the ethical and social responsibilities that STEM professionals have to society. This chapter provides an overview of current literature on the topic, including the state of educational interventions to develop responsibility attitudes among STEM students.

Many scholars and policymakers suggest that educational interventions in college are key to addressing deficiencies regarding professional ethics and social responsibility within STEM communities (e.g., NAE 2009, 2016). Ethics education in undergraduate and graduate STEM degree programs is ideally a method for familiarizing students with the obligations that they might have as future professionals. Some envision, for example, that such education will prevent or mitigate the chance that students will engage in unethical behavior in the workplace. However, traditional approaches to ethics education may have too narrow of a focus to address the full scope of the challenges facing STEM professionals.

Historically, ethics education in STEM programs usually highlighted issues at the microethical scale rather than broader, macroethical concerns. This trend is common in traditional approaches to engineering ethics (Herkert 2005). To illustrate the distinction: an issue closer to the microethical domain would be an individual engineer's obligation to tell the truth to a client, whereas a professional organization's stance on the morality of weapons development would be closer to macroethics. Assuming that professional ethics is even mentioned within a STEM degree program (which is not always the case), the focus tends to be on microethics, preventive ethics, and compliance with codes of ethics or the law. In response, some scholars advocate that a greater emphasis should be placed on aspirational ethics (Harris 2013; Huff and Barnard 2009; Pritchard and Pritchard 2006). In other words, the approach would not only involve teaching about harm avoidance, but also about the obligation "to do good" and cases of moral exemplars.

Along these lines, the ethical problems that the public perceives as emerging from STEM professions often go beyond the microethical level and range into the broader impacts that science and engineering have on society. There are many questions, for example, that the public and other stakeholders have about whether computing and other emerging technologies are being designed responsibly, especially considering the harmful consequences that many experience from such technologies. For instance, claims of harm have been articulated in relation to the use of Facebook in Myanmar (Toh 2021). Public concern about such issues is a key reason to examine the intersection of ethics and social responsibility education.

In this chapter, we highlight recent scholarship examining the concepts of personal and professional social responsibility, connections between conceptions of professional ethics and social responsibility, and a variety of factors that may influence social responsibility development in undergraduate students. Furthermore, we discuss what could comprise a future research agenda on social responsibility education.

2.2 A Renewed Interest in Ethics and Social Responsibility Education

There has been a renewed interest in integrating ethics and social responsibility content in STEM education. The reasons for this are varied, but studies indicating that STEM students might not be developing a genuine sense of concern for the public's well-being are a contributing factor. Several years ago, Erin Cech's work drew specific attention to engineering education (Cech 2014). She described a "culture of disengagement" – a widespread mindset that encourages bracketing or compartmentalizing of supposedly "irrelevant" (Cetina 1999) social and ethical issues when engaging in engineering practice in favor of technical concerns alone. What can follow from this is the belief that ethical and social issues are not, by definition, engineering problems. Cech (2014) argues that this mindset is furthered by three ideological concepts common in engineering: depoliticization, sociotechnical dualism, and meritocracy. Collectively, these three concepts, she suggests, contribute to a culture where engineers might dismiss their responsibility to help address societal problems.

This type of "disengagement" from ethical or other societal issues may be visible in other areas of STEM as well. Many academic programs do not require discussion of ethical issues relevant to their discipline and treat ethics as an issue external to their field. Similar attitudes are visible in the realm of Big Tech, where a "move fast and break things" mentality has been pervasive (Lidow 2019). This mentality can include a purposeful ignorance of "externalities" such as the societal impacts emerging from technology's development and implementation. In effect, this mindset and approach can foster a de facto or even active rejection of notions of social responsibility. Intertwined with the notion that social responsibility and ethics are outside of a STEM professional's purview is the commonly held belief that technology is "value neutral" (Morrow 2014), a belief that on many occasions is an attempt to insulate designers or their employers from the consequences of the technology they have created.

In response to these and other related concerns, many scholars, academic institutions, professional organizations, and funders of research are advocating for a sincere dedication to ethics and social responsibility education (Grosz et al. 2019). "Ethics across the curriculum," for instance, is one method tried by academic institutions to introduce students to ethics through multiple courses within their program of study. Professional organizations have responded to recent calls to focus on ethics as well. For instance, the Institute of Electrical and Electronics Engineers (IEEE) has dedicated resources to examining the macroethical dimensions of intelligent systems (Adamson and Herkert 2022), while the Association for Computing Machinery (ACM) revised its ethics code in 2018 and hosts an ongoing conference series on Fairness, Accountability, and Transparency (Gotterbarn et al. 2018). The Mozilla Foundation (2017) has also supported the development of ethics pedagogy for the undergraduate computer science curriculum.

Of course, this is not the first time that STEM communities have drawn attention to the need for ethics education. In prior decades, nanotechnology (Eosco et al. 2014) and the human genome project (Dolan et al. 2022) generated much interest in ethics. Yet, the most recent round of interest in ethics education has emerged due to many factors closely tied to the digital age. This includes growing pessimism about social networks and other computing technologies (e.g., Zuboff 2023), along with questions about the role of STEM professionals who are involved in developing technologies that could harm the public (e.g., Benjamin 2019). During this era, ethics education can play a key role in challenging assumptions about the "value neutrality" of technology, especially considering that individuals and groups can be exploited with relative ease and on a global scale with the computing devices that STEM professionals design. Overlapping with calls for increasing ethics education, scholars have also argued that more attention should be given to the importance of social responsibility (Bird 2014). In fact, this book chapter comes at the conclusion of our research team's multi-year research project examining social responsibility development among undergraduate STEM students.[1] A key motivation underlying social responsibility research, including our own, is to shed light on how students develop – or fail to develop – their personal and professional social responsibility attitudes.

2.3 What Is Social Responsibility?

Educators and others have put forward many proposals to bolster education on social responsibility (Gangone 2022). Yet, what exactly does the phrase refer to? While "social responsibility" seems intuitive, it is a difficult concept to define precisely. Also, disentangling it from the related concept of *corporate* social

[1] Project details, publications, and research tools are available at: https://serve-learn-sustain.gatech.edu/institutional-transformation-project

responsibility, which we do not cover here, is crucial, given that the latter concept has a narrower and somewhat different scope. Furthermore, whether and how social responsibility relates to ethics is not always clear. Yet, at the broadest level, social responsibility as a concept can be understood with respect to its constituent terms. Namely, as described by Bobo (1991), who contrasts social responsibility against traditional individualistic American values, it is *social* as opposed to individualistic, implying a collective duty. One social responsibility-oriented psychometric instrument, the Scale of Service-Learning Involvement (SSLI), based on Delve et al.'s (1990) service-learning model, notes the focus on pursuing "the common good rather than…personal advancement" (Olney and Grande 1995).

The concept of social responsibility also implies a sense of obligation to act in certain ways (and perhaps to hold certain beliefs or attitudes) so as to foster the social good. Early approaches to measuring social responsibility, for example, reveal a focus on a "sense of obligation to the group" (Gough et al. 1952). Thus, social responsibility is particularly relevant as both a pedagogical and professional-oriented concept because it is action-guiding. That is, while facets of it pertain to "mere" ethical beliefs and attitudes, it is also tied to behavior. Social responsibility is, therefore, often conceived of as a "value orientation" which "motivates certain actions" including "prosocial, moral, and civic behaviors" (Wray-Lake and Syvertsen 2011).

However, several intricacies reveal that social responsibility is more complicated than it initially appears, both in terms of its grounding and implications. First, a variety of normative ethical frameworks can serve as foundations for social responsibility. Wray-Lake and Syvertsen (2011) describe it as "rooted in democratic relationships...and moral principles of care and justice," invoking theories of both care and justice perhaps influenced by Gilligan (1993) or Rawls (1971). This grounding is also critical to the theoretical foundations of Canney and Bielefeldt's Professional Social Responsibility Development Model (PSRDM) (Canney and Bielefeldt 2015a, b, c). The PSRDM directly influences our own research on social responsibility development of STEM students (Schiff et al. 2021; Kreth et al. 2022). More specifically, the PSRDM delineates three realms that comprise social responsibility: (1) Personal Social Awareness, (2) Professional Development, and (3) Professional Connectedness. These realms influenced our research approach in terms of how we sought to examine connections (or lack thereof) between the development of personal social responsibility and professional social responsibility. Details about the PSRDM and how it influenced our research approach are provided in the "Efforts to Examine Social Responsibility in STEM" section.

Yet, it is possible to ground social responsibility within still other normative ethical frameworks. For instance, consequentialism could serve as its foundation (Skorupski 1995). Alternatively, it could be grounded in deontological thinking under the notion that moral prescriptions should be universalizable to all rational agents, or by virtue ethics as the manifestation of a desirable virtue (Pettit 2000), or in still other frameworks.

These disparate groundings do not necessarily entail incompatibility, as it is possible to align practical ethical considerations without sharing the same normative grounding. Yet, other considerations of key importance emerge, including *to whom* one is responsible. This question asks who (or what) "has standing" to be morally relevant with respect to one's own social responsibility. Even putting aside purely self-interested or egoistic notions of moral responsibility, a long history of philosophical thought has considered to what extent individuals are responsible for others, potentially including one's family, community, nation, ethnic or religious groups, the entirety of humanity, or non-human animals (e.g., Jaggar 1995; Greene 2014; Singer 1981). Further, if there are special obligations to be partial (or especially interested) with respect to one's family, local community, members of a shared social group or vulnerable sub-population, one's nation, and so on, this may delimit the responsibility to work on behalf of *other* groups (Friedman 1991). Thus, the recipients of one's responsibility, depending on the normative grounding under consideration, can range drastically.

A third key facet is the question, "*For what* am I responsible?" Social responsibility, certainly in the twenty-first century, is often cast with respect to aspects of social, economic, political, or environmental justice and injustice. Looking back, Bobo (1991) describes social responsibility as involving "a cluster of beliefs that endorse limitations to economic inequality, an obligation to meet the basic needs of all people in society, and a duty to redress unfair societal inequality" and to do so as part of "a responsibility of the social collective." Conceptualizations like this might align closely with notions of civic duty, implicitly focusing attention within national borders. A review of even older approaches to understanding social responsibility (for example, the questions developed for Gough et al.'s (1952) Personality Scale for Responsibility instrument) reveal that attitudes and actions associated with social responsibility vary widely. For example, questions in the instrument surround civic activities like voting and paying one's taxes, home-based activities like "taking care of one's aging parents," as well as more preliminary efforts like taking time to "find out about national affairs," and even broader concerns like worrying about "the rest of the world." Further, moral duties may include positive (to seek to actively benefit) and negative (to refrain from doing harm) obligations. As such, there is tremendous variation in the types of emphases and specific prescriptions that could emerge in relation to social responsibility, including variation over time and place.

From our perspective, the components contained within a given usage of "social responsibility," the extent of the implied obligations, and associated trade-offs are not always defined nor justified directly. To a great extent, how scholars define social responsibility depends on the cultural context as well as their underlying theoretical assumptions. Examining the questions in survey instruments that scholars select and the literature they decide to cite can provide a working conceptualization of what social responsibility means to them. Ultimately, though, we do not seek to resolve questions about the nature of social responsibility but note that the concept is dynamic and resists a singular and precise specification.

2.4 Personal and Professional Social Responsibility

Adding complexity to social responsibility's definitional challenges are the varying subtypes of the concept, including *personal* and *professional* social responsibility. Notions of personal and professional social responsibility overlap to some degree, but in important ways, they are distinct concepts. While both personal and professional social responsibility have macroethical facets to them, personal social responsibility refers to a general sense and awareness of societal problems. Professional social responsibility involves a context and action specific to one's own vocation and vocational skills.

Providing a definition for *professional social responsibility* may be an easier challenge than the broader root concept, but similar questions to those mentioned previously nevertheless apply. Is one responsible only for their own actions as a professional, those of their organization, or their profession as a whole? Are they responsible only to their clients, their local communities, national and international publics, or, say, the environment? It is worth noting here that at least some professional codes of ethics, a formal but incomplete articulation of one's professional obligations, have recently included obligations tied to macroethical issues. For example, when IEEE revised its Code of Ethics in 2020, it added a new obligation "...to improve the understanding by individuals and society of the capabilities and societal implications of conventional and emerging technologies, including intelligent systems" (IEEE 2020). What follows is one's professional social responsibilities can expand and become more complex over time.

Our research examined the distinction between personal and professional social responsibility in terms of which factors may contribute to each of their development in undergraduate students (Schiff et al. 2021). In addition, we mapped on to these two concepts the distinction between microethics and macroethics (see Fig. 2.1). While the connection between the microethical dimensions of personal and professional social responsibility is rather direct and intuitive, the connection between their macroethical counterparts may be more elusive. For example, notions of honesty from one's personal life can straightforwardly carry over to their behavior in the workplace. However, it may be a more difficult intellectual hurdle to determine how one's personal views about their broader ethical obligations to society (e.g., reducing homelessness) translate to professional contexts (e.g., software design). This connection, or lack thereof, between personal and professional social responsibility at the macroscale is something that STEM educators need to pay special attention to, especially since many students will have limited familiarity with the profession they plan to enter.

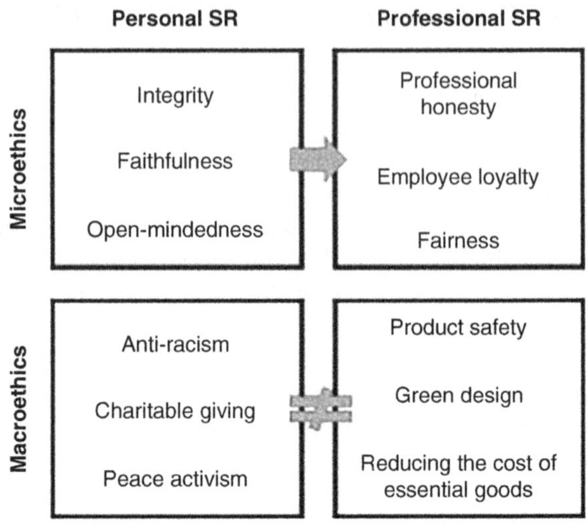

Fig. 2.1 Mapping social responsibility, microethics, and macroethics. (Adapted from Schiff et al. 2021)

2.5 Efforts to Examine Social Responsibility in STEM

While many definitions of social responsibility are possible, the options narrow considerably when the focus is constrained to a specific context. Here, we primarily examine recent work on social responsibility as it pertains to STEM students and professionals. A key advancement in this realm occurred with the development of the PSRDM by Canney and Bielefeldt (2015a, b, c). They describe social responsibility as a sense "of obligation to help others as both a person and a professional, with a special focus on helping disadvantaged or marginalized populations…both a value orientation and as a guiding principle for taking action" (Canney and Bielefeldt 2015a, p. 415). Their theoretical model and associated survey instrument, the Engineering Professional Responsibility Assessment (EPRA), help to assess social responsibility development among engineering students (Canney and Bielefeldt 2016).

The PSRDM emerged through the integration of three preexisting theoretical models: Schwartz's Altruistic Helping Behavior Model (Schwartz 1977; Schwartz and Howard 1982), Ramsey's framework for social responsibility in scientific decision-making (Ramsey 1989, 1993), and the Service-Learning Model (Delve et al. 1990). The Altruistic Helping Behavior Model describes moral development in individuals and its subsequent effect on the willingness to help others, corresponding to personal social responsibility in the PSRDM. Ramsey's work discusses the obligation that scientists have to be conscientious of social needs when making professional decisions. The Service-Learning Model describes a cycle of professionals applying their professional skills to address problems in society, then developing greater conscientiousness and willingness to act; this view draws analogies to physical exercise and is related to concepts of self-efficacy (Bandura 1977).

Canney and Bielefeldt's team has pursued various lines of inquiry at the intersection of engineering and social responsibility. Among their key findings is that engineering students often have lower social responsibility attitudes at graduation than when entering college (Bielefeldt and Canney 2016a). Our research team has been strongly influenced by the PSRDM framework and adapted their EPRA survey instrument to study social responsibility attitudes across STEM fields (Schiff et al. 2021; Kreth et al. 2022). The adapted survey, the Generalized Professional Responsibility Assessment (GPRA), was used to examine social responsibility development in undergraduate students (e.g., Erwin et al. 2018). The GPRA enables cross-disciplinary comparisons; among the goals tied to its use is to better understand the effect of extra-curricular activities on social responsibility development (Schiff et al. 2021).

Within our research, we sought to identify influences and inhibitors within STEM undergraduate education. For example, we found that students pick their field of study largely because of intellectual interest rather than concerns about ethical practice or impact (Schiff et al. 2021). This suggests that educational institutions may have challenges to overcome in order to cultivate student concern about professional social responsibility. That is, if students separate their ethical concerns (e.g., inequality) from their professional development (e.g., building technical skills) and conceive of these aspects of their identity along distinct paths, deliberate efforts may be needed to bridge this personal-professional divide. In addition, our research sought to investigate similarities and differences between the development of personal and professional social responsibility. Mapping that distinction onto the one between microethics and macroethics could be important for working in the realm of ethics and social responsibility education.

2.6 Ethics and Social Responsibility Education

Thus far, we have discussed ethics and social responsibility as though they are interrelated concepts. Yet, the connection between them is not always clear, in part because scholars have sharply varying definitions for each respective concept. Of course, the term "ethics" on its own is subject to countless definitions. Similarly, social responsibility, which we sought to highlight previously, has many associated definitional nuances. Some might perceive social responsibility as one type of ethical consideration (i.e., social responsibility as subsumed within ethics), whereas others may think these domains overlap but are in some ways distinct from one another. For instance, under some perspectives, social responsibility might include social and political considerations that go beyond the realm of ethics.

Something that can shed light on the similarities and differences between conceptions of ethics and social responsibility would be delineating what educators envision as the respective learning objectives associated with each topic. Some common goals for ethics education are increased moral sensitivity and improved moral judgment, aims often influenced by the work of Narvaez and Rest (1995) and

Bebeau et al. (1999). Another common goal of ethics education is to lessen the chance that students will perform bad behaviors in their future professional role. More generally, the purpose of ethics education could be to nurture people so that they become virtuous citizens (Narvaez 2005). Moreover, the type of educational offering can be an important factor. For instance, a traditional ethics course focusing on philosophical foundations can have different learning objectives than a professional ethics course. In terms of the latter, Martin et al. (2021) outline the multiplicity of learning objectives in engineering ethics education, including increasing student knowledge of professional codes of ethics or conduct.

Another method for examining overlaps between ethics and social responsibility centers on the particular educational activities or interventions that are intended to foster their development. The use of role playing scenarios is a common approach in ethics courses. For many years, professional ethics courses, including in medical fields and engineering, have embraced case studies as a method for highlighting ethical decision-making complexities. A large variety of interventions have been used in engineering ethics courses (Hess and Fore 2018).

A key goal of social responsibility education is tied to fostering awareness of the particular challenges plaguing society in one's time and context, such as environmental sustainability, social justice, and inequality. Within STEM education, social responsibility themes emerge when examining issues such as privacy and data, the power of technology companies, the role of algorithms in society, and the treatment of underrepresented groups in the workforce. In recent times, social responsibility education is increasingly being tied to learning about and making progress towards the sustainable development goals put forward by the United Nations (2021).

Efforts to nurture social responsibility development frequently seek to connect students to a specific community (Bielefeldt and Canney 2016a, b; Ward and Wolf-Wendel 2000). This can take the form of service learning, or, more broadly, community engagement. That is, such activities often take place in settings outside of a classroom and aim to build relationships with a local group. In that sense, community engagement more likely involves the practice or fulfillment of one's social responsibilities as compared to a more traditional, inwardly focused form of ethics education (Bielefeldt and Canney 2016a, b).

Along these lines, Bielefeldt and Canney (2016a, b) suggest that learning through service, which combines service-learning and co-curricular community service, can be a valuable approach for teaching social responsibility. Some common learning-through-service methods include working with community partners and reflecting on the service experiences. Pedagogical approaches like problem-based learning can involve challenging students to disentangle real-world, ill-structured problems through community engagement and extra-curricular activities (e.g., Engineers Without Borders), which can be accompanied by other educational activities such as mentoring and self-reflection (Hess and Fore 2018; Wittig 2013).

In our own research, we sought to identify, through surveys and interviews, which curricular, co-curricular, and extracurricular community engagement activities students took part in and draw connections between those activities and changes in social responsibility attitudes. The activities included attending a course within or outside of their primary discipline that contained material on community engagement or social responsibility topics, listening to campus speakers on the topic, and participating in projects that support local communities. Many of the activities had at least some affiliation with the Serve-Learn-Sustain initiative on our campus, which seeks to help students learn how to "create sustainable communities".[2]

During student interviews, we learned about the importance of family upbringing, the influence of peers, current events especially related to politics and racial (in) justice, as well as limitations associated with STEM coursework (Schiff et al. 2021). We found indications that increased community engagement is associated with growth in social responsibility attitudes over time, even controlling for pre-college social responsibility attitudes. Thus, attention to community engagement – whether through coursework or other venues – could increase student social awareness, and when associated with a student's disciplinary focus, may foster professional social responsibility.

2.7 Conclusion: Knowledge Gaps and Future Research Opportunities

As research into social responsibility development in STEM is a relatively new field, many knowledge gaps and opportunities for future research are in this realm. For example, it is important to consider what factors could influence the quality of community engagement experiences and their impact on social responsibility development. From our own study findings, reported in Schiff et al. (2021), students' personal and professional social responsibility seem to be strongly influenced by interpersonal interactions, particularly interactions with peers. Other components of quality and impact include the extent to which community engagement activities are collaborative, reflective, and connected to a student's discipline. Drawing from our own research experience and the literature, we identify below key research questions and knowledge gaps that could be examined during future inquiries.

Rather few studies, beyond Cech (2014) and Bielefeldt and Canney's work (Bielefeldt 2021; Bielefeldt and Canney 2016a, 2016b; Canney and Bielefeldt 2015a, b, c), focus on long-term social responsibility development within the education system, let alone after students leave college and enter the workforce. A standard approach to assessing ethical or social responsibility development makes use of a pre-post analysis of an experimental course or module. While such an approach can certainly be worthwhile and provide support for the efficacy of the particular intervention, it rarely answers the question of whether the change within a student persists over time. That is, relatively little is known about the longitudinal effects of ethics-related educational interventions during college. This problem intensifies, given the challenge of disambiguating the many influences students experience within the curriculum from

[2] For more information about Serve-Learn-Sustain at Georgia Tech, refer to: https://sls.gatech.edu/

non-curricular activities and pre-college attitudes. In particular, we suggest that more attention is needed on the extent to which co-curricular and non-curricular activities over the course of undergraduate education work together to shape student perceptions of their professional ethical obligations to the public.

Another key gap, as alluded to above, is that little is known regarding how social responsibility development during undergraduate education influences future professional behavior and practice. In general, research within the realm of ethical and social responsibility development focuses on undergraduate students; a key driving factor being that ethics education requirements are more likely to be in place at the undergraduate level. Yet, one should not assume a strong connection between at-graduation attitudes and future professional behavior. Additional longitudinal studies that follow students after they become working professionals would help close this knowledge gap. Along related lines, future studies could address questions on the extent to which social responsibility attitudes influence career choices post-graduation. Furthermore, whether and how professional social responsibility changes and develops within STEM professionals is another important but underexplored area of inquiry.

We suggest that more of a focus should be placed on whether and how professional social responsibility, especially at the macroethical level, develops within STEM students. Universities may need to take the step of actively encouraging faculty and departments to include in the STEM curriculum discipline-specific education and activities directly tied to cultivating professional social responsibility. Rigorous investigations should continue to evaluate which specific interventions foster social responsibility development. For instance, what (if any) curricular interventions can improve social responsibility attitudes? The same is true for extracurricular activities intended to foster social responsibility. Universities invest significant resources into student organizations, local and international volunteer projects, and many other activities that they hope will contribute to social responsibility development. Yet, the extent to which these activities achieve specific goals tied to ethical and social responsibility development could use the support of a fuller empirical foundation.

Acknowledgements This chapter is based on work supported by the National Science Foundation Cultivating Cultures for Ethical STEM Program (Award #1635554). Any opinions, findings, conclusions, or recommendations expressed in this chapter are those of the authors and do not necessarily reflect the views of the National Science Foundation.

Bibliography

Adamson, Greg, and Joseph Herkert. 2022. Addressing intelligent systems and ethical design in the IEEE code of ethics. In *Codes of ethics and ethical guidelines*, ed. Kelly Laas, Michael Davis, and Elisabeth Hildt, 145–159. Springer.

Allyn, Bobby. 2022, January 12. Former Theranos CEO Elizabeth Holmes to be sentenced on Sept. 26. *NPR*. https://www.npr.org/2022/01/12/1072612059/former-theranos-ceo-elizabeth-holmes-to-be-sentenced-on-sept-26.

Bandura, Albert. 1977. Self-efficacy: Toward a unifying theory of behavioral change. *Psychological Review* 84 (2): 191–215. https://doi.org/10.1037/0033-295X.84.2.191.

Bebeau, Muriel J., James R. Rest, and Darcia Narvaez. 1999. Beyond the promise: A perspective on research in moral education. *Educational Researcher* 28 (4): 18–26. https://doi.org/10.3102/0013189X028004018.

Benjamin, Ruha. 2019. *Race after technology: Abolitionist tools for the new Jim code*. Polity.

Bielefeldt, Angela R. 2021. Intersectional complexities of race/ethnicity and gender in engineering students' professional social responsibility attitudes. In 2021 ASEE Virtual Annual Conference Content Access.

Bielefeldt, Angela R., and Nathan E. Canney. 2016a. Changes in the social responsibility attitudes of engineering students over time. *Science and Engineering Ethics* 22 (5): 1535–1551.

———. 2016b. Relationships between religion, spirituality, and socially responsible engineering. *Engineering Studies* 8 (1): 66–90.

Bird, Stephanie J. 2014. Social responsibility and research ethics: Not either/or but both. *Professional Ethics Report* 27 (2): 1–4.

Bobo, Lawrence. 1991. Social responsibility, individualism, and redistributive policies. *Sociological Forum* 6: 71–92.

Canney, Nathan E., and Angela Bielefeldt. 2015a. A framework for the development of social responsibility in engineers. *International Journal of Engineering Education* 31 (1B): 414–424.

———. 2015b. Differences in engineering students' views of social responsibility between disciplines. *Journal of Professional Issues in Engineering Education and Practice* 141 (4): 04015004.

———. 2015c. Gender differences in the social responsibility attitudes of engineering students and how they change over time. *Journal of Women and Minorities in Science and Engineering* 21 (3).

———. 2016. Validity and reliability evidence of the engineering professional responsibility assessment tool. *Journal of Engineering Education* 105 (3): 452–477. https://doi.org/10.1002/jee.20124.

Cech, Erin A. 2014. Culture of disengagement in engineering education? *Science, Technology, & Human Values* 39 (1): 42–72.

Cetina, Karin Knorr. 1999. *Epistemic cultures: How the sciences make knowledge*. Cambridge, MA: Harvard University Press.

Delve, Cecilia I., Suzanne D. Mintz, and Greig M. Stewart. 1990. Promoting values development through community service: A design. *New Directions for Student Services* 1990 (50): 7–29.

Dolan, Deanne Dunbar, Sandra Soo-Jin Lee, and Mildred K. Cho. 2022. Three decades of ethical, legal, and social implications research: Looking back to chart a path forward. *Cell Genomics*: 100150.

Eosco, Gina M., Meghnaa Tallapragada, Katherine A. McComas, and Merrill Brady. 2014. Exploring societal and ethical views of nanotechnology REUs. *NanoEthics* 8 (1): 91–99. https://doi.org/10.1007/s11569-014-0192-z.

Erwin, Alexandra, Jason Borenstein, Wendy Newstetter, Colin Potts, and Ellen Zegura. 2018. Undergraduate STEM students and community engagement activities: initial findings from an assessment of their concern for public well-being. In ASEE Annual Conference & Exposition 2018.

Friedman, Marilyn. 1991. The practice of partiality. *Ethics* 101 (4): 818–835.

Gangone, Lynn M. 2022. *Educators and social responsibility: What this means to informed citizenry*. Washington, DC: AACTE.

Gilligan, Carol. 1993. *In a different voice: Psychological theory and women's development*. Cambridge, MA: Harvard University Press.

Gotterbarn, D.W., Bo Brinkman, Catherine Flick, Michael S. Kirkpatrick, Keith Miller, Kate Vazansky, and Marty J. Wolf. 2018. *ACM code of ethics and professional conduct*. Association for Computing Machinery.

Gough, Harrison G., Herbert McClosky, and Paul E. Meehl. 1952. A personality scale for social responsibility. *The Journal of Abnormal and Social Psychology* 47 (1): 73.

Greene, Joshua. 2014. *Moral tribes: Emotion, reason, and the gap between us and them*. Penguin.

Grosz, Barbara J., David Gray Grant, Kate Vredenburgh, Jeff Behrends, Hu Lily, Alison Simmons, and Jim Waldo. 2019. Embedded EthiCS: Integrating ethics across CS education. *Communications of the ACM* 62 (8): 54–61.

Harris, Charles E. 2013. Engineering ethics: From preventive ethics to aspirational ethics. In *Philosophy and engineering: Reflections on practice, principles and process*, ed. Diane P. Michelfelder, Natasha McCarthy, and David E. Goldberg, 177–187. Dordrecht: Springer.

Herkert, Joseph R. 2005. Ways of thinking about and teaching ethical problem solving: Microethics and macroethics in engineering. *Science and Engineering Ethics* 11 (3): 373–385.

Herkert, Joseph R., Jason Borenstein, and Keith Miller. 2020. The Boeing 737 MAX: Lessons for engineering ethics. *Science and Engineering Ethics* 26 (6): 2957–2974.

Hess, Justin L., and Grant Fore. 2018. A systematic literature review of US engineering ethics interventions. *Science and Engineering Ethics* 24 (2): 551–583. https://doi.org/10.1007/s11948-017-9910-6.

Huff, Chuck, and Laura Barnard. 2009. Good computing. *IEEE Technology and Society Magazine* 28 (3): 47–54. https://doi.org/10.1109/MTS.2009.934158.

Institute of Electrical and Electronics Engineers. 2020. *IEEE code of ethics*. Piscataway.

Jaggar, Alison M. 1995. Caring as a feminist practice of moral reason. In *Justice and care*. Routledge.

Kennedy, Brian, Alec Tyson, and Cary Funk. 2022. Americans' trust in scientists, other groups declines. *Pew Research Center*, 15 Feb 2022. https://www.pewresearch.org/science/2022/02/15/americans-trust-in-scientists-other-groups-declines/

Kreth, Quintin, Daniel S. Schiff, Jeonghyun Lee, Jason Borenstein, and Ellen Zegura. 2022. Social responsibility attitudes among undergraduate computer science students: An empirical analysis. In 2022 ASEE Annual Conference Minneapolis.

Lidow, Derek. 2019. Mark Zuckerberg and the tech world still do not understand ethics. *Forbes*, 11 Mar 2019. https://www.forbes.com/sites/dereklidow/2019/03/11/mark-zuckerberg-and-the-tech-world-still-do-not-understand-ethics/?sh=4305db035386

Martin, Diana Adela, Eddie Conlon, and Brian Bowe. 2021. A multi-level review of engineering ethics education: Towards a socio-technical orientation of engineering education for ethics. *Science and Engineering Ethics* 27 (5): 1–38.

Morrow, David R. 2014. When technologies makes good people do bad things: Another argument against the value-neutrality of technologies. *Science and Engineering Ethics* 20 (2): 329–343. https://doi.org/10.1007/s11948-013-9464-1.

Mozilla Foundation. 2017. *Responsible computer science challenge*. https://foundation.mozilla.org/en/what-we-fund/awards/responsible-computer-science-challenge/

Narvaez, Darcia. 2005. The Neo-Kohlbergian Tradition and beyond: Schemas, expertise, and character. *Nebraska Symposium on Motivation* 51: 119–163.

Narvaez, Darcia, and James Rest. 1995. The four components of acting morally. *Moral Behavior and Moral Development: An Introduction* 1 (1): 385–400.

National Academy of Engineering. 2009. *Ethics education and scientific and engineering research: What's been learned? What should be done? Summary of a workshop*. Washington, D.C.: National Academies Press.

National Academy of Engineering, Infusing Ethics Selection Committee. 2016. *Infusing ethics into the development of engineers: Exemplary education and programs*. Washington, D.C.: National Academies Press.

Olney, C., and S. Grande. 1995. Validation of a scale to measure development of social responsibility. *Michigan Journal of Community Service Learning* 2 (1) http://hdl.handle.net/2027/spo.3239521.0002.104.

Patel, Prachi. 2015. Engineers, ethics, and the VW scandal. *IEEE Spectrum* 25.

Pettit, Philip. 2000. Non-consequentialism and universalizability. *The Philosophical Quarterly* 50 (199): 175–190.

Pritchard, Michael S., and Michael S. Pritchard. 2006. *Professional integrity: Thinking ethically.* Lawrence: University Press of Kansas.

Ramsey, John. 1989. A curricular framework for community-based STS issue instruction. *Education and Urban Society* 22 (1): 40–53.

———. 1993. The science education reform movement: Implications for social responsibility. *Science Education* 77 (2): 235–258.

Rawls, J. 1971. *A theory of justice*. Cambridge, MA: Harvard University Press.

Schiff, Daniel S., Emma Logevall, Jason Borenstein, Wendy Newstetter, Colin Potts, and Ellen Zegura. 2021. Linking personal and professional social responsibility development to microethics and macroethics: Observations from early undergraduate education. *Journal of Engineering Education* 110 (1): 70–91. https://doi.org/10.1002/jee.20371.

Schwartz, Shalom H. 1977. Normative influences on altruism. In *Advances in experimental social psychology*, vol. 10, 221–279. London: Elsevier.

Schwartz, Shalom H., and Judith A. Howard. 1982. Helping and cooperation: A self-based motivational model. In *Cooperation and helping behavior*, 327–353. London: Elsevier.

Singer, Peter. 1981. The concept of moral standing. In *Ethics in hard times*, The Hastings Center series in ethics, ed. Arthur L. Caplan and Daniel Callahan, 31–45. Springer. https://doi.org/10.1007/978-1-4684-4022-5_2.

Skorupski, John. 1995. Agent-neutrality, consequentialism, utilitarianism… a terminological note. *Utilitas* 7 (1): 49–54.

Toh, Michelle. 2021. Facebook sued for $150 billion over violence against Rohingya in Myanmar. *CNN Business*, 7 Dec 2021. https://www.cnn.com/2021/12/07/tech/facebook-myanmar-rohingya-muslims-intl-hnk/index.html.

United Nations. 2021. SDGs Youth Leadership and Social Responsibility Education. Accessed September 14, 2022. https://sdgs.un.org/partnerships/sdgs-youth-leadership-and-social-responsibility-education-0.

Ward, Kelly, and Lisa Wolf-Wendel. 2000. Community-Centered service learning: Moving from doing for to doing with. *American Behavioral Scientist* 43 (5): 767–780.

Wittig, Ann. 2013. Implementing problem based learning through engineers without Borders student projects. *Advances in Engineering Education* 3 (4): n4.

Wray-Lake, Laura, and Amy K. Syvertsen. 2011. The developmental roots of social responsibility in childhood and adolescence. *New Directions for Child and Adolescent Development* 2011 (134): 11–25.

Zuboff, Shoshana. 2023. The age of surveillance capitalism. In *Social theory re-wired*, 203–213. Routledge.

Chapter 3
Developing an Ethics Credential for Undergraduate STEM Majors

Alexandra Bradner and Rebecca A. Bates

Abstract STEM faculty are encouraged to incorporate ethical, social, and historical content into their undergraduate STEM courses. This is a challenge, for there is more than enough foundational material, and interdisciplinary content can introduce a steep learning curve for students and faculty. As part of the NSF-funded Fall 2020 STEM Futures Education Project (https://serc.carleton.edu/stemfutures/about.html), we presented a plan for developing 1–2-day ethics modules that STEM faculty can easily incorporate into their courses and that STEM departments can use to craft ethics credentials that suit their purposes. Each module would have a materials list with readings, videos, and podcasts; a series of discussion questions; 2–3 interactive small group activities; and a list of prompts for an essay assignment. Some modules would have suggestions for community-based and project-based learning, and all would employ learner-centered and inclusive pedagogies to create a more inclusive and ethical culture in STEM undergraduate education. The modules empower STEM faculty with limited expertise in the humanities to deliver this new content responsibly, especially at 2- and 4-year institutions that need more resources to support team teaching.

Keywords Responsible research · Ethics · STEM · Inclusive pedagogy · Ethics education

A. Bradner (✉)
Department of Philosophy, Kenyon College, Gambier, OH, USA

R. A. Bates
Department of Integrated Engineering, Minnesota State University, Mankato, MN, USA

E. Hildt et al. (eds.), *Building Inclusive Ethical Cultures in STEM*, The International Library of Ethics, Law and Technology 42,
https://doi.org/10.1007/978-3-031-51560-6_3

3.1 Introduction

Undergraduate degree programs in science, technology, engineering, and mathematics (STEM) typically require more courses and prerequisites than those in the social sciences, arts, and humanities. Lab and practicum courses meet for long periods, 2–3 h a week or more. And most STEM degree programs recommend that students add supplementary courses in cognate disciplines, like statistics and computer science. Thus, STEM concentrators only have a little room in their schedules to explore interests outside their demanding academic track. This fact, combined with existing incentive structures, makes it challenging for STEM students to take courses in the humanities and social science of STEM, courses which enable them to think critically, historically, socially, and ethically about what they are learning and what they will eventually do with their degrees. Professors share this predicament. Doctoral training in the STEM disciplines rarely involves a deep dive into the history, philosophy, literature, religion, sociology, anthropology, and art of science. Later on, when new STEM professors are designing their first syllabi, they tend to stick closely to the materials they once taught as teaching assistants under the direction of their thesis advisors. Almost every STEM course focuses on transmitting foundational knowledge to students clearly and thoroughly. Reflective conversations between the STEM and non-STEM disciplines are rare because colleges and universities are organized divisionally, and there is already too much material to cover in one's classes within every single division.

This chapter imagines a flexible undergraduate ethics credential that STEM departments can offer students *within* their major or minor. Our model allows departments to tailor the credential to their degree programs, resources, and learners' needs. We first developed this model as part of the Fall 2020 STEM Futures initiative, a week-long, national design sprint funded by the National Science Foundation (DUE-1935479): "Workshop on the Substance of STEM Education" and organized by Arizona State University and Carleton College. In this workshop, small teams of pedagogical experts were charged with the task of developing curricular products that could prepare citizens to better "understand, discover, develop, and implement innovative and principled solutions to complex, STEM-infused problems in a rapidly changing environment."[1] The organizers believe "the citizens of tomorrow" will need more than mere foundational knowledge of chemical, biological, mathematical, etc., principles: "They will need creativity, ingenuity, and the ability to work collaboratively. And they will need to understand the broader social and the ethical contexts within which we live and work" (*Ibid.*). These citizens of tomorrow will fully represent the demographics of our nation, and our STEM

[1] "Stem Futures: the Future Substance of STEM Education Project," Stem Futures, Science Education Resource Center at Carleton College, last revised October 7, 2022, accessed November 4, 2022, https://serc.carleton.edu/stemfutures/overview.html

curricula need to be inclusive to meet their needs as learners. We must engage and motivate students of all backgrounds.

Twenty-five design sprint teams, assembled from 105 science educators from 53 institutions and 29 states, were asked to watch three preparatory webinars and develop a curricular innovation that met three associated desiderata. The curricular projects had to: (a) teach learners to view their STEM work within a broader human context, i.e., through a social, historical, or ethical lens; (b) incorporate the latest pedagogical research in the meta-cognition of learning, i.e., emphasize active-, student-centered learning, collaboration, and creativity; (c) retain the traditional goal of teaching students the foundational principles of the STEM disciplines. Teams had 4 days to prepare a presentation with a rationale for their curricular innovation and 2–3 sample assignments. Interested readers can find descriptions of the many curricular products here: https://stemfutures.education.asu.edu/curricular-products/.

Our team, composed of a leader in integrated engineering and a philosopher of science with pedagogical expertise, decided to produce two projects, a science writing concentration for a liberal arts college and the more portable ethics credential discussed in this chapter.

3.2 The Rationale

STEM work has exceptional momentum. Scholars are often in a race for accomplishment and priority. To distinguish themselves in publication and grant work, they must doggedly look for opportunities in the form of gaps in our existing knowledge and unsolved puzzles. In this race for professional respect, there needs to be more time or expertise available to help scientists reflect, in a humanistic way, on the long-term ethical consequences of their work and career choices. Such reflection is crucial, as we have seen recently and historically. You can invent an effective vaccine without a fair way to distribute it. You can create cancer therapeutics but fail to anticipate that they will cause birth defects. You can build an image search engine but later discover that it is imbued with racist bias. As John Horgan writes in *Scientific American* (Horgan 2013), "[I]t is precise because science is so powerful that we need the humanities now more than ever." To prevent needless suffering, STEM learners must be able to assess the social impact of their work; STEM learners must be able to communicate the complexity of that work to the general public; and, more personally, STEM learners must be able to assess their own high-stakes career choices carefully and confidently. In the study of science, students learn how to generate evidence for their beliefs by designing controlled experiments, crafting rule-governed proofs, and using those beliefs effectively to solve real-world problems. But it is through the study of the humanities that students learn how to question those evidentiary practices and determine whether they are good or bad. As Duke University political scientist Michael Allen Gillespie (Gillespie 2010) writes, "While science tells us how things work and thus opens up the possibilities for manipulating our world in countless ways, it does not tell us what to do with this power." One might say, after Kant, that studying the humanities without science can be speculative, but studying science without the humanities can be oblivious.

More specifically, studying history can teach us about the hidden external, social, and structural factors that influence scientific discovery and justification. What seems, at present, to be the upshot of a thoroughly rational process might appear, in a more historical light, entirely contingent. Studying philosophy teaches us how to interrogate concepts, question presuppositions, and generate alternative possibilities. Studying literature teaches us how to understand and empathize with people we do not know. Through studying religion, we learn about the objects, events, and traditions to which human beings attach meaning, the things without which we cannot endure. And we might add to this list sociology, anthropology, economics, and art. STEM learners should study it all—on this, there is widespread agreement. But STEM departments cannot transform into satellite humanities, social science, and art departments. The faculty do not have the requisite training, and the curriculum has no room to add whole courses. This was our puzzle. For our design sprint curricular project, we asked ourselves how STEM departments might begin incorporating humanities content into their degree programs quickly, cheaply, and responsibly.

3.3 Challenges

Two central concerns emerged through our series of cross-divisional conversations between a philosopher and engineer: (i) The philosopher emphasized that institutions already have on-site experts who study the social impacts of STEM work: philosophers, historians, and sociologists of science. STEM faculty interested in incorporating humanist content into their courses should not dabble or steal their colleagues' thunder. Instead, STEM departments should find a way to respect and leverage their institution's existing expertise. (ii) The engineer emphasized that STEM learners need to consider these broader questions within the context of their STEM courses instead of just taking separate courses in humanities departments. It is only within the STEM context that students can fully engage their foundational knowledge and entertain questions of personal ethics (such as how best to follow their chosen vocation in the field) and professional ethics (e.g., responsible conduct of research or RCR). Our translation goal was to value what other disciplines already do and know and find pathways to incorporate that knowledge both within and outside STEM.

3.4 Standalone Modules

We proposed the development of a collection of stand-alone ethics teaching modules that STEM faculty could adapt for use within their existing courses. We envision three content categories. Notably, the modules of each category would be produced by experts in their respective fields.

1. **Professional ethics/RCR:** These modules would engage questions of professional ethics or RCR. Readings and cases assigned within these modules would cover ethical questions that arise within human and non-human animal research, informed consent, data management and reproducibility, authorship and priority disputes, peer review, collaboration, conflicts of interest, mentor/mentee responsibilities and relationships, incivility and bullying within laboratories and the professional culture, implicit bias, and workplace harassment. The modules could be produced by professionals working within the National Science Foundation, the National Institutes of Health, the Environmental Protection Agency, the Department of Energy, and the central professional organizations for the various STEM disciplines, such as the American Chemical Society, the American Society for Engineering Education, and the Mathematical Association of America, among many others. One example of RCR modules developed by experts is the CITI training platform, whose modules are widely used by institutions to educate researchers. Still, they are less commonly used as components of undergraduate courses and come at a cost (The CITI Program 2023).
2. **History, philosophy, and sociology of science:** These modules would engage the canonical literature in the history, philosophy, and sociology of science and would be authored by scholars with doctorates in the humanities or social sciences. Topics covered would include: aspects of the scientific method, including discovery, hypothesis formation, and prediction, confirmation, and explanation; scientific rhetoric and peer review; instrumentation; the theory-ladenness of observation; the old demarcation problem (science vs. pseudoscience); the new demarcation problem (productive vs. pernicious values in science); theory choice and the structure of scientific revolutions; the sociology of scientific knowledge; history of the philosophy of science, including Aristotle's four causes, Francis Bacon's inductive method, the Mill-Whewell debate, logical empiricism, Michel Foucault, Bruno Latour, and feminist philosophy of science; and historical case studies, including the Copernican revolution, the discovery of oxygen, the emergence of the theory of natural selection, and the development of the hydrogen bomb, among so many others. Modules in this category would also cover lessons in bioethics and medical ethics, which might include material on reproductive ethics (abortion, genetic testing, germline gene therapy, and human enhancement); macro allocation and micro allocation of healthcare resources; tissue and organ donation; elective surgery and refusal; problems with the very concept of sexual orientation; assisted dying; suicide; immortality and the sanctity and meaning of life; and biases and injustice in medical research (which might address cases such as the Tuskegee Syphilis Study and the harvesting of HeLa cells from Henrietta Lacks, for example).

3. **Vocation and personal decision-making:** These modules would enable students to think about their professional trajectories in the company of similarly placed students and faculty members. The modules describe a series of high-impact, community-engaged activities in which students would meet with mentors engaged in local and regional STEM institutions which could offer site-specific, trustworthy career advice to nearby undergraduates. Mentors might include local faculty, alums, community members, and industry professionals. Activities might include interviews, student-led podcasts, community activism, site visits and reports, job shadowing, etc. For instance, a student interested in emergency medicine might arrange to shadow a hospital chaplain for a week and report to the class after the experience. A student interested in structural engineering might meet with a local civil engineer, conduct an interview, and produce a podcast.

The modules should possess the following features:

- The modules should quickly adapt to suit several local courses, degree programs, or co-curricular organizations.
- Each module should take 1–3 class sessions to teach. Faculty should be able to incorporate the modules into an existing STEM syllabus without disruption.
- The modules should be leveled to accommodate both introductory and advanced courses.
- An expert with relevant credentials should develop each module.
- The modules should be open educational resources (OER), free and accessible to faculty who teach undergraduates, perhaps housed at the Online Ethics Center for Engineering and Science (https://onlineethics.org/).
- The lesson of each module should be teachable for STEM faculty with either minimal or extensive experience in the humanities. The lesson should be easy to prepare, with a minimal learning curve.

More broadly, our collection of ethics teaching modules is designed to satisfy the three constraints established by the STEM Futures organizers:

- **The foundational knowledge constraint:** Though humanistic content appears at intentional points throughout the semester or degree program, STEM content remains the focus of the course or major/minor. Students should leave each course prepared to take the next step in their degree progression and be able to pass whatever national exams are supported by their STEM degrees.
- **The humanistic context constraint:** The whole point of this project is to consider novel ways in which we might bring humanistic ways of thinking and humanities content to STEM education. Students should be thinking about how to extrapolate their work's moral consequences and social significance, connect their present studies to past discoveries, and communicate their work to a general audience that often lacks even the most basic STEM education.
- **The meta-cognitive constraint:** Faculty should adopt learner-centered teaching practices and minimize passive, professor-centered teaching practices, like lecturing, which students struggle to retain. Such active-learning practices might

include in-class debates; film discussions; creative assessment options, like the writing and performance of podcasts; team papers and other small group projects; practitioner shadowing; community observation; volunteer activities; interviews; or reading groups.

The modules are there to serve as raw materials. Each school, department, or faculty member can decide how to use the modules toward some credential. A faculty member might add three modules to their introductory chemistry course. An environmental science department might require its majors to complete six modules at any point by the end of their senior year. A department chair might ask the campus registrar to add a special designation on student transcripts for STEM courses that contain four or more modules. An extra-curricular student group for women engineers might meet once a month to work through a module. The flexibility enables each site to add humanistic content to its curriculum in a way that suits local resources and needs and provides meaningful credentials for the students, the institution, and their stakeholders.

3.5 Learning Outcomes

Our proposed credential has eight learning outcomes that could be adopted or adapted to meet local needs.

- Learning Outcome 1: Identify expectations for responsible conduct of STEM research as codified in United States laws, regulations, and policies and the historical motivation for ethical standards in research.
- Learning Outcome 2: Recognize and differentiate ethical theories such as utilitarianism, deontology, virtue ethics, standpoint theory, ethics of care, evolutionary ethics, divine command theory, and ethical egoism. This list of Western philosophies can and should be complemented by Eastern, African, and Indigenous American philosophies.
- Learning Outcome 3: Describe processes for personal decision-making related to STEM activities, including, but not limited to, how one plans to determine whether to serve the public, through governmental work, or the private sector, through industrial work and how one intends to determine the appropriate balance of work and caregiving responsibilities.
- Learning Outcome 4: Apply earlier learning in an active, engaged mode through (a) discussing issues related to responsible conduct of research, historical motivation for ethical standards, and processes for personal decision-making, (b) describing how practicing scientists and engineers use ethical codes to guide behavior and actions, or (c) interacting with community representatives through volunteer or learning activities such as participation on an institutional review board (IRB) meeting, shadowing a hospital chaplain, or interviewing a practicing engineer.

- Learning Outcome 5: Integrate individual STEM knowledge with foundational knowledge about ethics and ethical decision-making by addressing ethical issues associated with a current topic in a student's field. For example, an engineering student might connect their study of aerosols and filtration in class to the broader social issue of public health and mask mandates. A computer science student might consider ways that choosing a data set to build a facial recognition model might result in different recognition rates for people with varying skin tones, which has implications when used (Buolamwini and Gebru 2018).
- Learning Outcome 6: Incorporate diverse perspectives and consider stakeholders that may be unrecognized or otherwise marginalized in an ethical decision-making process.
- Learning Outcome 7: Appreciate the gray areas of ethical decision-making, science, and engineering in spaces that need answers.
- Learning Outcome 8: Imagine and describe alternative possibilities and explore avenues to support and extend each student's work in STEM and ethical decision-making. For example, students might identify relevant periodicals that cover ethical decision-making in STEM and aim to keep up with those outlets, or students might identify and join local resources and communities of practice that routinely engage in ethical discussions on STEM subjects.

3.6 A Sample Module: Quality-Adjusted Life Years (QALYs)

We produced a sample ethics teaching module for the STEM Futures project on the just distribution of limited medical resources. (Please see Appendix.) This module would be classified under the second content collection, "history, philosophy, and sociology of science," and was developed by a philosopher who teaches bioethics/medical ethics regularly. The module could take anywhere from one to six class sessions to prepare and would be appropriate for inclusion in an infectious disease, pharmacology, biochemistry, or intro biology course. The modules we envision would have the following features:

- Suggested readings are selected and organized into 3 days to help STEM instructors plan their class time. But instructors can cut down the days spent on the module or enhance the lesson with additional high-impact activities.
- Three-to-four discussion questions are provided for each reading. Each set has been tested with live students to ensure it is appropriate for the allotted class time.
- Two active-learning assignments and a traditional essay prompt are provided to assist the STEM instructor in thinking about assessment.
- Author credit and contact information are provided in case the STEM instructor has questions for the module's author.

We suggest that every humanities teaching module, before it is posted to the Online Ethics Center and incorporated into the classroom, be reviewed and approved by a panel that includes two humanities or social science PhDs, two STEM faculty members with teaching expertise, and two student STEM concentrators. It is crucial that the material provided "works." In other words, developers should not create these modules *a priori* from their office desk chairs. Experienced classroom teachers with expertise in the covered content area should write the modules.

3.7 Other Example Activities

The suggested traditional and more engaged learning activities below support the development of the three knowledge areas and can be easily adapted to address the module content areas.

Foundational Knowledge

1. Webinars, documentaries, thought-provoking films (e.g., *Dear Scientists*,[2] *Coded Bias*[3]), videos of scientists and engineers discussing why and how they are a scientist/engineer
2. Lectures, e.g., bioethics and the distribution of health care costs; scientists studying social phenomena (e.g., https://mappingprejudice.umn.edu/)
3. Readings, e.g., brief readings that define ethical frameworks or readings that support a specific topic (e.g., the list in Appendix)

Meta Knowledge

1. Conversations (online or face-to-face): Topics from the webinars, lectures, and readings can be discussed in various ways. Some examples include processing case studies; following an ethical framework to make low-stakes or high-stakes personal decisions; debating a work of fiction with a similar decision-making analysis; reviewing redacted IRB applications, particularly for student-led research (asking, for example, "Would the group approve this project?").
2. Setting up observations: Participating in Institutional Review Board deliberations, shadowing practitioners, interviewing practitioners. Preparing guided questions for observations and interviews. Incorporating perspectives and awareness of diverse populations (considering, for example, "Whose voices are included?")
3. Setting up volunteer activities, while acknowledging the expertise of both volunteers and community members to avoid casting the volunteer as "savior." Pulling materials from local service-learning resources.

---

[2] Dear Scientists. Directed by Ioanna Semendeferi, 2013.

[3] Coded Bias. Directed by Shalini Kantayya, 7th Empire Media, 2020.

Humanistic Knowledge

1. Bring in storytellers: Invited speakers from various perspectives (to support connections to equity, social justice, and inclusion). Address issues of both personal and professional decisions.
2. Assign readings or movies that focus on story or narrative and weave together personal and professional decisions (include links to others' work in these areas). Suggested stories and discussion prompts can be found in these (and other) papers: Bates 2011; Sleezer and Bates 2020; Summet and Bates 2020.
3. Create an interview process: Modeled on pre-created interviews with scientists and engineers, have a class consider what questions they would ask to define a set of questions, have students interview practitioners, return to discuss a variety of answers, and then reflect in writing on how they might answer these questions in 5 years.

3.8 STEM Faculty Assessment of Humanities Assignments

One of the challenges of this proposal is how STEM faculty are suddenly made responsible for class content outside of their fields of expertise. Regarding the grading of student work, the faculty learning curve could be especially problematic. Students will only take this broader, humanistic material seriously if it is possible to receive a passing grade. If students are guaranteed an A or B on their humanities and social science papers (interviews, volunteer work, etc.), this vital content will become the "fluff" of the course. STEM faculty must become confident in their ability to assess excellent versus failing humanities and social science assignments, so they can produce meaningful feedback to students who do not yet grasp the thought styles the department and instructor hope to inculcate by teaching the modules.

One potential solution would be to offer STEM faculty targeted training in assessing written work in the humanities. We will produce an hour-long faculty development video: "How to grade an ethics paper: a resource for STEM faculty." The tutorial would include information that could be used by STEM faculty to assign, mentor, and evaluate written assignments in the humanities. Critical information would consist of the following:

- How to write a paper prompt that guides students toward an outline for their paper, works within the required page length and deadline, tests students' knowledge of the course materials, and points students toward developmental resources, like on-campus writing centers and faculty office hours.
- How to introduce a paper prompt in class.
- How to help students develop a thesis and a paper outline in an office hour meeting.
- A discussion of the pros and cons of grading rubrics and a few samples
- What to look for in a humanities paper

- How to mark up the paper: what kind of targeted feedback to provide in-line and what sort of narrative feedback to provide at the end of the paper.
- A 15-min section in which a humanist grades a paper in real time for the viewer

3.9 Extension to an Inclusive Degree

Several months after the STEM Futures project, a group interested in expanding STEM education along the lines of the same three principles came together to develop the structure for a new bachelor's degree program with an interdisciplinary curriculum at Minnesota State University, Mankato to create a more inclusive degree that increases access to STEM careers and career advancement. Unlike the proposal above, which provides one-to-three-day humanistic modules for STEM faculty to incorporate into their STEM courses, this program's interdisciplinarity operates at the curriculum level. Using a similar "design sprint" process, the group created a flexible interdisciplinary studies degree with twin cores of STEM foundational knowledge and humanistic knowledge (supported partly by National Science Foundation funding, DUE-1843775). The working group included representatives from multiple STEM disciplines, the field of philosophy, the field of technical communications, and a partner 2-year college.[4]

The 120-credit degree requires an active and multi-disciplinary project-based capstone experience, student-proposed and advisor-approved STEM coursework (15–29 credits), and humanities and social sciences (15–29 credits). Required portfolio development asks students to perform reflection that draws connections across their topics. Prospective students for this program include (a) students with articulated career goals, such as science communication, who could benefit from a 4-year path that develops, from the very first semester, contextual knowledge and cross-divisional skills; (b) students who come to the program later, struggling to choose a specific STEM major and looking for ways to build breadth into their STEM curriculum; and (c) students transferring from 2-year colleges with associate's degrees in applied science, who have taken many technical credits, but now want to focus on humanistic knowledge and the contextualizing of their foundational knowledge. Many non-traditional students who have applied for science degrees from 2-year colleges need a 4-year degree to advance in their careers. The flexible degree increases timely access to 4-year degrees for many students who would otherwise have a more extended pathway because their applied science credits can be used toward this degree.

The plural pathways of the program empower students to choose the character of their curriculum. All pathways require students to take at least five courses in STEM and at least five humanities and social science disciplines. In both their STEM and

[4] MinnPolynators Curriculum Working Group (D. Armfield, B. Bektas, M. Diab, M. Minicozzi, D. Rogalsky, P. Schumacher, D. Sharlin, R. Sleezer, B. Williams), Minnesota State University, Mankato, June 2021.

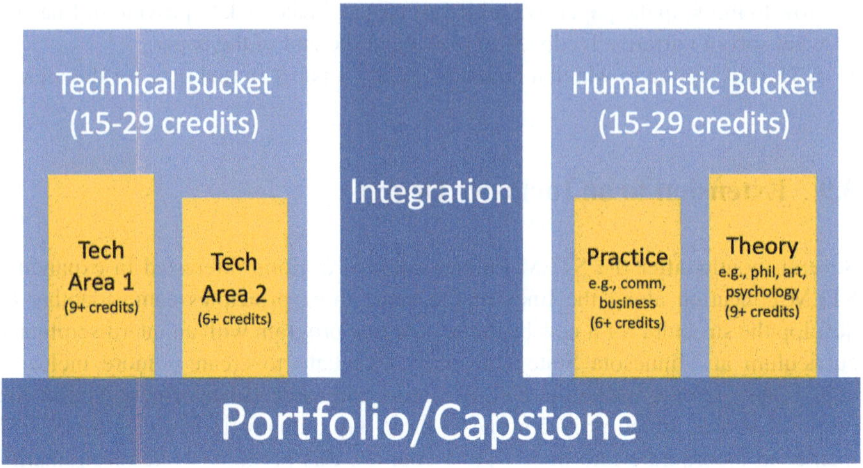

Fig. 3.1 The BS degree requires students to choose courses that cover at least two STEM topics and credits addressing theoretical and practical aspects of humanistic knowledge. Portfolio and project-based capstone credits allow students to integrate their understanding. (MinnPolynators Working Group)

humanistic courses, students will pursue work that is both practical and theoretical. As soon as possible, students assemble a portfolio in which they reflect upon their work in the STEM and humanistic realms, drawing connections across the divide. Then, throughout both semesters of the senior year, students complete a capstone in which they design and complete a real-world STEM project with humanistic components (see Fig. 3.1). Like the module proposal above, this degree program takes advantage of existing institutional expertise in the social sciences, humanities, and STEM. In addition to providing an opportunity for a degree that crosses disciplinary boundaries, the structure allows for student autonomy in topics, creating a 4-year degree that includes students previously excluded from earning a 4-year STEM degree.

3.10 Conclusion

Building an inclusive and ethical culture in STEM requires knowledge from multiple fields. It is more likely to be successful when information from the humanities and social sciences is clearly shown to be relevant for STEM students and faculty. The modules and degree described here can leverage existing resources to build new learning opportunities, provide flexible access to degrees and training, and move towards aspirations of well-informed and effective graduates who can address the complex problems facing humanity. Future work includes seeking resources to develop and disseminate modules, with the aim of better including experts in the humanities and social sciences in the STEM conversation.

Appendixes

Appendix: QALY (Quality-Adjusted Life Years) Module for STEM Faculty Integrating Bioethics Content into Their Courses

This is a guide for STEM faculty who would like to incorporate content about the fair distribution of rare healthcare resources into one of their courses. Faculty can pick and choose from the materials below, to suit their own educational context. Teaching and discussing all of the readings will likely take three 90-min class sessions. If you only have time for one class period of material, assign the readings from "Day two."

Day One: John Stuart Mill's Utilitarianism

Mill, John Stuart, "Chapter 2: What Utilitarianism Is," *Utilitarianism,* edited by Jonathan Bennett (2017), https://www.earlymoderntexts.com/assets/pdfs/mill1863.pdf.

This online edition of Mill's text has been edited for undergraduates by Jonathan Bennett. You can ask students to read a short excerpt, which introduces Mill's theory (pages 4–8) or a longer excerpt, which both introduces Mill's theory and considers objections and replies (pages 4–18).

Helpful Supplements

Crash Course Philosophy #36: "Utilitarianism"

https://www.youtube.com/watch?v=-a739VjqdSI&t=520s

Wireless Philosophy video "Ethics: Utilitarianism, Part 1: What is Utilitarianism?" https://www.youtube.com/watch?v=uvmz5E75ZIA (4:31 min)

Wireless Philosophy video "Ethics: Utilitarianism, Part 2: Problems for Utilitarianism"

https://www.youtube.com/watch?v=uGDk23Q0S9E (6:22 min)

Discussion Questions

1. Utilitarians want to maximize pleasure. What kind of pleasure?
2. A utilitarian would distribute health care resources on the basis of the consequences of that distribution. What else might we consider, other than the consequences of the distribution, when we're trying to determine how to distribute a scarce resource?
3. What might be some of the negative consequences of always acting with the aim of maximizing your society's happiness?
4. Is suffering always something to be avoided?

Day Two: The QALY Approach to Distributing Rare Health Care Resources

Williams, Alan, "The Value of QALYs," *Health and Social Services Journal*, 1985.

This selection is from the paper that introduces QALYs (quality-adjusted life years), a utilitarian mechanism for distributing scarce healthcare resources.

Harris, John, "QALYfying the Value of Life," *Journal of Medical Ethics*, vol. 13, no. 3 (September 1987), pp. 117–23. https://jme.bmj.com/content/medethics/13/3/117.full.pdf.

This paper objects that the QALY view discriminates against older people and people with disabilities. Utilitarians are interested in maximizing the amount of pleasure writ large, but do not care about how that pleasure is distributed among metaphysical persons.

Discussion Questions

1. Imagine you are a policy maker. What kinds of healthcare policies would maximize QALYs?
2. Can you think of a case in which a scarce healthcare resource distribution according to QALYs would not be fair?
3. Should a society save as many lives as possible? Or, should a society save as many lives as it can cheaply or economically save?

Day Three: From the Philosophical to the Personal

Hill, Chris, "The Note," in *Willing to Listen, Wanting to Die*, edited by Helga Kuhse, (Penguin Books: Ringwood, Victoria, 1994), pp. 9–17.

An affecting open letter about one person's decision to end his life, given the quality of his life.

"Breathing Lessons: The Life and Work of Mark O'Brien," directed by Jessica Yu (1996).

An Oscar-winning documentary about a journalist and poet living with an iron lung. Here are some reviews:

https://en.wikipedia.org/wiki/Breathing_Lessons:_The_Life_and_Work_of_Mark_O%27Brien
https://www.imdb.com/title/tt0115753/
https://variety.com/1997/tv/reviews/breathing-lessons-the-life-and-work-of-mark-o-brien-1117341375/

Discussion Questions.

1. Why are the Hill essay and the Yu film affecting?
2. Compare the Hill essay and Yu film to the Harris and Alan essays. Did the reading of the Hill and the watching of "Breathing Lessons" change how you felt about the two scholarly essays?
3. Can you think of a case in which quantity of life would be more valuable than quality of life? Can life be valuable in itself, regardless of its quality? If so, what is the ground of this value? From where does this intrinsic value derive?
4. From what does the value of a human life derive, according to Jessica Yu, the director of "Breathing Lessons?" According to Hill?

Assessments

In-class activity

"Teaching Quality-adjusted Life Years (QALYs)," an assignment by Dr. Bonny Parkinson, Senior Research Fellow, Macquarie University Centre for the Health Economy (MUCHE), Australia. https://www.mq.edu.au/__data/assets/pdf_file/0011/787079/QALYs-exercise-instructions.pdf

High-impact learning

Ask students to interview someone more than 75 years of age about the best year of their life. Here are some possible interview questions:

- What was the best year of your life and why?
- If you had a time machine, would you return to that year? Why or why not?
- If you could remain at that age forever and relive that year again and again, would you choose to do so? Why or why not?
- When you were younger, did you think your quality of life would improve or decline in old age? Why?
- Throughout the last 10 years, has your quality of life improved or declined? How so?
- What do you think an outsider, i.e., anyone who is not you, would say about your quality of your life? Would they have a different perspective from you on the quality of your life?

Paper assignment

Critically assess *one* of Harris's objections to the QALY view. In a 4–6-page paper, begin by summarizing Harris's argument for the objection under consideration. Your summary should be a scholarly one that uses for support relevant quotations from the text. Next, present a back-and-forth conversation that starts with an objection to Harris's objection. Why do you think Harris's initial objection is unreasonable? Now, how would Harris defend himself against your response to his objection? And, finally, how would you respond to that defense?

References

Bates, R.A. 2011. *AI & SciFi: Teaching writing, history, technology, literature, and ethics.* Paper presented at 2011 ASEE Annual conference & exposition, Vancouver, BC. https://doi.org/10.18260/1-2—17433. https://peer.asee.org/17433

Buolamwini, J., and T. Gebru. 2018. Gender shades: Intersectional accuracy disparities in commercial gender classification. In Proceedings of the 1st conference on fairness, accountability, and transparency, proceedings of machine learning research 81, 77–91. http://proceedings.mlr.press/v81/buolamwini18a/buolamwini18a.pdf

Gillespie, M.A. 2010. *Science and the humanities, On the human.* The National Humanities Center. 18 Jan 2010. https://nationalhumanitiescenter.org/on-the-human/2010/01/science-and-the-humanities/. Accessed 4 Nov 2022.

Horgan, J. 2013. Why study humanities? What I tell engineering freshmen. *Scientific American*, 20 June 2013. https://blogs.scientificamerican.com/cross-check/why-study-humanities-what-i-tell-engineering-freshmen/. Accessed 4 Nov 2022.

Sleezer, R., and R.A. Bates. 2020. *Ethical development through the use of fiction in a project-based engineering program.* Paper presented at 2020 ASEE virtual annual conference content access, Virtual On line. https://doi.org/10.18260/1-2%2D%2D34586. https://peer.asee.org/34586

Summet, V.H., and R.A. Bates. 2020. *Science fiction as an entry point for ethical frameworks in engineering and computer science education.* Paper presented at 2020 ASEE Virtual Annual Conference Content Access, Virtual On line. https://doi.org/10.18260/1-2%2D%2D35180. https://peer.asee.org/35180

The CITI Program. 2023. *Research, ethics, and compliance training | CITI Program.* Accessed 28 Feb 2023. https://about.citiprogram.org/

Chapter 4
Ethics Education in Engineering and Technological Institutes in India: Challenges and Looking Forward

Reena Cheruvalath

Abstract In India, many engineering and technological institutes have incorporated ethics into their curriculum. Ethics or professional ethics is added as an elective course separated from the mandatory subjects in these institutes. There needs to be more than the existing codes of ethical conduct for the instructors and the learners at the higher education level to prepare them for ethical decision-making. Learners lack a foundation in the subject, as 'ethics' is not taught as a core subject in schools in India. Moral education offered at the school level in India focuses only on human values and cultural expectations, not the discipline of 'ethics.' The examples in the textbooks hardly invoke critical thinking skills among learners. This lack of ethics education at the lower level makes it hard for teachers to impart ethics at the higher level. In addition, there are sociocultural challenges in imbibing the concept of 'ethics' in the minds of students. Reflecting on one's decision-making process is necessary to improve one's ethical sensitivity. Integrating critical thinking in terms of analyzing issues from multiple perspectives and logical reasoning in terms of constructing inductive or deductive arguments while inferring ethical judgment as part of the teaching pedagogy can address these challenges. Instructional methods, which make the learners participate in the teaching-learning activities focusing on multiple perspectives, enable them to look at an issue from various perspectives. Improving meta-moral cognitive skills help the students to reflect on their strengths and weaknesses in their ethical decision-making.

Keywords Moral education · Critical thinking · Logical reasoning · Meta-moral cognitive skills · Ethical decision-making

R. Cheruvalath (✉)
Birla Institute of Technology and Science Pilani Goa, Zuari Nagar, Goa, India

E. Hildt et al. (eds.), *Building Inclusive Ethical Cultures in STEM*, The International Library of Ethics, Law and Technology 42,
https://doi.org/10.1007/978-3-031-51560-6_4

4.1 Introduction

As science and technology invite issues related to human safety, and conflicting values with its progressive and new inventions, deliberating about the ethical components of these issues becomes very significant. Due to the complex nature of ethics, it is often difficult to decide what is right and what is wrong. The primary reason for the complexity is the availability of many alternative solutions for an ethical issue. Choosing the best solution for the complicated ethical issues arising in a professional context is not easy for a layperson. This invites the need to develop ethical decision-making skills for future professionals. Most business organizations in India have developed ethical codes of conduct to ensure an ethical environment at the workplace. For example, the companies such as WIPRO, Tata steel, etc., have their ethical codes of conduct for their employees. WIPRO's ethical codes proclaim integrity, honesty, and accountability.

A few engineering and technological educational organizations in India provide educators similar ethical codes of conduct.[1] The University Grants Commission (UGC 2010), a statutory body set up by the Department of Higher Education, Ministry of Education, Government of India, prescribes a code of professional ethics for college and university teachers. Though these codes do help the teachers familiarize the do's and don'ts in the profession, they are insufficient to prepare a professional to be a better ethical decision-maker. These guidelines consist of teachers' responsibilities toward students, colleagues, administrators, non-teaching staff, guardians of students, and society. In summary, the codes mention the need to follow the rules and regulations and behave in accordance with societal and cultural expectations. It might not guide them to find an ethical decision when they are in an ethically conflicting situation, such as either extending deadline for assignment submission to support a student who has depressive symptoms or strictly following the institute guidelines to submit the assignment on time and fail the student if they miss the deadline. Consequently, they may fail to transfer ethical decision-making skills to their students. Furthermore, the guidelines do not contain any codes of ethics for students in higher education.

The All-India Council for Technical Education (AICTE), another statutory body under the Department of Higher Education, provides guidelines for students who pursue engineering and technological education to absorb universal human values. The council suggests group discussions about universal human values as part of their induction program. AICTE prescribes including professional ethics and value education in the third semester of the undergraduate program. The course mainly focuses on value education, self-improvement, and the need to build a good relationship with others. Supporting this, the New Education policy 2020 makes it mandatory to a holistic value-based education for technical education.

[1] For example, https://www.iitb.ac.in/sites/www.iitb.ac.in/files/Faculty-Handbook-2021.pdf) as well as for the students (https://iitpkd.ac.in/sites/default/files/2019-07/StudentCodeofConduct.pdf

The policy draft (Section- P4.5.14. accessed from https://www.education.gov.in/en/nep-new) suggests adding Panchatantra stories in ethics classes. The Panchatantra are a collection of animal fables written in the Sanskrit language and was translated from Sanskrit to Pehlevi in the sixth century BC (Buchthal 1941). It is translated into many languages at present. Although it is a fact that these stories inculcate values among students, these stories 'often have one-dimensional characters with rewarding good characters and punishing evil characters' (Khanna 2015) and thus, critical analysis becomes difficult. As per the policy, it is mandatory to have a 1-semester foundation course on universal human values. The prescribed textbooks for the course also explain the need to uphold various human values and maintain good relationships. However, there is no discussion about the discipline of 'ethics' in these textbooks. For example, the authors write: "Developing the right understanding about oneself and the rest of reality through self-exploration and realization of the inherent co-existence, harmony, and self-regulation at various levels in existence is seen to be the real basis for imbibing universal human values and ethical human conduct" (Gaur et al. 2010, 17). There is a tendency to integrate culture and religion to teach ethics. However, researchers point out that theory-based ethics learning is ineffective in the present Indian context (Gangopadhyay et al. 2020).

Indian higher education system is divided into mainly two categories: private and public. Under these categories, public institutions are fully funded by the government. The public institutions include Central universities, state universities, Indian Institute of Technologies (IITs), and the private institutions consist of Deemed universities (which are autonomous) and private universities (some are partially funded, and some are unaided). IITs and Birla Institute of Technology and Science Pilani (BITS Pilani), one of the Deemed universities, are some of the premier institutes in engineering and technology. The traditional public universities, such as central and state universities, and the private colleges and universities registered under these universities have full-fledged programs in the areas of Humanities and Social Sciences (HSS) courses.

On the other hand, in engineering and technological institutes, the major focus is on engineering and science programs, though a few offer full-fledged Bachelor of Arts (BA) and Masters of Arts (MA) programs in the areas of HSS. For example, the Indian Institute of Technology Hyderabad (Development Studies, Cultural Studies, etc.) and IIT Madras have programs. However, they do not offer BA/MA programs in Philosophy.

In India, many technological and engineering institutes have incorporated ethics into their curriculum. These courses aim to sensitize the students about the ethical dilemmas they might face in their professions. Ethics or professional ethics is offered as an elective course separated from the mandatory subjects in these institutes. In addition, a few science departments offer courses like bioethics, medical ethics, environmental ethics, etc. These courses are often offered as non-graded courses or as electives with a grade. As per the websites of some of the premier institutes, professional ethics and social responsibility are included as subjects in their syllabus for engineering and science undergraduates. The primary objectives are to "sensitize them in various aspects of life such as rights and wrongs, parental expectations, respecting diversity in society," etc. It is meant to develop "foundations and skills for lifelong learning such as, "understanding one's goals of life; its impact on one's relationship with the society & environment, time & stress management, 'Thinking' as a skill - enhance the ability to think." Students enrolled in these programs should get a satisfactory grade in these ethics courses, and the course needs to be repeated until they get an 'S' grade which indicates 'satisfactory performance'. The minimum credit required for the course is 1 or 2. In a few institutes, either professional ethics is a graded and mandatory course with three credits or one of the elective courses offered. The course content of Professional Ethics in BITS Pilani includes Three philosophical theories (Utilitarianism, Deontology, and Virtue ethics), Professional responsibilities, Ethical codes of conduct of various business organizations, ethical issues in Employer-Employee relationships, Whistle Blowing, Business ethics, Management Ethics, Corporate Social Responsibility, Working women and issues, Affirmative action and discrimination, Dismissal, Occupational Health and Safety, Engineering ethics, and Computer ethics. It is a 3-credit elective course.

Though there are benefits to not letter grading a course, this practice can often lead to poor learning attitudes and behaviors (McMorran et al. 2017). It is doubtful whether the students take the course seriously or not. Furthermore, not every student needs to choose the course when the course is offered as an elective. The teaching methods used are case discussions, workshops, debates, etc. (https://web.iitd.ac.in/~hegde/pesr/lab.htm). Some of the institutes conduct workshops as part of the course, which is interactive sessions with the students under the guidance of a resource person (For example, IITs and BITS Pilani). The author uses real-life examples from the profession to teach ethics with group discussions, as the real-life examples help the students to put themselves in the shoes of the professionals involved in the case. A case situation is given with two alternatives: either action is ethically right or ethically wrong. Students are asked to provide reasons for their ethical judgments, and every group is asked to find out what the problems are with the reasons provided by the other group members. This also makes the students think from multiple perspectives and find out the weaknesses in their own judgments. Nonetheless, imparting ethics education in India's science, engineering, and technological institutes is challenging.

4.2 Engineering and Technology and Ethics Education: Challenges

The challenges to imparting ethics education include logistical factors such as socio-cultural, the availability of qualified teachers, curriculum-related issues, and lack of foundational knowledge in ethics. Since most teachers are not trained to make ethical decisions, they cannot transfer the same to their students. Ethical decision-making is a skill, and to make the right ethical decision, a professional must collect all the facts and be able to analyze them critically. A professional needs critical thinking and logical reasoning skills to collect all the facts related to the issues, critically analyze them, and choose the best solutions, avoiding biases. Critical thinking skills consist of 'inquisitiveness, open-mindedness, seeking and offering clear reasons, being alert for alternatives, analyzing the arguments, deducing and judging deductions, making, and judging inductive inferences and arguments' (Ennis 2015). Logical reasoning in this context means the skills to construct arguments using inductive or deductive logical reasoning. [Examples for deductive and inductive arguments are given at the end of this chapter].

Discriminating biased information from the unbiased is not easy when one is not trained to do so. For instance, if an instructor supports the biased information that the physicians who belong to the socially disadvantaged group are less effective compared to their counterparts who belong to the advantaged groups in India, students will not be able to realize the limitations in the above judgment. An instructor should be able to make the students think that the quality of a professional is not assessed based on group membership but on educational qualifications, skills, and experience. As different 'unequal' social strata in terms of 'caste,' (Gupta 2005), which is a sub-sect of religion that exists in India, the teachers need to make the students reflect on the logic behind constructing such biased judgments about professionals. Lower and higher education institutions play an essential role in this regard as they can train students.

Sociocultural factors play a role in imbibing the concept of 'ethics' in students' minds. A study among children showed that socioeconomic status influences their moral judgment. The lower the socioeconomic status, poor the skill in making moral judgments (Tripathi and Misra 1979). People might give different reasons for their ethical judgment due to their cultural diversity in the Indian context (Shweder 1990). Thus, setting up a common criterion based on religious beliefs or values for assessing an action, whether ethically right or wrong, does not work. In India, there is a trend to treat religion, spirituality, and ethics as related subjects (Radhakrishnan 1923; Raju 1982), though there are criticisms of this view (Mohapatra 2019). Research findings support that in India, many people believe that religion plays a major role in promoting ethical conduct in different domains of life (https://www.gov.uk/research-for-development-outputs/religions-ethics-and-attitudes-towards-corruption-in-india)*. The Indian researchers propagate the need for laying foundations using the socio-cultural and religious framework for the development of individual and social moral principles with Vedic principles (Srivastava et al. 2013). Socioeconomic and cultural factors of an individual affect their evaluation of situations with ethical dilemmas and, thereby, their moral judgment (Arutyunova et al. 2016). That is to say, people construct moral decisions based on socio-conventional rules. For instance, as per the social convention, in India, a mistake made by a female is less tolerable than a similar mistake made by her male counterpart. Some people believe these rules are less or entirely breakable, and their ethical decisions are based on these beliefs (Caravita et al. 2012). As a collective society (Mehrishi 2015), an individual's role in deciding what is right or wrong in Indian society is undermined. For example, a collective society emphasizes the interdependence between people and collectivistic values (Arutyunova et al. 2016); thus, people make decisions in conformity with the group or social norms. Often socio-religious and cultural factors influence our reasoning and information processing. Studies support that religiosity and reasoning are negatively related (Daws and Hampshire 2017), and people with high extrinsic religious orientation often have poor critical thinking skills (Mathew 2008). The association between religious or spiritual aspects and the concept of morality can also be seen in school textbooks. Related issues will be discussed in the next section. The need is to improve the learners' ethical sensitivity, which is presently lacking in the existing curriculum. Researchers believe that since care and interdependence are part of the Indian culture, teaching ethics education focusing on socio-cultural aspects helps to improve ethical sensitivity (Gielen 2016).

In many Indian universities, Departments of HSS or the Department of Philosophy offer ethics courses. In engineering and technological institutes in India, the HSS department co-exists with the science departments, such as physics, chemistry, biology, mathematics, and engineering departments, such as mechanical, chemical, electronics engineering, and computer science. This coexistence allows students and instructors to obtain knowledge from different disciplines and fields and aims to promote interdisciplinary research. However, only a few professionals who belong to these disciplines approach issues from perspectives other than their disciplines. A learner with a science or engineering background needs to gain the skills to think from the perspectives of an engineer or social scientist or vice versa. To manage interstitiality, there should be a connection between students and teaching structures (Lindvig et al. 2019). One solution to avoid this issue might be that faculty members from the engineering or science discipline can be invited to explain the engineering part in an ethics class. Both faculty members can explain ethical issues that arise while designing or manufacturing in a group discussion mode. Though many departments coexist, more than academic interaction between faculty members and students is required. There is a privation of recognition to understand the need for this kind of interaction (Subrahmanian et al. 2018). Furthermore, engineering or technological institute students expect to learn a pure disciplinary subject, i.e., how to build a rocket or robot, etc., and not necessarily humanities and social science subjects.

Second, as 'ethics' is offered as a separate subject or elective, the course objectives of STEM ethics are often not met. Students study it either to satisfy a graduation requirement or for a better grade. Getting a good grade can never be a criterion for assessing the ethical skills of a person. Receiving an A or B grade in ethics does not necessarily indicate that a student is ethical. They might have learned various ethical theories and the basic concepts of ethics. However, they might not have learned how to apply the knowledge in real-life situations or might not have learned the skills to find out the limitations in their ethical judgment, which help them be ethical. In addition, while teaching core science and engineering subjects, teachers do not discuss the ethical implications of it. The author has interviewed a few faculty colleagues who teach engineering subjects to understand whether they discuss ethical implications in their regular classes related to the technology they teach. For instance, when an instructor discusses Computer-Aided Design (CAD) technologies, they might not discuss the harmful use of these technologies. Only a very few discuss it, and many of them are not aware of the need for discussing ethics in their classrooms. A respondent (X) states that "we do not get time to finish our syllabus, and thus, explaining the ethical implications is not possible." Another respondent (Y) states, "I never thought about it before, as we are not trained in the subject, but I can discuss the ethical implications next time."

Z states that "I do not know, and I never discussed ethics in my class." As a result, students lack the skills to identify the ethical issues which arise in their specific disciplinary engineering subjects. For example, while designing a machine, some questions arise: whether it is user-friendly, culturally friendly, and gender-neutral, whether it ensures the occupational safety of the user, what ethical considerations need to be checked, and so on. Discussing a few cases in the classrooms is not sufficient to understand the depth of the ethical implications of technology use. As per the available syllabus, learners are mainly concerned about its cost-effectiveness rather than safety.

There are discussions on 'who should teach ethics?' Johnson (1994) and Fiesler et al. (2020) provide two arguments: (a) philosophers and social scientists who are trained in ethics are experts in the area and thus, they should teach and (b) teaching ethics by scientists themselves help the students to understand that science has social implications. Incorporating social science and humanities with engineering subjects helps the students understand these disciplines' merits as reflections on society and human conditions (Kaur 2005). Involving instructors from various disciplines is another solution (Grosz et al. 2019). For instance, some universities involve faculty members from science and engineering and HSS disciplines while designing an interdisciplinary course, such as engineering and computer ethics. A course in Engineering ethics helps engineering students identify the various ethical issues while designing, developing, testing, or manufacturing a product. For instance, whether the design of an engineering product respects the culture of consumers or is it gender neutral? Is there any harm in using the product? How to solve those issues? and so on. A course in computer ethics, for instance, encourage students to check when they develop software whether it accommodates disabled person or people who are old and so on. However, often these courses are taught by only one instructor due to a dearth of interdisciplinary activities among the faculty members from the different departments or due to their busy work schedules or lack of faculty members.

Encouraging interdisciplinary research projects will help promote interdisciplinary teaching. However, the funding provided by the various agencies in India needs to be increased, and there is a lack of funding from external agencies (Sonetti et al. 2020). For instance, the major funding agencies in India for science and technology are the Atomic Energy Regulatory Board (AERB), Department of Biotechnology (DBT), Aeronautics Research and Development Board (ARDB), Defence Research & Development Organisation (DRDO), Council of Scientific & Industrial Research (CSIR), Department of Science & Technology (DST), etc. However, these agencies providing funding in the area of social science subjects are very minimal. Though these agencies add guidelines to show how the proposed project is useful for society, the focus is more on technology than the latter. The major funding agencies in the areas of Humanities and Social Sciences are the Indian Council of Social Science Research (ICSSR), the University Grants Commission (UGC), and the National Council of Educational Research and Training (NCERT), etc.

Since in engineering and technological institutes, HSS subjects are offered as electives; students still need to receive a degree in these subjects. Accordingly, these subjects are likely considered as having 'lower status,' and so too teachers who teach them. The faculty members who teach humanities and social sciences in engineering and technological institutes feel powerless or ignored (Ravinder 2005), leading to less job satisfaction. The teachers complain that only a few students are enthusiastic about learning the subjects at a deeper level, like those who study these subjects as a core discipline and regularly participate in classroom activities. When students are interested in the subjects, they participate in the teaching-learning process (Hidi and Harackiewicz 2000). Students' interest galvanizes their learning experience (Harackiewicz et al. 2016), and it is a powerful motivational process (Renninger and Hidi 2016). For instance, the author has noticed that students are interested in discussing any social issue that negatively affects them, such as affirmative action. They also participate in discussions if the issue questions the established social norms, say gender neutrality, in the context of India. Some students give the reason for enjoying the HSS subjects is that it is easy to receive good grades in these subjects compared to their core disciplinary subjects. Subsequently, instructors with a Ph.D. in HSS subjects prefer to teach in state or central universities where philosophy and ethics are offered as core subjects, unlike in engineering and technological institutes. As a result, students in the former institutes are interested in the subject and show interest in knowing it in depth when compared to students who study the course as an elective.

Many state and central universities in India offer ethics as part of philosophy and applied ethics, subjects that are taught at the graduate and postgraduate levels. These syllabi include engineering ethics, computer ethics, bioethics, medical ethics, environmental ethics, professional ethics, etc. However, learners in these courses rarely get exposure to engineers, scientists, or medical professionals to understand their perspectives. Moreover, the focus is mainly philosophical, and the students need to gain advanced knowledge of science or engineering subjects. When they become teachers, they may lack the skills to apply these philosophical principles to science and engineering disciplines effectively. The teachers need the skills to make their students reflect on their ethical judgment rather than byhearting the principles. A few Ph.D. scholars who prefer to work in these specific areas (for example, in engineering ethics) contribute to teaching these subjects. However, they may prefer to teach in an institute where Philosophy and ethics are offered as core subjects. Likewise, though HSS departments exist in many engineering and technological institutes, only a few institutes offer graduate or postgraduate programs in these core subjects. In addition, only a few engineering and technological students pursue higher education in these fields. This also leads to a need for more qualified faculty members to teach ethics in technical institutes. At times, the subject of 'ethics' is taught by engineering or science faculty members with no proper orientation in philosophy or ethics. This leads to the class discussions focusing on the contents of the major engineering or science disciplines rather than ethics. The learners are familiar with cases discussed in the text or reference books. However, they do not learn to identify ethical components in issues or situations. This makes it difficult for the learners to apply knowledge from the classroom to professional situations. After graduation, a professional often faces new situations not necessarily discussed in the classrooms and, thus, fails to address and solve them effectively.

Besides these issues, teachers face difficulty imparting ethics to engineering and science students in the classroom (Cheruvalath 2015; Martin et al. 2021). One of the primary reasons is the students' lack of foundational knowledge of ethics. They come to the college with a preset notion of 'ethics' or 'morality' which is deeply rooted and confirmed by their culture. Generally, the word ethics is understood by the learners as do not cheat, do not copy, etc. In an introduction class, when the instructor asks what their understanding of the term is, a few state that 'ethics is concerned with moral principles, and right things prescribed by the society or religion.' They also come with the mindset that they are already ethical and have nothing new to learn. As the students in higher education lack foundational knowledge in the area of ethics, it poses a significant challenge for the teachers to convince them that ethics is a discipline and essential. In India, many parents believe that a child's future lies in a medical or engineering degree (Gupta 2020; Wilson 2011; Nambissan 2009). They train their children from 11 to 12 for engineering entrance examinations and are sent to the coaching centers very early. Since 'ethics' is not part of these entrance examinations, the subject does not get proper attention at this stage either.

4.3 Ethics and School Curriculum in India

What we learn in the early years of education often provides the foundation for what we learn at a higher educational level. Children learn ideas of right and wrong from their family members, schools, peers, culture and religion, etc. Accordingly, everyone follows their criteria to assess what is right and wrong. From first to fifth standards in India, children mainly learn English, Hindi, or a regional language and mathematics. Some syllabi include environmental science, computer science, general knowledge, etc., as additional subjects. Though 'ethics' is not taught as a separate subject, there are schools that provide moral education, usually titled 'moral science.' The textbook content focuses on value education. The book contains traces of values of friendship, courtesy, etc. A few examples are explained in the following paragraphs.

A moral science course syllabus often focuses on social virtues and becoming a good citizen. In general, Moral Philosophy is considered a Moral science that consists of human conduct, motives related to that conduct, and the aims of which it ought to be directed (Blakey 2006). Indian moral science education is based on the cultivation of moral and spiritual values (Mefodeva et al. 2016). For example, one syllabus includes the importance of life, love, doing righteous action without worrying about the results (the theme comes from the Hindu religious text- Bhagavat Gita), compassion towards patients or people with illness, respecting parents and teachers, having a cordial relationship with neighbors, the relationship between an individual and the society, to be in unison with the world, development, importance of employment, importance of having a cordial relationship between teachers and students, the art of communication, need for following virtues, and the relationship between the development of science and the living conditions of human beings. The concepts are narrated through stories that are imaginary and in the form of advice. These stories are based on various religious texts, and the pedagogy used to teach these contents does not allow learners to reflect on their thoughts or beliefs.

For instance, a story used in a moral science class (seventh standard Text book in one of the states in India) goes like this:

> There was a saint who was resting under a tree. Two birds in the tree disturbed his sleep, and he cursed the birds, and the birds were burned to death. The saint was very proud of his power. Then he visited a house to collect his alms. The lady in the house said, "Please wait for a while until I give food to my husband and children." The saint got angry and asked, "Do you know who I am?" The lady replied, "Do you have any plan to char me to death as you did to the birds with your power? He was surprised and became respectful.

Small children develop or confirm their belief that a saint can be powerful in the above manner, though it is factually not true. Stories like this contain unscientific information, so no critical analysis of the contents is likely. Content analysis shows that the contents of the textbooks are socio-cultural, religious, and spiritual, not based on the philosophical analysis of rights or wrongs. In addition, a few school teachers mentioned that in ninth and tenth standards, the class hour for moral science is used for other subjects such as science or mathematics as they consider the latter to be more important than the former. The point here is that though moral science exists in schools, it might not be taught with vigor.

Some schools impart value education through a program called the 'Awakened Citizen Program' (https://kvsangathan.nic.in/awakened-citizen-programme), offered for 3 years at the secondary level. This program focuses on the enhancement of the inner strength of students. The course's primary objectives in the first year are: to enable the students to live in harmony, seek perfection by identifying their excellence, uphold the sacred traditions and culture of India and democracy, and develop solutions to complex social challenges. In the second year, the course focuses on developing personality, contributing to the world, improving self-confidence, facing challenges, courageously acting with honesty and integrity, and considering the impact of one's actions. In the third year, students learn to explore their potential in the classroom with friends, family, nation, and employment. Apart from this, how to make ethical choices, help others, concentrate, and understand other human beings as equals irrespective of gender, caste, religion, etc., and positive engagement is also taught. There are practical sessions to apply and inculcate these above concepts. These sessions provide motivation to improve oneself.

Similarly, traces of moral education can be seen as part of a few stories in English textbooks at various levels in India. Stories in literature are often used to inculcate moral education in children and youth (Engelen et al. 2018; Willows 2017; Fraustino 2014). A few school syllabi provide a list of unethical practices in sports as part of physical education. When we analyze the curriculum of higher standards related to ethics or moral education, only a few traces can be seen in English textbooks or the social sciences.

In summary, there needs to be an in-depth discussion of ethics at the school level. There needs to be more discussion happening at the pre-university level that arouses students' curiosity about rights and wrongs in real-life scenarios. The training the learners get in their school does not prepare them for identifying ethical components in the issues they face, as they lack clarity in what 'ethics' means. Teachers and parents believe that children can be taught ethics when they become 15 years old or above so that they can understand the concept well. However, by that time, they might have developed the notion of rights and wrongs, and it is difficult to change their moral standards. This indicates the need for rigorous training at the school level to develop ethical sensitivity among students.

4.4 The Way Forward: Improving Meta-moral Cognitive Skills for Better Ethical Decision Making

The author uses case study discussions and debates as a teaching method in the professional ethics class for engineering and science students. One or two cases are given for each topic, and students are asked to discuss them in small groups. After they complete the discussion among their group members, students are asked to share their views, and other group members are invited to find out their weaknesses. Apart from that, documents that contain Indian laws related to topics such as whistle-blowing or corporate social responsibility are given, and they are asked to find out the flaws in these laws. As teaching ethics to adults is challenging, one way to improve ethical decision-making is by focusing on the ethical decision-making *processes*. There are various ways to improve ethical decision-making processes. Any decision-making process contains many stages. To make a better decision, it is important to identify the strength and weaknesses of one's moral judgment, called meta-moral cognitive skills (Cheruvalath 2015). Meta-moral cognitive skills are a combination of critical thinking and logical reasoning skills. In the case of an ethical decision-making process, the following steps are involved:

- To identify the ethical component involved in the issue.
- Formulate the ethical issue clearly
- Choosing the option: whether an action/behavior is ethically right or wrong.
- Provide reasons for choosing the specific option.
- Identifying criticisms for those reasons
- Identifying counter criticisms.
- Constructing an argument using logical reasoning: Deductive or Inductive

The above steps help a learner identify whether the ethical decision they have taken is logical. Students need basic knowledge about philosophical theories such as utilitarianism, deontology, virtue ethics, etc. These theories provide guidance for making an ethical decision. The learner examines the consequences, assesses the action, and whether any values conflict. A learner needs ethical decision-making skills to apply these theories and make better decisions. Meta-moral cognitive skills assist in apprehending the strengths and weaknesses of one's ethical decision-making skills. A better ethical decision depends on a learner's skills to identify reasons, criticisms, and counter-criticisms for one's own decisions. Providing reasons, criticisms, and counter-criticisms makes the decision-maker reflect on one's decision and improves their critical thinking skills. To do so, they need all the information related to the issue. While analyzing, if a learner misses any vital information, the ethical solution will be a failure.

The author provides an exercise that can be done during the class.

For example:

A college student 'X' sends emails to other students from 'Y's account without the permission of Y, just for fun.' Is that right? Why?

In this case, there are two alternatives: Yes, it is right, and No, it is wrong. A learner can choose one option and provide reasons.

The reasons for the options can be:

It is not correct because:

(a) Accessing someone else's account without the account holder's permission is like accessing someone else's physical property without the owner's permission.
(b) The account holder's privacy is at stake.
(c) Entering someone else's property without the owner's permission, even for fun, is ethically and legally wrong.

The criticisms for the above reasons can be:

(a) It is not like accessing someone else's physical property, as it is not tangible.
(b) The hacker is not taking any private data from the other's account.
(c) X is not entering Y's property as X only uses an email account.

The counter criticisms can be:

(a) An email account is one's property; thus, accessing it without permission is like accessing a physical property though it is not tangible.
(b) Private data will be available to X when accessing Y's email account; it does not matter whether X uses it.
(c) Y's email account is Y's property.

Summaries of applying theories in the given case are shown below:

1. Utilitarianism:

   What are the consequences of hacking someone else's account?

(a) Good result: X has fun
(b) Bad result: Y will be sad or angry. Y's private email contents are exposed. Hacking others' accounts is illegal, so X will be punished if caught.

If we weigh the result, bad consequences are more than good consequences, so X is not right.

2. Deontology:

What is X's intention: to have fun
What is X's duty as a student: To study, to respect others' privacy
What is X's maxim: I hack another student's email whenever I want to have fun.

Is it universalizable: No, if everyone starts hacking each other's accounts for fun, it will create chaos. Y gets upset when his account is hacked, and Y can also hack X's account. There is an inconsistency.

Thus, X's action is not right.

3. Applying logical reasoning, students are asked to construct arguments for their own ethical decision or judgment using deduction and induction.

For example,

(a) Deductive argument (Valid. It follows the form of one of the rules of inference-Modus Ponens): There are nine Rules of inference which are elementary valid argument forms (Copy 2010). Modus Ponens is of the form:

$p \rightarrow q$
p
$\therefore q$

Accessing others' accounts for fun without permission cannot be justified.
X has accessed Y's account for fun without permission.
Therefore, X's action cannot be justified.

(b) Inductive argument (Strong): In an Inductive argument, the conclusion probably follows from the premises. If the probability is high, it is called a strong inductive and the if the likelihood is low, it is called a weak inductive argument (Hurley 2012).

Many instances show that hackers are punished. Thus, X will probably be punished for unauthorized access.

Every learner in the class can do the above exercise individually, or it can be done as a group. When everyone or every group discusses their outcomes in the class, it helps other students to absorb how to approach an issue from multiple perspectives.

The above examples show different approaches to solving an ethical dilemma. Learner trained with such standards can improve their meta-moral cognitive skills and, thereby, their ethical decision-making skills. The above methods can be adopted while teaching ethics to engineering and technology students to strengthen the ethical decision-making processes.

4.5 Conclusion

In conclusion, there is a need to integrate ethics with the school curriculum and with every significant disciplinary subject taught in engineering and technological institutes. Teaching the subject collaboratively with philosophers, scientists and engineers helps meet the course's objectives. Incorporating the subject as only electives does not allow the learners to understand the ethical implications of various engineering and technological issues. To learn to make better ethical decisions, an individual needs critical thinking and logical reasoning skills. In addition, the instructional methods introduced above, which make the learners participate in the teaching-learning activities focusing on multiple perspectives, improve their critical thinking skills. Enabling professionals to look at an issue from various perspectives prepares them to make better decisions. Using technology casually without considering the consequences on society and the environment is a disaster. Understanding the ethical implications of one's decisions and their effects on society can ignite the growth and development of society and, thereby, a nation in a sustainable manner. India's premier engineering and technological institutes that mold future leaders should educate them on 'ethics' appropriately for a sustainable world.

References

Arutyunova, K.R., Y.I. Alexandrov, and M.D. Hauser. 2016. Sociocultural influences on moral judgments: East-west, male-female, and young–old. *Frontiers in Psychology* 7: 1334. https://doi.org/10.3389/fpsyg.2016.01334.

Blakey, R. 2006. *History of moral science*. Edinburgh: Kessinger Publishing.

Buchthal, H. 1941. Indian fables in Islamic art. *Journal of the Royal Asiatic Society* 73 (4): 317–324. https://doi.org/10.1017/S0035869X00097604.

Caravita, S.C.S., S. Giardino, L. Lenzi, M. Salvaterra, and A. Antonietti. 2012. Socio-economic factors related to moral reasoning in childhood and adolescence: The missing link between brain and behavior. *Frontiers in Human Neuroscience* 6: 262. https://doi.org/10.3389/fnhum.2012.00262.

Cheruvalath, R. 2015. Teaching ethics to engineering students in India: Issues and challenges. In *Contemporary ethical issues in engineering*, ed. S.S. Sethy, 121–132. Hershey: IGI Global.

Copy, I.M. 2010. *Symbolic logic*. New Delhi: PHI Learning.

Daws, R.E., and A. Hampshire. 2017. The negative relationship between reasoning and religiosity is underpinned by a bias for intuitive responses, specifically when intuition and logic are in conflict. *Frontiers in Psychology* 8: 2191. https://doi.org/10.3389/fpsyg.2017.02191.

Engelen, B., A. Thomas, A. Archer, and N.V.D. Ven. 2018. Exemplars and nudges: Combining two strategies for moral education. *Journal of Moral Education* 47 (3): 346–365. https://doi.org/10.1080/03057240.2017.1396966.

Ennis, R.H. 2015. Critical thinking: A streamlined conception. In: *The Palgrave handbook of critical thinking in higher education*, ed. M. Davies, and R. Barnett. New York: Palgrave Macmillan. https://doi.org/10.1057/9781137378057_2.

Fiesler, C., N. Garrett, and N. Beard. 2020. What do we teach when we teach tech ethics?: A syllabi analysis. In *SIGCSE '20: proceedings of the 51st ACM technical symposium on computer science education*, 289–295. https://doi.org/10.1145/3328778.3366825.

Fraustino, R.L. 2014. The rights and wrongs of anthropomorphism in picture books. In *Ethics and children's literature*, ed. L.R. Fraustino. London: Routledge.

Gangopadhyay, K., R.T. Jacob, and R.S.S. Jayanth. 2020. How effective is theory-based learning of ethics in India? *IIM Kozhikode Society & Management Review* 9 (1): 55–61. https://doi.org/10.1177/2277975219889778.

Gaur, R.R., Sangal, R., and Bagaria, G.P.A. 2010. *Foundation Course in Human Values and Professional Ethics*, 17. Excel books: New Delhi.

Gielen, J. 2016. Education in care ethics: A way to increase palliative care awareness in India. *International Journal of Ethics Education* 1: 15–24. https://doi.org/10.1007/s40889-015-0003-6.

Grosz, B. J., D.G. Grant, K. Vredenburgh, J. Behrends, L. Hu, A. Simmons, and J. Waldo. 2019. Embedded ethics: Integrating ethics broadly across computer science education. *Communications of the ACM* 62(8): 54–61. arXiv:1808.05686 http://arxiv.org/abs/1808.05686

Gupta, D. 2005. Caste and politics: Identity over system. *Annual Review of Anthropology* 34: 409–427. https://doi.org/10.1146/annurev.anthro.34.081804.120649.

Gupta, A. 2020. Heterogeneous middle-class and disparate educational advantage: Parental investment in their children's schooling in Dehradun, India. *British Journal of Sociology of Education* 41 (1): 48–63. https://doi.org/10.1080/01425692.2019.1660142.

Harackiewicz, J.M., J.L. Smith, and S.J. Priniski. 2016. Interest matters: The importance of promoting interest in education. *Policy Insights From the Behavioral and Brain Sciences* 3 (2): 220–227. https://doi.org/10.1177/2372732216655542.

Hidi, S., and J.M. Harackiewicz. 2000. Motivating the academically unmotivated: A critical issue for the 21st century. *Review of Educational Research* 79: 151–179. https://doi.org/10.1007/s11948-021-00333-610.2307/1170660.

Hurley, P.J. 2012. *A concise introduction to logic*. Belmont: Wadsworth Publishers.

Johnson, D. 1994. Who should teach computer ethics and computers & society? *Computers and Society* 14 (5): 6–13. https://doi.org/10.1016/0268-4012(94)90078-7.

Kaur, R. 2005. Locating the humanities and the social sciences in institutes of technology. *Sociological Bulletin* 54 (3): 412–427. http://www.jstor.org/stable/23620616.

Khanna, M. 2015. The Panchatantra stories – An embodiment of management wisdom. *IIBM's Journal of Management Research* 1 (2): 20–24.

Lindvig, K., C. Lyall, and L.R. Meagher. 2019. Creating interdisciplinary education within monodisciplinary structures: The art of managing interstitiality. *Studies in Higher Education* 44 (2): 347–360. https://doi.org/10.1080/03075079.2017.1365358.

Martin, D.A., E. Conlon, and B.A. Bowe. 2021. Multi-level review of engineering ethics education: Towards a socio-technical orientation of engineering education for ethics. *Science and Engineering Ethics* 27: 60. https://doi.org/10.1007/s11948-021-00333-6.

Matthew, K. 2008. *The impact of religious schema on critical thinking skills*. All graduate theses and dissertations. 10. https://doi.org/10.26076/e355-8ca7.

McMorran, C., K. Ragupathi, and S. Luo. 2017. Assessment and learning without grades? Motivations and concerns with implementing gradeless learning in higher education. *Assessment & Evaluation in Higher Education* 42 (3): 361–377. https://doi.org/10.1080/02602938.2015.1114584.

Mefodeva, M.A., A.V. Fakhrutdinova, and R.R. Zakirova. 2016. Moral education in Russia and India: A comparative analysis. *The Social Sciences* 11 (15): 3765–3769.

Mehrishi, K. 2015. *The India collective: What India is really all about?* KW Publishers.

Mohapatra, P.K. 2019. *An applied perspective on Indian ethics*. Singapore: Springer.

Nambissan, G. 2009. The Indian middle classes and educational advantage: Family strategies and practices. In *The Routledge international handbook of the sociology of education*, ed. M.W. Apple, S.J. Ball, and L.A. Gandin, 285–295. London: Routledge & Francis.
Radhakrishnan, S. 1923. *Indian philosophy* (I): 34–41. London: George Allen and Unwin.
Raju, P.T. 1982. *The Western and the Indian philosophical traditions, Indian philosophy: Past and future*, 88. New Delhi: Motilal Banarsidass.
Ravinder, K. 2005. Locating the humanities and the social sciences in institutes of technology. *Sociological Bulletin* 54 (3): 412–427. https://doi.org/10.1177/0038022920050306.
Renninger, K.A., and S. Hidi. 2016. *The power of interest for motivation and engagement*, 2016. New York: Routledge.
Shweder, R. 1990. Cultural psychology, what is it? In *Cultural psychology: Essays on comparative human development*, ed. J.W. Stigler, R. Shweder, and G. Herdt. Cambridge: Cambridge University Press.
Sonetti, G., O. Arrobbio, P. Lombardi, I.M Lami, I.M, & S. Monaci, 2020. "Only social scientists laughed" Reflections on social sciences and humanities integration in European energy projects. *Energy Research & Social Science* 61: 101342. https://doi.org/10.1016/j.erss.2019.101342.
Srivastava, C., V. Dhingra, A. Bhardwaj, and A. Srivastava. 2013, January. Morality and moral development: Traditional Hindu concepts. *Indian Journal of Psychiatry* 55 (Suppl 2): S283–S287. https://doi.org/10.4103/0019-5545.105552.
Subrahmanian, E., T. Odumosu, and J.Y. Tsao. 2018. *Engineering a better future interplay between engineering, social sciences, and innovation*. Cham: Springer.
Tripathi, D.K.M., and G. Misra. 1979. Development of moral judgement in Indian children. *Psychologia* 22: 164–169.
UGC. 2010. https://www.ugc.gov.in/oldpdf/pub/report/5.pdf
Willows, A.M. 2017. Stories and the development of virtue. *Ethics and Education* 12 (3): 337–350. https://doi.org/10.1080/17449642.2017.1356664.
Wilson, C. 2011. The social transformation of the medical profession in urban Kerala: Doctors, social mobility, and the middle classes. In *Being middle-class in India: A way of life*, ed. H. Donner, 139–161. New York: Routledge.
Workshop held on 28th–29th January 2010 at the University of Hyderabad, Andhra Pradesh, India. 5*.

Chapter 5
Embedding Moral Reasoning and Teamwork Training in Computer Science and Electrical Engineering

Raquel Diaz-Sprague and Alan P. Sprague

Abstract During fall 2018 and fall 2019, we designed a highly interactive 4-class period ethics and teamwork minimodule and embedded it in an upper-level design course in Electrical & Computer Engineering at the University of Alabama at Birmingham (UAB). In the first two class periods, we covered ethical principles, moral theories, and the pillars of morality: honesty, fairness, and reciprocity. In class, students watched segments of video lectures by Harvard's Michael Sandel and Emory University's Frans De Waal. Small group discussions followed the video lectures. Student teams led teamwork demonstrations via in-class games during the last two class periods. Strong student engagement was observed.

Previously, in the fall of 2016 and spring of 2017, we instigated prosocial software design and development in the UAB Computer Science capstone course, which requires a team project. Three out of seven student teams developed prosocial app concepts. Additionally, from 2019 to 2021 we led the UAB Ethics in Action: Art or App Design Challenge, which encourages college students nationwide to develop prosocial app concepts or artworks for class projects. In the fall of 2021, students at three different institutions submitted award-winning entries. We urge faculty leadership to encourage a prosocial approach to software design. Research on developing moral reasoning algorithms is needed.

The authors' interdisciplinary research on detecting cyberbullying and nastiness in social media and their history of collaboration in university outreach projects and Computer Science and Engineering education and research are also described.

Keywords Ethics and teamwork · Interactive minimodule · Moral reasoning · Prosocial app · Ethics in action

R. Diaz-Sprague (✉) · A. P. Sprague
University of Alabama at Birmingham, Birmingham, AL, USA
e-mail: diazspra@uab.edu; sprague@uab.edu

E. Hildt et al. (eds.), *Building Inclusive Ethical Cultures in STEM*, The International Library of Ethics, Law and Technology 42,
https://doi.org/10.1007/978-3-031-51560-6_5

5.1 Introduction

Albert Einstein once said that "We should take care not make the intellect our god – it has powerful muscles, but no personality" (BrainyQuote n.d.). Science and technology tell us what is, but not "what should be" to promote societal benefit and foster human flourishing.

In 1961, President Kennedy set the goal of a safe moon landing by the end of the decade by saying:

> We set sail on this new sea because there is new knowledge to be gained and new rights to be won, and they must be won and used for the progress of all people. For space science, like nuclear science and technology, has no conscience of its own. Whether it will become a force for good or ill depends on man (Azquotes n.d.).

If science and technology becoming a force for good or ill depends on man, moral STEM leadership, cultivating ethical cultures, and embedding moral reasoning and a prosocial approach in STEM education could lead students to think about, discussing, designing, and building "what should be."

Ethics is, therefore, an essential component of STEM education. However, while most engineering educators agree on the importance of incorporating ethics into the curriculum, there needs to be a consensus on how ethics should be taught and who should teach it. Some alternatives are standalone courses, discussions or lectures about professional responsibility, modules on professional responsibility, or case studies. Case studies are a popular tool for teaching engineering ethics (Fiesler et al. 2020; Skirpan et al. 2018).

At Harvard University, Barbara Grosz has led the creation of Embedded EthiCS – the most comprehensive approach for integrating ethics into the Computer Science curriculum (Grosz et al. 2019). Embedded EthiCS "makes ethical reasoning integral to Harvard's computer science education with a distributed pedagogy that introduces ethics directly into courses across that curriculum." It works by embedding philosophers - postdocs or advanced graduate students – into courses to teach modules that explore ethical issues raised by course materials (Harvard University n.d.).

ABET, the Accreditation Board for Engineering and Technology, requires that students at accredited schools of Engineering or Computer Science show competence in ethics and teamwork (ABET 2023). ABET accreditation requirements describe training results by accredited schools rather than prescribe training. The ABET guidance was interpreted at the authors' previous institution as ensuring that ethics training and teamwork skills be listed as learning objectives in at least one course. However, implementing course objectives was discretionary, up to the individual professor teaching the course. Faculty are often pressed for time to cover required course materials and may feel unprepared or unqualified to teach ethics and/or teamwork as they may not have much training on those topics. Time constraints and the absence of clear guidelines on how such issues as ethics and teamwork should be covered can make the course objectives on those topics discretionary at best and perfunctory at worst. Ethics training for the Capstone course had consisted of as much as 1/3 of the course or as little as a single lecture in the course syllabus. Teamwork training was generally not included in the syllabus. Absent specific guidance, students tend to apply their work habits and moral code – virtue ethics – or lack thereof, to the team project, usually a software design project required by the course. In 2016–2017, the authors observed several teamwork failures and some instances of student misconduct in the Capstone course. This served as motivation to seek funding to develop a minimodule covering ethics and teamwork training, not as theoretical or esoteric topics, but as constructive information with practical applications essential for collaborative behavior needed for completing a team project.

5.2 History of the Authors' Collaboration

We are a husband-and-wife team of multidisciplinary backgrounds who have collaborated extensively, primarily in university outreach projects. Raquel led the Women in Science Day Program at The Ohio State University (OSU) for many years. Alan, our son, Kevin Sprague, and daughter, Susan Sprague, were critical partners in the logistics and communications for the acclaimed program, which each year brought hundreds of gifted and talented middle school girls from all-over Ohio to participate in women-led workshops on STEM and medical fields. Similarly, we have collaborated to run the annual High School Programming Contest, led by Alan, hosted by the University of Alabama at Birmingham. In 2016 we received a grant from the National Security Agency (NSA) to increase female participation in the coding contest and established a Grace Hopper Award for female contestants.

Since the early 2010s we have collaborated in Computer Science and Engineering education research. Given Alan's expertise in Data Mining, our research initially focused on analyzing social media posts to detect insults and working to neutralize the threats and harms of cyberbullying. Nine teenagers suicides were linked to cyberbullying on the social networking site Ask.fm in 2012. In Ask.fm, members ask questions to other members. Many of these "questions" are actually insults and may be posed anonymously. After Ask.fm became infamous for the suicides attributed to its influence; we analyzed thousands of Ask.fm posts and presented a paper: "Exploring Roadblocks to Deter Cyberbullying," at a conference on Computer Ethics & Philosophical Enquiry (Diaz-Sprague and Sprague 2014). We are coauthors, with Computer Science researchers at the University of Houston, in published research that uses Natural Language Processing techniques for analyzing teen social media posts to detect invective and nastiness in teen social media communications (Samghabadi et al. 2019).

In 2015 Raquel became a Visiting Scholar at the UAB Computer Science Department, where Alan was a full Professor. In fall 2016, they jointly taught the CS Capstone course (CS 499). It was in teaching CS499 that it became evident that embedding ethics and teamwork training in the Capstone course was necessary. Also, in CS499, we encouraged students to develop apps to combat cyberbullying and to consider other prosocial software designs for their required team project.

Three out of seven teams (42%) developed such apps. Two apps were designed to flag profanity in messaging posts; one offered an alternative, more polite wording. Particularly memorable is a thoughtful app called "Moral Reminders," developed by a team of African students. It reminded the user to offer appreciation or uplifting African proverbs to people around them three times a day. Through these projects, we recognized the potential to empower students' insights, cultural background, imagination, and creativity to design and create technologies about "what should be." Faculty should encourage students to work on course-required projects to emphasize potential societal benefits. We submit that students can be instrumental in cultivating ethical cultures and be a force for good through software design.

In 2017 and 2018, the authors were awarded unrestricted grants from the UAB Center for Teaching and Learning, undertaking a campus-wide effort to support faculty proposals for its theme "Learning in a Team Environment" (UAB Reporter 2017). After receiving approval from the Institutional Review Board, we developed and implemented a minimodule on Ethics and Teamwork. The minimodule was designed to be embedded in an upper-level UAB Electrical & Computer Engineering Department course. We were invited to teach the minimodule in the upper-level Electrical Engineering course (EE485/EE585) in fall 2018 and again in fall 2019. The minimodule consisted of four class periods, accounting for 15% of the course grade.

5.3 Minimodule Design

The minimodule on ethics and teamwork had to be substantive and succinct. It was designed to be implemented within four 75-min class periods. We divided the minimodule into two segments and allowed a 2-week intermission between the two. The first two class periods were devoted to a refresher on ethical principles, moral theories, and moral reasoning. In class, students watched short video lectures by masterful professors. On Day 1, they watched Lecture 1 of Harvard's Michael Sandel video lecture on "Ethics: What's the Right Thing to Do" (Sandel n.d.). On Day 2, students watched Emory University's distinguished primatologist and ethologist Frans de Waal on "Moral Behavior in Animals" (De Waal 2010).

After each video lecture, students gathered in small groups to participate in guided discussions. They worked to provide answers to questions posed to them by the instructors and offered reflections both individually and as a group. There was strong student interest – discussions sometimes continued in the hallways after class ended.

Our highly interactive minimodule was modeled after a 12-h minimodule on Medical Communication with Latino Patients which Raquel designed and taught for several years at The Ohio State University medical school. Typically, students learn best by being actively engaged with the subject matter. A recent survey of 1250 undergraduates published in the April 4, 2023 edition of Inside Higher Ed states: "Students say they typically learn and retain information best in an interactive lecture – somewhere between a traditional lecture and a high-intensity active learning environment."

Why did we choose the De Waal video? Given that the minimodule had to deliver basic teamwork concepts: fairness, cooperation, and social responsibility in a short period, we chose Frans De Waal TED talk (16 min), which shows that animals such as primates have social norms that help them live harmoniously and work collaboratively within their social groups. They demonstrate moral emotions such as empathy, offer consolation to those in distress, and seek reconciliation after fights. Most memorably, the video shows that capuchin monkeys have a strong sense of fairness and vehemently and decisively reject unfairness. Honesty, fairness, and reciprocity are pillars of morality, essential for teamwork.

After the first two class periods, students were given readings and assignments on teamwork. They were instructed to work in small groups to prepare to present either a PowerPoint on "Ethics and Teamwork" or to select a game involving collaboration and practice it to present it to the class. The teams were advised that their peers would judge them and that a team would be selected as the winner of the teamwork class competition. After a 2-week intermission, on Days 3 and 4 of the minimodule, student teams presented games that required planning, effective communication, and teamwork. Most teams chose to present a game. Only one team, composed of three graduate students, presented a PowerPoint. Their presentation included quotes about teamwork – what it is, why it matters – and a couple of slides reflecting on how the keen sense of fairness and collaborative behaviors demonstrated by monkeys and other animals in De Waal's lecture had awed and inspired them.

The games the students presented ranged from simple word games to complex card tower building to guiding blindfolded team members through a maze with only a few verbal clues. Each student team demonstrated its chosen game to the class and invited the other teams to play the game within the allotted time. The students were intensely engaged. The course had a friendly competitive atmosphere. Also, on Day 4, students took a comprehensive quiz on the topics covered and filled out the VALUE rubric on teamwork and evaluation forms (VALUE Rubric n.d.). Students performed well in the examination and showed teamwork in action through the games. Individual and team grades were given, and the minimodule accounted for 15% of the grade for the course. We received appreciation from the students, the faculty teaching the course, and the department chair.

We conclude that a minimodule, conceived and organized as described here, effectively conveys the fundamentals of ethics and teamwork in four class periods. We report that the minimodule embedded in an upper-level Electrical Engineering and Computer class (EE485/EE585) at UAB was effective, engaging, and even enjoyable. Particularly noteworthy is the effectiveness of providing ethics and teamwork instruction via short (< 25 min) video lectures by renowned professors followed by small group discussions. We found the students' interest and engagement deeply rewarding. We encourage faculty at other institutions to consider the minimodule approach to teaching ethics. (Diaz-Sprague and Sprague 2019) (Fig. 5.1).

Fig. 5.1 Ethics in action logo

5.4 Ethics in Action: Art & App Challenge

To further instigate students to use their creativity and imagination to build apps or create artworks to promote everyday ethics and cooperative behaviors, in the fall of 2019, we established a novel competition at UAB – the Ethics in Action: Art & App Challenge. It is a challenge for students to create artwork or design apps or app concepts promoting civility, altruism, conflict de-escalation, and other prosocial behaviors. The rationale was that if every student has a smartphone and there is an app for almost anything they need – but not one for moral guidance – why not challenge students to develop some type of "moral app" for their software design projects? Such an app could help students improve their interpersonal skills and make moral decisions in everyday situations. The app could help manage anger, promote civility, reject vulgarity, de-escalate conflict, etc. Eventually, a new line of software products - prosocial apps or software for the soul – could be introduced into software design courses (Diaz-Sprague and Sprague 2021). The instructions and requirements for participation in the Ethics in Action Challenge were described in the announcement and in the Ethics in Action Entry Form published in the School of Engineering News 2020 and 2021.

The initial competition was in the fall of 2019. It was limited to UAB students competing for $1500 in monetary prizes, funded by our grants. That year, a senior majoring in Art at UAB won an honorable mention award for his "Helping Hands" drawing, which shows little monkeys behaving altruistically. In spring 2020, the UAB Department of Electrical and Computer Engineering again sponsored the Ethics in Action: Art or App Design Challenge, supported by a UAB Center for Teaching and Learning grant. The authors were listed as leaders of the contest. The announcement on the college website indicated that the goal of the Ethics in Action Challenge was "to ignite students' imaginations, inspiring them to show through drawings, digital artworks or an app concept that ethics in action, cooperation, honesty, building a respectful and diverse community and caring for our planet can happen in big or small ways."

To expand the competition beyond the UAB campus, in an announcement in early 2021, the 2021 Ethics in Action Art or App Design Challenge sponsored by the Electrical and Computer Engineering Department was opened to college students at any accredited institution nationwide (School of Engineering News 2021). Three universities participated. We are grateful to the UAB Electrical & Computer Engineering Department, the UAB College of Engineering, and the University of Alabama at Birmingham for giving a home to the Ethics in Action: Art or App Design Challenge. Our grants expired in 2021. We hope to continue this effort in the future.

In the fall of 2021, we received the following entries:

Digital art: "The Race to Justice", submitted by a Math & Science senior, Auburn University.

Digital art: "Conversion", submitted by a Computer Science senior at Auburn University.

Art entry: "Misconceptions", submitted by a Design junior at the University of Cincinnati.

App concept: "Montgomery 1960 Experiential Learning Mobile Platform", submitted by a graduate student at Purdue University, who did the project while an undergraduate at the University of Tennessee at Knoxville.

Three jurors judged the entries: a UAB Associate Professor in the Art Department, the Art Director at a national bath products company, and an Adjunct Instructor at Phoenix College. All the entries were deemed meritorious. All received honorable mention awards. Particularly worthy of recognition is the "Montgomery 1960 Experiential Learning Mobile Platform," designed to put participants in the "virtual shoes" of the Alabama State College students who, in 1960 – at significant risk to themselves – organized sit-ins at segregated lunch counters. The Experiential Learning Mobile Platform was initially conceptualized in the Augmented Reality Interactive Storytelling app (ARIS). It is part of an ongoing project: The Montgomery 1960 Project - developed under the leadership of Professor Karen D. Boyd at the University of Tennessee at Knoxville (Montgomery 1960 Project n.d.).

5.5 Coding Competitions Abound: in Most, Speed Is Critical

There are numerous high school or college computer programming competitions at local, regional, national, and international levels. The International Collegiate Programming Contest (ICPC), sponsored by the ICPC Foundation, is the oldest, largest international programming competition. It is open to college students worldwide and draws participants from 111 countries. It focuses on the skills and speed of programming to solve complex algorithmic problems (International Collegiate Programming Contest n.d.).

Coding skills and speed are necessary, but more is needed to lead students to conceptualize and develop technologies that may improve societal well-being. In our limited research, we have not come across college-sponsored contests that work in teams to design technologies to address complex social issues, tackle local or global concerns, or have as a stated goal to create programs or apps to benefit people's lives, mitigate harms or to offer tips for moral reasoning and ethical decision-making. Further research and activity are needed in these areas. We submit that students are willing and able to use their knowledge and skills in team projects to contribute toward a better world. Leadership is needed.

The Microsoft Imagine Cup does have a prosocial, beneficent goal. Creating programs to improve people's lives is the competition's goal. The Microsoft call for participation in the Imagine Cup states, "reimagine technology solutions in a competition designed to help you make a difference", and "Create something that matters" (Imagine Cup n.d.). The first place winner in 2022 was a team from Saudi Arabia and India, studying in Germany, that created ExoHeal. This modular exoskeletal hand rehabilitation device provides rehabilitation exercises to people with hand paralysis. The second and third place teams created software for children with learning disabilities and for people with hearing impairments (Imagine Cup Winners 2022).

The stated goal of the Congressional App Challenge (CAC) is "to inspire students from every corner of the country to explore STEM, coding, and computer science through hands-on practice" (Congressional App Challenge n.d.). The CAC is open to students in high school or earlier. Teams of up to four students may participate if their representative in Congress has decided to sponsor teams and organize the judging of the resulting apps. Teams have 4 months to turn in their apps. The CAC could very well add a prosocial goal to the competition. Currently, the CAC does not mention improving people's lives as a desired goal for the students' apps. Still, a few of congressional districts' winning apps do have beneficent goals, such as "Meals on Wheels" – which helps drivers to organize their routes, and "A Smarter Insulin Pump" – which aids in optimizing insulin dosage for a given patient (Congressional App Challenge Winners 2019).

There are many other programming competitions where programming speed is critical for success. Many have a restrictive time limit, for example, 3 h. Four such competitions are described below.

1. For the Facebook Hackercup, each annual competition consists of 5 rounds of 44–72 h each. In each round, several problems are stated and need to be solved (Facebook Hacker Cup n.d.).
2. Topcoder runs a half dozen worldwide contests annually (Topcoder Competitive Programming n.d.). Some contests are: given requirements for a program, produce technical specifications for it; given technical specifications for a program, produce code; and testing and case writing. Countries having the most winning teams are China and Poland.
3. Google runs three or so competitions yearly, HashCode, Code Jam, and Kick Start (Google's Coding Competitions n.d.).
4. UAB runs the Alabama High School Programming Contest. It typically attracts 50 students from across the state – they are given six problems to solve within 3 h (Department of Computer Science Outreach n.d.).

5.6 Why Do Prosocial Technologies Matter?

The purpose of a college education should be not only to provide students with job-related skills and knowledge but also help them to grow intellectually and acquire social responsibility, ethical decision-making, and leadership skills leading to human flourishing. Colleges can choose to prepare students to be socially conscious global citizens by challenging them and encouraging them to develop prosocial app concepts, artworks, games, or other prosocial projects. As we observed in our experience teaching the Capstone course, where teams usually choose to do a game, which is relatively simple to program, a sizable number of students (42% of the class) decided to take the extra challenge of working to conceptualize, design, and write a program to develop a "prosocial app" – something that they had not done before. Allowing students to step out of their comfort zone enhances their intellectual growth and leads to a more profound satisfaction. According to Nicomachean ethics, doing the right thing for its own sake, even if it's more challenging, leads to lasting happiness – eudaimonia – and human flourishing.

Our nation's problems are staggering, and we don't have the tools to combat them. We need more significant input in Computer Science from the social sciences and the humanities to develop the prosocial technologies of the future. We need the talents, skills, and ingenuity of every educated man or woman to address our nation's problems, particularly the antisocial – sometimes even vile and abhorrent – use of the internet. We need more female participation in the computing workforce, especially women of color in technological fields. According to Fernandez and Wilder (Fernandez, J., and J.A. Wilder. 2020. Communications of the ACM 63 (8): 18–21), women account for only 26% of the computing workforce, but only 1% of the computing force is occupied by Latinas.

Current technological tools are being used for harm. We have technologies spreading disinformation, cyberbullying, hate speech, conspiracy theories, election interference, etc. As of 2020, there were more cell phones and smartphones than humans on the planet. There could be moral-guidance apps embedded in the phones. As computing-based technologies continue to advance and amaze us, we live in a time of hope as well as fear, an age of advanced knowledge and abysmal ignorance. Intolerance, incivility, antisocial behavior and noncooperation are commonplace. Abusive language, untruths and invective permeate the internet eroding civil discourse, respect for the truth, the rule of law, and democracy itself. Our nation is slipping in quality of life through virulent polarization, chaos, and violence. We may be lacking something important in our educational system – perhaps it is a solid grounding in the humanities, advanced language skills, training on ethics and values, and social responsibility. Our nation's students need guidance to apply their skills to projects that serve not just their private interests but the public interest as well. We should be able to have an educated citizenry that lives in harmony with all kinds of diversity and shows the utmost respect for the dignity of each person. It's hoped that an army of students coding for kindness, coding for caring, and developing prosocial technologies can help build a happier America.

5.7 Envisioning a Happy Future for America

> Speaking at Amherst College in October, 1963, President Kennedy said: I look forward to a great future for America –
>
> a future in which our country will match its military strength with our moral restraint,
> its wealth with our wisdom,
> its power with our purpose.
> I look forward to an America which will command respect throughout the world,
> not only for its strength,
> but for its civilization as well.

References

ABET. 2023. *Accreditation criteria*. https://www.abet.org/accreditation/accreditation-criteria/. Accessed 14 Apr 2023.

Azquotes. n.d. *John F. Kennedy*. https://www.azquotes.com/author/7900-John_F_Kennedy/tag/science. Accessed 22 Jan 2023.

BrainyQuote. n.d. *Albert Einstein quotes*. https://www.brainyquote.com/authors/albert-einstein-quotes/. Accessed 22 Jan 2023.

Congressional App Challenge. n.d.. https://www.congressionalappchallenge.us/. Accessed 22 Jan 2023.

Congressional App Challenge Winners. 2019. https://www.congressionalappchallenge.us/2018-winners. Accessed 22 Jan 2023.

De Waal, F. 2010. *Moral behavior in animals*. https://www.ted.com/talks/frans_de_waal_moral_behavior_ in_animals. Accessed 22 Jan 2023.

Department of Computer Science Outreach. n.d.. https://www.uab.edu/cas/computerscience/outreach/. Accessed 22 Jan 2023.

Diaz-Sprague, R. and A. Sprague. 2014. Exploring roadblocks to deter cyberbullying. In Computer Ethics & Philosophical Enquiry (CEPE) conference. Paris. June 22–24.

Diaz-Sprague, R., and A. Sprague. 2019. Tying ethics and teamwork in engineering education: some results and observations. In Annual conference of the association for practical and professional ethics, Baltimore, MD, February 28–March 2.

———. 2021. Towards an ethics in action: App design challenge as a tool in STEM ethics education. *IEEE Ethics*.

Facebook Hacker Cup. n.d.. https://www.facebook.com/codingcompetitions/hacker-cup. Accessed 21 Jan 2023.

Fernandez, J., and J.A. Wilder. 2020. TECHNOLOchicas: a critical intersectional approach shaping the color of our future. *Communications of the ACM* 63 (8): 18–21.

Fiesler, C., N. Garrett, and N. Beard. 2020. What do we teach when we teach tech ethics?: A syllabi analysis. In SIGCSE'20: Proceedings of 51st ACM technical. symposium on computer science education, 289–295.

Google's Coding Competitions. n.d. https://codingcompetitions.withgoogle.com. Accessed 4 Apr 2023.

Grosz, B.J., D.G. Grant, K. Vredenburgh, J. Behrends, L. Hu, A. Simmons, and J. Waldo. 2019. Embedded EthiCS: Integrating ethics across CS education. *Communications of the ACM* 62 (8): 54–61.

Harvard University. n.d. *Harvard works to embed ethics in the computer science curriculum*. https://news.harvard.edu/gazette/story/2019/01/harvard-works-to-embed-ethics-in-computer-science-curriculum/. Accessed 22 Jan 2023.

Imagine Cup. n.d.. https://imaginecup.microsoft.com. Accessed 2 Apr 2023.

Imagine Cup Winners. 2022. https://imaginecup.microsoft.com/en-us/winners#2022. Accessed 22 Jan 2023.

International Collegiate Programming Contest. n.d.. https://icpc.global. Accessed 2 Apr 2022.

Montgomery 1960 Project. n.d.. https://1960.csctw.org. Accessed 14 Apr 2023.

Samghabadi, N., S. Maharjan, A.P. Sprague, R. Diaz-Sprague, and T. Solorio. 2019. Detecting nastiness in social media. In Proceedings of the first workshop on abusive language online.

Sandel, M. n.d. *Harvard University's justice with Michael Sandel.* http://justiceharvard.org/. Accessed 21 Jan 2023.

School of Engineering News. 2020. *Ethics in action.* https://www.uab.edu/engineering/home/news-events/school-of-engineering-news/ethics-in-action. Accessed 22 Jan 2023.

———. 2021. *Announcing the ethics in action art or app design challenge 2021.* https://www.uab.edu/engineering/home/news-events/school-of-engineering-news/announcing-the-ethics-in-action-art-or-app-design-challenge-2021. Accessed 21 Jan 2023.

Skirpan, M., N. Beard, S. Bhaduri, C. Fiesler, and T. Yeh. 2018. Ethics education in context: A case study of novel ethics activities for the C.S. classroom. In SIGCSE'18: Proceedings of the 49th ACM technical symposium on computer science education, 940–945.

Topcoder Competitive Programming. n.d.. https://www.topcoder.com/community/competitive-programming. Accessed 4 Apr 2023.

UAB Reporter. 2017. https://www.uab.edu/reporter/resources/learning-development/item/7678-6-grants-awarded-to-promote-teaching-innovation. Accessed 1 Apr 2023.

VALUE Rubric. n.d.. https://www.aacu.org/initiatives/value-initiative/value-rubrics. Accessed 14 Apr 2023.

Part II
Introduction: How the Socio-political Context Influences STEM Ethics Education

As STEM ethics education addresses the standards and norms of research and practice and the values held in society, it is always influenced by its sociocultural and socio-political context. The challenges and goals of a society frame both the topics, ethical frameworks utilized, instructional methods, and moral values emphasized. In addition, the economic conditions are critical concerning the power actually to achieve innovation.

These are factors that touch each and every approach toward STEM ethics education. STEM ethics education in the United States, for example, has been shaped by Responsible Conduct of Research (RCR) education, and codes of ethics adopted by professional organizations such as the National Society for Professional Engineers (NSPE 2019), which are often used in teaching professional ethics. In contrast, European countries tend to focus more on societal aspects and implications of STEM. A liberal Western perspective and Western sociocultural and socio-political values shape both approaches. While this Western perspective towards STEM ethics education has been influential, there are other perspectives on STEM ethics, and other paths towards including ethics in STEM education.

This section explores STEM ethics education and the socio-political and economic context of ethics education in South Africa, Ghana, Eastern Europe, the United States, and China. In each chapter, the authors explore how STEM education can be adapted to include a reflection on ethics in STEM and how educational approaches in STEM ethics can be and have been developed. They reflect on how socio-political forces shape the values emphasized in STEM ethics, how instructional methods can be changed to meet student needs and understanding, and how to meet communities' societal challenges and needs.

Without a doubt, the social environment where future STEM professionals will work plays an enormous role in the values emphasized in STEM education. In Laetus Lategan's chapter, "A Framework for STEM Ethics Education in South Africa: Holding Values Paramount," he provides an orientation of the current state of STEM education in South Africa, with a particular focus on agricultural science, engineering and applied technologies, health sciences, and mathematics. After emphasizing the critical role ethics education plays in helping researchers and technologists meet the challenges and needs of communities – both globally and in the developing world – the author provides a detailed analysis of ethics education's role in South African higher education. Lategan highlights the key principles and values emphasized by the curriculum for these disciplines and provides seven observations on how to better orient ethics education. Finally, he provides a framework for STEM ethics education that holds these values as paramount and provides students with a deep and nuanced understanding of how these principles apply to their professional practice in the academic and social environment they will practice.

Catastrophic events like the COVID-19 pandemic helped illustrate some of the faultlines in the global economy and local healthcare provision as the limits of technological solutions to these kinds of global challenges became obvious. F. K. Abagale and M. A. Akudugu's chapter, "Ethics Education in Science, Technology, Engineering, and Mathematics (STEM) in Africa: A Reflection on the Successes, Failures and the Way Forward in the Era of a Global Pandemic" discusses how the COVID-19 pandemic exposed the vulnerability of human populations and how science and technology helped to somewhat alleviate these vulnerabilities during this time. With this in mind, the authors discuss how STEM education in Africa can better integrate ethical considerations to help students in these disciplines meet sustainable development goals, and better position African communities to weather future challenges. In their chapter, the authors advocate for a paradigm shift from the over-focus on access to STEM education to including ethics in the curriculum as a way to ensure that this kind of education does not widen the inequality gap, but rather narrows it. After providing an overview of current ethics education in STEM and the exodus of skilled African professionals to other countries, the authors look at some of the challenges African countries face and provide suggestions to help surmount these challenges. This includes creating centers of excellence at existing African universities, continued investment and improvement in digital technologies that can open educational opportunities to more students, improving links with technological hubs, and increasing links with the private sector. The authors conclude with how lessons from the COVID-19 pandemic helped illustrate the power of collaborative global learning and reflect on how these lessons could help strengthen STEM ethics education.

The ensuing chapters further stress the influence of the sociopolitical context for ethics education in STEM. They describe political and cultural factors play an important role in shaping the focus and uniformity of STEM ethics education. Is it necessary for professionals globally to endorse a unified set of ethical values, or should regional differences exist? How should we bridge potential divides between different approaches to STEM ethics?

In the chapter "Ethics Education in STEM in Eastern Europe, Moral Development or Professional Education?" by Aive Pekur, the author discusses the history of ethics education in STEM fields in former Soviet countries. During the Soviet era, there was a unified curriculum for teaching all subjects, including philosophy, ethics, and STEM. As countries in this area gained independence, different STEM ethics approaches emerged. The author seeks to answer two questions in this chapter: first, what is the global recognition and acknowledgment of ethics in the former

Soviet region? Second, what are the differences in the understanding of core concepts in STEM ethics in the Western world compared to the former Soviet region? The author describes the results of an empirical study of articles on STEM education from the 23rd International Conference on Interactive Collaborative Learning, Educating Engineers for Future Industrial Revolutions (ICL2020), and discusses possible approaches for bridging the divide between local and global approaches to STEM ethics.

Lina Wei's chapter, "Engineering Ethics Education in China: Development, Promoters, and Challenges for the Future," provides a history of government efforts to integrate ethics into engineering educational programs and the important role ethics plays in a country where the engineering sector continues to grow and be a key part of China's development plan. With the explosion of infrastructure projects, expanding collaboration with global partners, and the growing need to pay closer attention to these projects' social and ethical implications, there is increasing awareness that ethics education needs to step up to meet these needs. Wei provides a detailed overview of key developments in engineering education in China in the past 20 years and discusses many new developments in engineering ethics education stimulated by the Chinese government. She then highlights some of the unique characteristics of these developments and how engineering ethics is being promoted to the wider engineering community and the public.

In the United States, due to several recent social movements such as Black Lives Matter and an expanding understanding of how issues such as systemic racism impact individuals' daily lives, there has been a significant push to both bring more traditionally marginalized and underrepresented groups into STEM fields, and help universities, departments, and instructors adopt more inclusive practices to help minoritized students succeed and not feel as if they have to assimilate or repress their identity and culture. Karina Vielma, in her chapter, "Building Ethical Awareness Using Culturally Relevant Practices in STEM Departments," speaks to this need by providing a detailed discussion of culturally relevant education, how educators can use this approach to build connections between students and course content through this approach, and by providing some examples of what these practices might look like in a STEM department.

Reference

National Society for Professional Engineers. 2019. Code of ethics. Accessed 28 Mar 2023. https://www.nspe.org/resources/ethics/code-ethics

Chapter 6
A Framework for STEM Ethics Education in South Africa: Holding Values Paramount

Laetus O. K. Lategan

Abstract This chapter discusses STEM ethics education in *South African* higher education. This discussion is not a systematic review of STEM ethics education in South Africa but rather an orientation of ethics education in STEM. This orientation confirms global views and identifies what ethical considerations are relevant to South Africa. The focus is on agricultural sciences, engineering and applied technologies, health sciences, and mathematics as representative of STEM disciplines. These disciplines are discussed in the context of the South African university typology, qualifications framework, learning outcomes, and national imperatives linked to higher education. South African authors' views and opinions on STEM ethics education are presented to contextualize the salient perspectives offered in this chapter. A literature search of South African voices in support of ethics education in STEM education confirms the importance and relevance thereof. Principles such as human dignity, autonomy, informed consent, respect for vulnerable persons, confidentiality, the absence of harm, maximum benefit, justice, and social responsibility should be evident in STEM ethics education. Based on these discussions and observations, a framework with six domains is presented for STEM ethics education.

Keywords Agriculture · Ethics education · Engineering · Health sciences · Mathematics · National qualifications framework

6.1 Orientation of the Chapter

STEM fields are commonly known as science, technology, engineering, and mathematics. These fields are often referred to as "hard sciences" as opposed to the humanities, social sciences, and management, which are labeled as "soft sciences."

L. O. K. Lategan (✉)
Central University of Technology, Bloemfontein, South Africa
e-mail: llategan@cut.ac.za

E. Hildt et al. (eds.), *Building Inclusive Ethical Cultures in STEM*, The International Library of Ethics, Law and Technology 42,
https://doi.org/10.1007/978-3-031-51560-6_6

Ethics and ethics education is usually associated with the latter academic disciplines as these disciplines are easily associated with values and norms. Although ethics, as a branch of philosophy, and applied ethics, such as business ethics, are effortlessly connected with the humanities, social sciences, and management, ethics is not limited to these academic disciplines only. Apart from the well-known examples from medical ethics, as professional and applied ethics, there is a growing awareness of ethics education in fields such as engineering, agriculture, and information technology. This ever-increasing awareness confirms the importance of ethics and ethics education in the STEM disciplines.

This chapter discusses STEM ethics education in *South African* higher education. The intention is not to present a systematic review of STEM ethics education but rather to identify the practices associated with STEM ethics education, what lessons can be learned, and what pointers can be offered to enhance STEM ethics education in South Africa and elsewhere. The analysis of STEM ethics education is based on the national Department of Higher Education and Training's (DHET) classification of academic disciplines. Relevant to this classification are agricultural sciences, engineering and applied technologies, health sciences, and natural sciences, including mathematics, as part of STEM disciplines.

The flow of the chapter first identifies some (global) observations from the practice of ethics and ethics education in STEM. This will be followed by an overview of the South African university typology, qualifications framework, learning outcomes, and national imperatives linked to higher education. Such an overview will provide context to a literature review of South African authors' views and opinions on STEM ethics education. From these perspectives, a framework consisting of eight dimensions is presented for STEM ethics education.

6.2 Introductory Comments

A random read of (global) literature on ethics and ethics education in STEM suggests that there is no shortage of opinions and perspectives, and neither is the importance of ethics and ethics education in STEM discarded. The opinions may differ in the depth of analysis and presentation, but there is global consent that ethics education in STEM is relevant and essential. Academic publications and public science communications suggest possible curriculum content and the positive impact that ethics education will have. The following examples can support these observations:

Chrispeels and Mandoli (2003) reflect on agriculture and agriculture ethics. They provide a basic definition of ethics by referring to ethics as "making choices." They continue to argue in favor of *utilitarian ethics* as the preferred approach to ethics: *the greatest good for the largest group*. Their view is complemented by Dyck (2017), also from an agricultural perspective. He remarks that ethics is an integral part of farming and that farmers and their associated industries should understand what morals inform their actions. The United Nation's Sustainable Development Goals (SDGs) also emphasize the importance of ethics. For example, Goal two,

"Zero Hunger", depends on food security and safety (SDG 2015). *Food security* is a challenge caused by, among others, genetic manipulation, climate change, and limited water resources. The burden is increased by many developing countries that are in no position to provide for their citizens' basic food needs. *Food safety* refers to food that is good for human and animal consumption. Food safety, nutrition, and security are inextricably linked and regarded as the basis for sustaining life and promoting health (World Health Organization 2022).

Health sciences without ethics is unthinkable. The Ethics Declaration of Helsinki (World Medical Association 2013), with its focus on respect for the individual, the right to make informed decisions, and recognition of vulnerable groups, are crucial for sustainable healthcare. The same sentiments expressed above can be attributed to engineering and applied technology. Varma argues in favor of human and social sciences within engineering to sensitize future engineers to the needs of humans and societies. She says: "The faith in and expectations of S&T [science and technology] need to be matched by a comparable level of understanding of the social and ethical dimensions of scientific and technological activities" (Varma 2000, p. 218).

Rycroft-Smith (2021) echoes the same sentiments about the importance of ethics in mathematics. His view is that ethics is not only required in how we *do* mathematics but also in the *application* of mathematics in decision-making, reporting, or predicting events. To his observation can be added the calculations for retirement planning or healthcare facilities required for a pandemic.

The importance of STEM ethics education is supported from a broader basis than academic disciplines. The COVID-19 pandemic reminded society again of protecting especially *vulnerable communities*. This is true for all societies, particularly developing countries, where more checks and balances may be needed to protect vulnerable communities. An effective way to address this concern is a discussion document at the seventh World Conference on Research Integrity. This document, known as the Cape Town Declaration on Fostering Research Integrity (Horn et al. 2022), promotes research integrity through fairness and equity in research, research contexts, environments, and collaboration. This document is important because it wants to protect researchers and communities from developing countries. In developing countries, marginalized societies and communities can easily be exploited when the richness of the data sampled by researchers from these societies or Indigenous Knowledge systems from these communities are not recognized or acknowledged. This happens when researchers' or communities' contributions are used without receiving credit for their input or when their contribution is ignored as not crucial due to the dominance of researchers from advanced communities.

STEM ethics education should also be viewed in the context of higher education as a global activity. We owe a big thanks to globalization which contributed to higher education without borders and the global scholar who can benefit from knowledge bases worldwide. Obviously, global higher education also depends on ethics and integrity of knowledge creation, knowledge transfer, and knowledge application. The Fourth Industrial Revolution (4IR) changes the way in which people are living and working. The disruptive technologies and trends associated with 4IR, such as the Internet of things, robotics, virtual reality, and artificial intelligence, necessitate an ethics orientation in the curriculum. This comment is informed by a burning discussion within 4IR about whether "machines" can behave responsibly. To what extent can machines operate based on morality, values, and care? In his book, "*Machines behaving badly: The morality of AI*", Walsh (2020) raises questions about the unintended consequences of Artificial Intelligence (AI). Two observations are important for the discussion on ethics in technology: who is building machines, and what can these machines do. The "*who*" raises the question if machines are portraying "characteristics" of their "builders"? The "*what*" raises the abilities of these technologies. Are these abilities ever considered?

All the above observations are relevant to the South African higher education system. For example, the South African Presidential Report on the Fourth Industrial Revolution supports the ethical and transparent use of new technologies (Republic of South Africa [RSA] 2020). This report identifies the South African response, strategy, and framework for 4IR. This report shows that ethics education must be included in discussing and developing 4IR technologies and practices (RSA 2020, pp. 31, 43, 44, 55). In fact, ethics should become part of a program on human capacity development (RSA 2020, p. 143). The importance of ethics education is also confirmed by the recent Doctoral Degrees National Report of the Council on Higher Education (CHE) (CHE 2022), the CHE's ongoing auditing of institutional practices and accreditation of programs (CHE 2021), and the valuable contributions from professional councils. To illustrate this point: The CHE's report on doctoral standards identifies ethical awareness in research and professional conduct as part of research ethics. Of note is that these aspects of research ethics are not only limited to the researcher but should also be evident when research is performed. Research subjects and communities are affected by how research is carried out (CHE 2022, p. 27).

Regardless of the fact that the value of STEM ethics education is recognized, ethics education in STEM is challenging. Steele and Brew (2012) distinguish between *controversial* and *problematic* ethics. Based on their distinction, it is controversial if the expectation is to present facts and to refrain from personal values in science. It is problematic because students may not professionally be well prepared to deal with ethical dilemmas. The meaning of this distinction for promoting ethics in the curriculum is that facts cannot be removed from a value system. The application of facts is based on value systems. It is equally challenging when an own value system is imposed on others without considering what is generally accepted as ethical values. A student can have a theoretical understanding of what ethics is but may need more experience to pass ethical judgment in a matter. In the curriculum, it is beneficial to encourage students to form an opinion but only by understanding what is regarded as good behavior. Also, students should have enough exposure to case studies to prepare themselves for being ethical in the workplace.

These random comments point toward the importance of ethics and ethics education in STEM. The obvious role of ethics is fivefold:

- Ethics enables the student to grow their ability to identify potential ethical dilemmas and develops skills to deal with these dilemmas.

- Ethics protects the integrity of science and the trustworthiness of results, recommendations, and applications.
- Ethics prepares students as future professionals and how they should behave in the world of work.
- Ethics guards the beneficiaries of science and scientific results, namely government, business, industry, and social communities, from science's potential wrongdoing and harmful practices.
- Ethics promotes multidisciplinary studies by adding human and social sciences perspectives to STEM disciplines.

For the South African part of the discussion, it is important first to understand the South African higher education landscape.

6.3 The South African Higher Education Landscape

There are 26 public universities in South Africa. The university system in South Africa consists of three typologies, namely (a) Traditional Universities, (b) Comprehensive Universities, and (c) Universities of Technology. The focus of *Traditional Universities* is a broad range of general formative and professional programs at undergraduate and postgraduate levels. The *Comprehensive Universities* offer the full spectrum of programs, including vocational, professional, and general formative programs at undergraduate and postgraduate levels. The Universities of Technology originate from previous Technikons (non-university higher education institutions in a former binary system) and, in some cases, through mergers with former universities. These universities offer a range of programs, primarily at the undergraduate level, that are vocationally and/or professionally orientated (see CHE 2020).

The Higher Education Act (RSA 1997a) regulates the three university types. This act created a single-coordinated higher education system. One of the act's objectives is for institutions and program-based education "to respond better to the Republic's human resource, economic and development needs" (RSA 1997a). It is expected of higher education institutions to live up to the values embodied in the White Paper on Higher Education, namely *human dignity, equality,* and *freedom* (RSA 1997b). The White Paper comments that the type of post-school education system required by the country is expected to adapt to "changes in technology, industry, population dynamics, and global trends" and contribute to economic development and life-long learning (RSA 1997b). Institutions are also expected to be responsive to, for example, the country's National Development Plan, 2030, the United Nations' SDGs, and the African Union's Agenda 2063. These comments all speak to the role that higher education institutions should play in developing South Africa's democracy and economy and being a part of global developments and demands. Evidently, this contribution requires much more than a strict focus on disciplinary programs and that STEM disciplines should be infused with an ethical view to contribute to the application of knowledge.

All education institutions share a common National Qualifications Framework (NQF). The National Qualifications Framework Act established a single integrated framework (RSA 2008a). The NQF has ten levels representing qualifications ranging from school, post-school, graduate, and postgraduate. NQF Levels 7–10 qualifications mostly represent university-based qualifications, with Level 7 as a three-year qualification, Level 8 as a 4-year or honors qualification, Level 9 as a master's qualification, and Level 10 as a doctoral qualification. At the university level, these qualifications run across six main scientific domains, namely Agricultural Sciences, Engineering and Applied Sciences, Health Sciences, Humanities, Natural Sciences, and Management and Social Sciences, representing numerous scientific fields. The South African Classification of Educational Subject Matter (CESM) manual (RSA 2008b) deals with the classification level of instructional programs.

The South African Qualifications Authority (SAQA) is the oversight body of the NQF and the custodian of its values and quality character. SAQA (2012, pp. 3–4) uses level descriptors that refer to the expected learning outcomes for academic and occupational qualifications. Ethics and professional practice are one of the cross-cutting level descriptors. The following ethics and professional practice level descriptors for NQF Levels 7–10 are relevant for this chapter:

- NQF Level 7: The student should be able to make ethical and professional decisions and justify the decisions and actions taken (SAQA 2012, p. 10).
- NQF Level 8: The student should be able to identify and address ethical matters (SAQA 2012, p. 11).
- NQF Level 9: The student should be able to make decisions affecting knowledge production or complex organizational or professional issues. The student should also be able to develop ethical standards for a specific context (SAQA 2012, p. 12).
- NQF Level 10: The student should be able to identify, address and manage emerging ethical issues. In addition, the student should also be able to advance ethical decision-making processes and monitor and evaluate the consequences of decisions (SAQA 2012, p. 12).

With this background of the South African higher education system, an overview of ethics education in STEM can be presented.

6.4 A Literature Review of Ethics Education in STEM in South Africa

A literature search of South African voices in support of ethics education in STEM confirms the importance and relevance thereof. A sample from Agricultural Sciences, Engineering and Applied Technologies, Health Sciences, and Mathematics substantiate this comment. A diverse body of literature resources was thoughtfully identified to inform this discussion. Following the literature review are some observations contributing to a framework for ethics education in STEM. The context for the discussion is offering ethics education in STEM disciplines at universities.

6.4.1 Agricultural Sciences

The South African *Agricultural Research Council* (ARC) has as core business animal sciences, crop science, impact and partnerships, and research and innovation (ARC 2022). The ARC's core business implies a wide range of agricultural sciences activities. A study of various universities' curricula suggests that Agricultural Sciences can include aspects such as food safety and security, treatment of animals, use of hazardous chemicals, the environment, sustainable development, technologies, agricultural extension, and management. (See Swanepoel et al. 2017 and RSA 2008b for an overview of activities associated with Agricultural Sciences.) Agricultural sciences are covered by the first category in the CESM code and exclude Agricultural Engineering (covered in Engineering Sciences categories) and Veterinary Medicine (covered in Health Professions and Related Clinical Sciences), and includes Animal Sciences, Food Sciences and Technology, Plant Sciences, Soil Sciences, Forestry, and Wood Sciences (RSA 2008b). The agriculture knowledge triangle comprises education, research, and extension and is farmer-centered (Swanepoel et al. 2017, pp. 25, 63ff).

Another way to profile Agricultural Sciences is to view it from an *agro-food value chain*. Production, processing, storage, trading, distribution, and consumption are usually associated with this value chain. The agro-food value chain represents rural and urban spaces and represents many industries and enterprises beyond the agricultural context. The agro-food supply chain extends to many academic disciplines and commercial and manufacturing institutions and organizations involved in this chain (Smidt and Jokonya 2022; Swanepoel et al. 2017).

Ethics is deeply embedded in this discipline, fields of study, and the agro-food value chain, as matters of *doing right* and *wrong* are inherently part of people's livelihood and well-being. For example, a report from the South African Cities Network (2015) identifies challenges linked with the value chain, including matters such as land tenure issues in South Africa, access to local and international profitable markets, and competitiveness with international markets. The significance of this observation is that participating in the agro-food value chain creates ethical challenges beyond food production, such as genetic manipulation of plants or animals.

In a commissioned study by the Academy of Science of South Africa (ASSAf), Swanepoel and ten other authors (Swanepoel et al. 2017) completed a survey of agricultural education and training (AET) in South Africa. From their research, it is evident that there must be a *paradigm shift* leading to graduates that "have to be entrepreneurs outside of and across international and local value chains; able to work effectively in systems with and as researchers, extension agents, farmers, and entrepreneurs; and adaptive enough to evolve with new demands and opportunities" (Swanepoel et al. 2017, p. 123). One outcome of the AET is "Cultivating the right set of values towards people and towards the complementarity between agriculture and the environment" (Swanepoel et al. 2017, p. 123). Consequently, the "implications of global and regional and national policies and value chains" (Swanepoel et al. 2017, p. 125) should be part of the curriculum.

The South African Council for Natural Scientific Professions (SACNASP), established by the Natural Scientists Act, 1982 (RSA 1982), makes a meaningful contribution to ethics. The Council's mission is "to establish, direct, sustain and ensure a high level of professionalism and ethical conscience among its scientists" (SACNASP 2022). The Act monitors the standard of education and training to ensure that prospective registrants meet the registration requirements.

6.4.2 Engineering

The Engineering Profession Act, 2000 (RSA 2000) mandates the Engineering Council of South Africa (ECSA) to register engineers, technologists, and technicians based on the accreditation of programs and qualifications.

ECSA (2022a, b) refers to engineering as "the practice of science, engineering science, and technology concerned with the solution of problems of economic importance and those essential to the progress of society". From this expectation of what engineers should be doing, it is evident that the engineering profession is much more than having a required knowledge basis. The "needs of society, sustainability, and the protection of the physical environment" guide the engineer when solutions are considered. According to applicable legislation, engineering work has to be ethical and conducted: "Effective, safe, and sustainable engineering work is founded on the competence and integrity of engineering professionals."

Gwynne-Evans et al. (2021) argue that the integration of ethics in engineering education should move beyond the curriculum design level. They advocate that ethics should be part of a program's accreditation and qualification standards. Ethics can be comprehensively integrated within a program through graduate attributes [i.e., skills, knowledge, and abilities that engineering graduates should attain beyond disciplinary content knowledge]. Such a perspective can move ethics from the *periphery* to the *program's center.*

From their article, it is evident that ethics is inherently part of engineering education and profession. This is based on the engineer needing more than technical skills when determining or addressing risk. Solutions should be presented in such a way that future generations can also benefit from current practices. This makes ethics an integral part of engineering education and profession.

An important observation is that ethics should be more than personal and include corporate and social participation to secure ethical behavior. In engineering ethics, it should consist of what is *implemented* (what is perceived and experienced by the participants) and what is *attained* (what can be measured).

They comment that a limitation within the ECSA's graduate attributes is that no specific mention is made of ethics. However, the implicit importance is echoed through the reference to "attitude". This creates the opportunity to make ethics more explicit in pursuing graduate attributes. In the ECSA documents, ethics is linked with knowledge, skills, values, and attitudes. This broad base approach to ethics paves the way to integrate ethics into the engineering curriculum.

An excellent example of ethics in the curriculum is offering a module in Philosophy and Ethics in the third and fourth years of study at Stellenbosch University. In this module, the focus is on "Applied ethics; the code of conduct for professional persons of the Engineering Council of SA (ECSA); case studies of typical situations from engineering practice, including the social, workplace and physical environments" (Stellenbosch University 2022). This orientation will prepare the student to deal with ethical challenges in the engineering practice and to be professional in performing work. The assumption is that this module will contribute to the expected graduate attributes and influence how knowledge of engineering disciplines is aligned with the expected ethical behavior.

6.4.3 Health Sciences

As expected, there is no shortage of ethics education within the health sciences. Programs in ethics education are evident at the graduate and postgraduate levels and cover medical, nursing, and additional medical qualifications.

The education focus is not limited to formal education as part of the curriculum. Although bioethics and medical ethics are well developed and evident in healthcare, an approach such as *medical humanities* can be regarded as a useful addition to enlarge the scope of ethics education in healthcare. In explaining what medical humanities is, Reid states (2014, p. 2), "medical humanities primarily as a critical intellectual space for reflexive contemplation of the power of medicine, the history of medicine and the cultural ways in which biomedical frameworks have been interpreted." The value of medical humanities in South Africa is that it promotes a deeper understanding of society's humanity and culture in health and healthcare. Such an understanding will extend appreciation for and understanding of different value systems. Ethics in health sciences can also be promoted through research ethics committees as part of the ethical/legal framework to promote stakeholder engagement in health research (Wilkinson et al. 2021) and how to do research during a pandemic. An appropriate example is a process of obtaining informed consent during COVID-19. De Vries et al. (2020) propose combining individual consent with delayed and proxy consent. The latter applies to people who may temporarily be unable to consent during a pandemic.

A valuable observation from health ethics education is the multi-disciplinary nature thereof. A relevant example from the University of Pretoria's Centre for Ethics and Philosophy of Health Sciences (CEPHS) can be cited. Scholars from within Health Sciences and associates from other disciplines, such as humanities, law, and philosophy, participate in the research activities of this Centre. They include themes such as Ethics of Mental Health, Ethics of Health Research, and Ethics of Pandemic Diseases. This Centre goes beyond the bioethics focus and is guided by "conceptual work and philosophy" in approaching health ethics. Of note is the comment that "health ethics is recognized in the full scope of health, thus inclusive of clinical ethics, research ethics, professional ethics and relevant to all areas of health including clinical assessment and diagnosis, treatment, prognosis, teaching and training, organizational and institutional management, and policy-making" (University of Pretoria 2022).

The National Health Research Ethics Council (NHREC), established under the National Health Act (RSA 2003), and the Health Professions Council of South Africa (HPCSA 2022) also provide direction on ethical issues.

Notable functions of the NHREC relevant to this chapter are:

- Set norms and standards for conducting research on humans and animals, including norms and standards for conducting clinical trials, and
- Advise the national department and provincial departments on any ethical issues concerning research (RSA 2003).

The Council, for example, enhances healthcare standards for training and promotes the discipline of the professionals registered with the HPCSA.

6.4.4 Mathematics

From reading some literature on diverse topics relevant to ethics education in mathematics, four observations provide an overview of challenges that are relevant to the South Africa context (Anderson and Le Roux 2017; Chiodo and Bursill-Hall 2019; Ernest 2018; Müller 2022; Taylor 2021; Vithal and Volmink 2005). Although these comments cannot be limited to mathematics education, it should be noted that these comments are raised in the context of mathematical education. Possible reasons for why these comments are raised in literature are that mathematics education relates directly to school education, and mathematics is seen as an abstract science that is difficult to link to ethics. The comments are:

- *Firstly*, the change from inclusive education informed by race to a democratic education recognizes human rights values, equity, and equality. While values are generally at the core of education, the shift in representative values contributed to a major change in the curriculum.
- *Secondly*, regardless of the importance of moving away from a culture-dominated value system to a citizen-driven value system, structural challenges within dysfunctional schools are evident and a major reason for concern. Mathematics education generally does not contribute to an education system that should contribute to economic advancement and nation-building.

- *Thirdly*, although the general expectation is to have broad-based values within education, in particular mathematics education, new pedagogical developments and curriculum design challenges cannot be ignored.
- *Fourthly*, regardless of the general agreement that all education should be embedded in values, is there an opinion that mathematics is a "pure science" informed by theory, not values.

To extend on these general comments. A comprehensive book on mathematics education in South Africa deals with values in mathematics education. In this book, edited by Vithal and Volmink (2005), the reference is made to values that are used interchangeably for ethics, sociatal aspirations, and the core of mathematics as a scientific discipline. Although values, similar to ethics, are not discussed in a standalone chapter, these concepts are embedded in various aspects of the curriculum. This is a crucial observation signaling that ethics should not be an add-on approach but rather a rooted approach.

Two other observations are that in the Apartheid curriculum, values were closely associated with religion, particularly the Christian religion. Christian National Education was an essential basis for education in South Africa (Naidoo 2005, p. 184). The downside of the approach was that there was a close link between the shared values of an ethnic group represented by policy and practice. The shift away from this approach was primarily not because of religion but rather ethnic representation. The preferred values in the post-Apartheid curriculum represent diversity, not only of people and their attitudes but also knowledge and skills to contribute to society. The new approach is to engage with a new society's values. Naidoo (2005, p. 199) says: "The time to be mere recipients of values from a higher authority was over. Teacher education needs to take its place as a focal point for changes in the education system." Vithal and Volmink (2005, p. 15) correctly argue in favor of "skills, knowledge, values, and attitudes that should be necessary for all South Africans". They should serve as an enabler to participate in society (Vithal and Volmink 2005, p. 16). This, in turn, will contribute to the reconstruction and development of the country (Brodie and Pournara 2005, p. 31). Diversity of representative values is unavoidable and contributes to building a democratic value system.

To sustain values in education is an enabling environment important. Such an environment is currently a challenge in our society. In responding to the challenges within the school system, Taylor (2021) argues that professionalizing the teachers' profession in terms of preparing future teachers and teaching as a profession may contribute to solving the challenges experienced in the school system. This recommendation implies that the educator should also support a value system to secure the effective transfer of ethics education to the learner.

6.5 Observations

At least seven observations can be made from the above references to ethics education in STEM. The observations are in no particular order:

- Ethics education in STEM should lead to applied knowledge dealing with moral problems in science, the world of work, and engagement. A fair interpretation will be to focus on identifying ethical challenges in the workplace and to deal with these challenges rather than having conceptual knowledge of what ethics is or how to deal with these challenges.
- It appears that the health disciplines often include philosophical underpinnings of ethics in their understanding of ethics more than other STEM fields. These fields tend to grow knowledge of and develop skills in the ethics education curriculum based on application, hence a more hands-on approach. This does not mean that a discipline such as philosophy is ignored when conceptualizing what ethics means for a specific discipline. All STEM fields can benefit from philosophical underpinnings and hands-on approaches to ethical challenges.
- Ethics education is not limited to the discipline content only. Still, it is also useful for the world of work (hence professionalism), stakeholders (government, business, and industry), and the public as end users of scientific and engineering results.
- Ethics education should also contribute towards fostering the values of democracy and result in growing responsible citizens.
- Students should not be passive recipients of ethics and values but should be able to critically engage with understanding the principles informing the desired ethical behavior.
- Ethics, as a graduate attribute, should not be added to the curriculum but integrated into the curriculum. Such an approach is still regarded as the best way to accommodate ethics in STEM education.
- Promoting ethics education requires more than the universities' input. Professional councils, advisory bodies, government, business, industry, and end-users have a shared responsibility to secure a reflective curriculum equipping the prospective employee with relevant knowledge supported by a skills basis that can be applied to prevent moral dilemmas and to deal with ethical challenges.

An applied ethics approach should be followed for ethics education in STEM disciplines. Applied ethics can be understood as practical ethics directed at the application of ethics to real-world problems. Such an orientation will enrich the understanding of making the right or the best choices. The four typical ethics questions that should be asked are:

- *Are we doing things right?* This question hints toward compliance and diligence.
- *Are we doing the right thing?* This question assesses if there was no better way to deal with a matter.
- *How can the common good be promoted?* This question points towards accepting responsibility and creating benefits.

- *What benefit is there?* This question reminds us to the purpose of ethics, namely to create value in human behavior.

Guiding principles for ethics education in STEM will promote principles such as human dignity, autonomy, informed consent, respect for vulnerable persons, confidentiality, the lack of harm, maximum benefit, justice, and social responsibility.

The next step in the discussion is presenting a STEM ethics education framework based on the integration of the above principles.

6.6 A Framework for STEM Ethics Education

Based on the previous paragraphs' discussions and observations, a framework for STEM ethics education is presented. From the discussions and observations, six domains for a framework can be identified. These domains capture the essence of what can be identified as a framework for STEM ethics education. These domains are:

- Domain 1: The core of STEM ethics education
- Domain 2: The role of applied ethics in STEM ethics education
- Domain 3: Professional and care ethics in STEM ethics education
- Domain 4: Basic ethics principles in STEM education
- Domain 5: The academic sphere of influence of STEM ethics education
- Domain 6: The social sphere of influence of STEM ethics education

A brief discussion of each domain will outline its significance.

Domain 1: The Core of STEM Ethics Education
STEM ethics is based on principles representative of science, the world of work, service delivery, and social engagement. Within academia is *knowledge,* the core activity of a university. In the creation (research), transfer (teaching), and application (engagement) of knowledge, the integrity of the knowledge activity should be linked to the STEM disciplines. As education is a common good, values such as public accountability, social responsibility, and respect for the university's value system, science councils, funding agencies, and the value systems of government, business, and industry should be honored. In the engagement with social communities, the principles of do no harm, protect vulnerability, accept responsibility, and create benefit and meaning should be the basis for engagement. The same approach should be in place in the engagement between academic teachers and students. The core of STEM ethics education revolves around values relevant to academia, the public, and private engagement.

This view resonates with the ethics and professional behavior learning outcome in NQF Levels 7–10.

Domain 2: The Role of Applied Ethics in STEM Ethics Education
This chapter discusses applied or practical ethics in STEM ethics education. Applied ethics refers to the application of ethical principles to real-life situations, either in

the classroom, laboratory, world of work, or engagement. Applied ethics gives guidance on how to deal with specific challenges in the profession or activity. A relevant description of applied ethics is to say that it is the *application of ethical principles through decisions and professional behavior in the workplace.*

Beauchamp (2005, p. 11) comments from a bioethics perspective that the value of applied ethics is its contribution to identifying and applying ethical principles to a given situation or activity. Of note is his comment that applied ethics is more a process than a final product for ethical consideration.

This view resonates with the ethics and professional behavior learning outcome in NQF Levels 7–10.

Domain 3: Professional and Care Ethics in STEM Ethics Education
In addition to the value of applied ethics, professional and care ethics should also be promoted.

Professional ethics embodies engagement with people either through the world of work, service delivery, or interaction with communities. Professional ethics refers to ethical behavior when someone engages with other people, structures, and systems. A more detailed overview of professional ethics points toward how someone engages with people, on what basis is the engagement with people, and the level of preparedness to engage with people. Any professional action and behavior are based on a solid knowledge basis of a subject. Departing from a solid knowledge basis creates confidence that people can trust someone's judgment. The absence of a knowledge basis reduces confidence in judgements and creates doubts if the challenge is adequately addressed. The superiority of knowledge has a downside, too, and, therefore, the potential to dominate people. Professional behavior is required when engaging with the significant "other".

Care ethics is essentially applied ethics based on relationship building between people through applying ethical principles in their engagement. A care ethics approach can add enormous value to STEM ethics education as it will assist in avoiding power relationships and domination. The challenge is to meet the significant other in their context respectfully.

This view resonates with the ethics and professional behavior learning outcome in NQF Levels 7–10.

Domain 4: Basic Ethics Principles in STEM Education
Beauchamp and Childress (2013) express the backbone of ethics principles as *beneficence, non-maleficence, autonomy,* and *justice.* Although these principles were identified for bioethics, they can be regarded as the basic principles for ethics behavior and activity regardless of the discipline. Irrespective of its value for ethics in general, more specific guidelines can be added either as an *extension* of these principles or as a *nuance* to these principles. Consideration can be given to guidelines such as human dignity, informed consent, respect for vulnerable persons, confidentiality, maximum benefit with minimum consequences, redress, and social responsibility.

This view resonates with the ethics and professional behavior learning outcome in NQF Levels 7–10.

Domain 5: The Academic Sphere of Influence of STEM Ethics Education
In general, (all) ethics education should influence the academic sphere's activities. There are three obvious spheres influenced by ethics education:

- **Academic citizenship:** The fostering of academic citizenship should make research and teaching relevant, directed at public and private markets and enterprises, grow social impact, create role models, partnerships, the next generation of academics and contribute towards sustainable development goals. Ethics education should also question if what we are doing is right and if we are doing the right thing.
- **Science:** No knowledge, either existing or new, can go without basic values such as the human rights agenda, responsible scholarship, quality and standards, and a competitive but value-adding edge.
- **Application:** Ethics education should prepare students for report writing and scientific authorship, data, and Intellectual property (IP) protection, and respect for communities participating in learning and research. In return, senior academics should also be trained on how to deliver postgraduate supervision and avoid all kinds of misuse of resources and abuse of students.

This view resonates with the ethics and professional behavior learning outcome in NQF Levels 7–10.

Domain 6: The Social Sphere of Influence of STEM Ethics Education
Education has only value when it contributes to a changing society. (All) ethics education can influence social spheres' activities based on the following:

Respectful engagement: Respect takes on various forms and applications. Towards people's respect will embody others' feelings, circumstances, and rights. The engagement should also lead to improving the well-being of society. This will include improved living conditions, human and environmental safety, and security. Through engagement, the inequalities in society should be reduced. The trademark of all engagement is that it can never harm people, structures, systems, and the environment. Engagement should open up new spaces and opportunities for marginalized people and communities.

Gatekeepers: Education involves people, systems and structures, their data, and IP. A gatekeeper system is necessary to avoid any exploitation of either people or systems and structures. The key to the gatekeeper's function is securing that permission and respectful engagement are in place. Ethical clearance is no traffic light allowing you to stop or go. It is an instrument to assist the researcher in being mindful of the consequences of the research process and what the results can be. Ethical research implies, by default, ongoing self-review: Am I still respecting the research subject and research community and honoring the values of my discipline?

This view resonates especially with the ethics and professional behavior learning outcome in NQF Levels 7–10.

The above framework with the identified dimensions can visually be presented in Fig. 6.1:

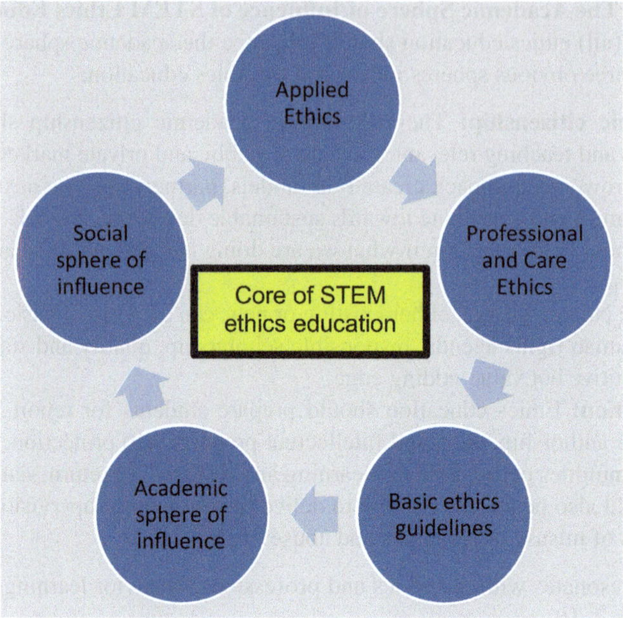

Fig. 6.1 Framework with dimensions of STEM ethics education

6.7 Pointers from the South African Discussion and Practices

The *scope* of this chapter is to identify the practice of ethics education in STEM in South Africa.

The *departure point* for the discussion is the assumption that there is a shared understanding of what ethics is. Based on this understanding is the working definition of ethics as the application of principles, norms, or values to a situation to either avoid moral dilemmas or to address existing or potential dilemmas.

This *assumption* on ethics is based on the acceptance that everyone knows what ethics is, although not all human behavior is reflective of ethical standards. The assumption opens the opportunity to grow ethical knowledge and practice but simultaneously questions whether ethics is a lived experience, especially in the curriculum. Therefore, the question is whether ethics is evident in the three main fields of universities' mission statements, namely research, teaching, and service, or merely an add-on to meet graduate attributes or satisfy professional councils and bodies.

The scope of the chapter was intentionally *delineated* to identify broad base practices instead of providing detailed curriculum evidence. The *outcome* of the literature search hints toward policy, systems, and curricula in place to support ethics education in STEM, but there needs to be more evidence of this learning outcome's success, if any.

While the *practice of ethics education* tends to be more of an applied nature, more evidence is needed to confirm the successful delivery of ethics education in STEM. The best way to deal with ethics education in STEM is to have a basic understanding of what ethics is and its role within a specific academic discipline and expose the student to as many opportunities as possible to grow ethical and professional experience in a future profession.

From a *curriculum* perspective, it is safe to say that ethics should be part of graduate attributes, move from an add-on approach to the center of an academic discipline, and be integrated into the three mission fields of university mission statements. Ethics should be seen as broader than social responsibility. Ethics should be *recognized* as a discipline with its own body of knowledge, interdisciplinary in nature and content, a balance between concept and abstract understanding, and useful knowledge relevant to the world of work and service delivery.

An evident need in the discussion of ethics education in South Africa is a public discussion on how universities promote responsible citizenship through ethics education in STEM.

6.8 Summary

This chapter focuses on STEM ethics education in South Africa. The discussion has three main aims, namely to (a) identify what the practices associated with STEM ethics education are, (b) what lessons can be learned, and (c) what pointers can be offered to enhance STEM ethics education in South Africa and elsewhere.

A random literature review indicates that:

- Ethics enables the student to grow the ability to identify potential ethical dilemmas and a basic know-how to deal with these dilemmas.
- Ethics protects the integrity of science and the trustworthiness of results, recommendations, and applications.
- Ethics respects the public trust in students' future professions and the professions' role and contribution to the world of work.
- Ethics guards the beneficiaries of science and scientific results, namely government, business, industry, and social communities, from science's potential wrongdoing and harmful practices.
- Ethics contributes to multi-disciplinary education based on STEM and Arts. The latter is a collective for human and social sciences.

A specific look at the South African higher education system identified three university typologies sharing with General Education and Further Education and Training a single qualifications framework, the National Qualifications Framework. Each qualification level has specific learning outcomes, of which ethics and professional practice are one of the cross-cutting level descriptors. A literature search of South African voices in support of ethics education in STEM education confirms the importance and relevance thereof. A sample from agricultural sciences, engineering and applied technologies, health sciences, and mathematics outlines the role of ethics in STEM education. Principles such as human dignity, autonomy, informed consent, respect for vulnerable persons, confidentiality, the lack of harm, maximum benefit, justice, and social responsibility should be evident in STEM ethics education.

Based on the discussions, a framework of six domains can be considered for ethics education in STEM. The domains are:

- Domain 1: The core of STEM ethics education
- Domain 2: The role of applied ethics in STEM ethics education
- Domain 3: Professional and care ethics in STEM ethics education
- Domain 4: Basic ethics guidelines in STEM education
- Domain 5: The academic sphere of influence of STEM ethics education
- Domain 6: The social sphere of influence of STEM ethics education.

References

Agricultural Research Council (ARC). 2022. Homepage. https://www.arc.agric.za/Pages/Home.aspx. Accessed 3 Nov 2022.

Anderson, A., and K. Le Roux. 2017. Toward an ethical attitude in mathematics education research writing. *Journal of Urban Mathematics Education.* 10: 74–94. https://doi.org/10.21423/jume-v10i1a303. Accessed 4 Nov 2022.

Beauchamp, T.L. 2005. The nature of applied ethics. In *A companion to applied ethics*, ed. R.G. Frey and C.H. Wellman, 1–16. Malden: Blackwell Publishing.

Beauchamp, T.L., and J.F. Childress. 2013. *Principles of biomedical ethics*. 7th ed. New York: Oxford University Press.

Brodie, K., and C. Pournara. 2005. Towards a framework for developing and researching group-work in mathematics classrooms. In *Researching mathematical education in South Africa: Perspectives, practices and possibilities*, ed. R. Vithal, J. Adler, and C. Keitel, 28–72. Cape Town: HSRC.

Chiodo, M., and P. Bursill-Hall. 2019. Teaching ethics in mathematics. *EMS Newsletter*. 2019–12. 38–41. https://doi.org/10.4171/NEWS/114/10. Accessed 4 Nov 2022.

Chrispeels, M.J., and D.F. Mandoli. 2003. Agricultural ethics. *Plant Physiology* 132: 4–9. https://doi.org/10.1104/pp.103.021881.

Council for Natural Scientific Professions (SACNASP). 2022. *Overview*. https://www.sacnasp.org.za/overview. Accessed 3 Nov 2022.

Council on Higher Education (CHE). 2020. *VitalStats. Public Higher Education 2018.* Pretoria. Accessed on 19 Nov 2022.

———. 2021. *A quality assurance framework (QAF) for higher education in South Africa.* Pretoria. Accessed 28 Jul 2022.

———. 2022. *National review of South African doctoral qualifications, 2020–2021*. National Doctoral Standards Report. Pretoria. Accessed 7 Jul 2022.

De Vries, J., T. Burgess, M. Blockman, and N.A.B. Ntusi. 2020. Research on COVID-19 in South Africa: Guiding principles for informed consent. *South African Medical Journal* 110 (7): 635–639. https://doi.org/10.7196/SAMJ.2020.v110i7.14863.

Dyck, T. 2017. Why the ethics of farming are more important than ever. *The Conversation*. 23 Oct 2017. https://financialpost.com/commodities/agriculture/why-the-ethics-of-farming-are-more-important-than-ever. Accessed 26 Jul 2022.

Engineering Council of South Africa (ECSA). 2022a. Homepage. https://www.ecsa.co.za/default.aspx. Accessed 22 Jul 2022.

———. 2022b. *What is engineering?* https://www.ecsa.co.za/engineeringsa/SitePages/What%20is%20Engineering.aspx. Accessed 3 Nov 2022.

Ernest, P. 2018. The ethics of mathematics: Is mathematics harmful? In *The philosophy of mathematics education today*, ICME-13 monographs, ed. P. Ernest. Cham: Springer. https://doi.org/10.1007/978-3-319-77760-3. Accessed 4 Nov 2022.

Gwynne-Evans, A.J., M. Manimagalay Chetty, and S. Junaid. 2021. Repositioning ethics at the heart of engineering graduate attributes. *Australasian Journal of Engineering Education* 26 (1): 7–24. https://doi.org/10.1080/22054952.2021.1913882.

Health Professions Council of South Africa (HPCSA). 2022. Homepage. https://www.hpcsa.co.za/. Accessed 22 Jul 2022.

Horn, L. S. Alba, F. Blom, M. Faure, E. Flack-Davison, G. Gopalakrishna, C. IJsselmuiden, K. Labib, J. Lavery, R. Masekela, T. Olomola, D. Schroeder, N. Simon, C. Van Zyl, S. Vasconcelos, R. Visagie, F. Kombe, N. Barsdorf, J. De Vries, D. Karatzas, G. Lescano, and P. Roxana Saner. 2022. Fostering Research Integrity through the promotion of fairness, equity and diversity in research collaborations and contexts: Towards a Cape Town Statement. In 7th World conference on research integrity. Preconference discussion paper. https://osf.io. Accessed 2 Aug 2022.

Müller, D. 2022. *Situating ethics in mathematics as a philosophy of mathematics ethics education.* https://www.researchgate.net/publication/358290928. Accessed 4 Nov 2022.

Naidoo, A. 2005. Pre-service mathematics teacher education: Building a future on the legacy of apartheid's colleges of education. In *Researching mathematical education in South Africa: Perspectives, practices and possibilities*, ed. R. Vithal, J. Adler, and C. Keitel, 183–232. Cape Town: HSRC.

Reid, S. 2014. The 'medical humanities' in health sciences education in South Africa. *South African Medical Journal* 104 (2): 109–110. https://doi.org/10.7196/SAMJ.7928.

Republic of South Africa (RSA). 1982. *Act 55 of 1982. Natural Scientists Act, 1982.*

———. 1997a. *Act 101 of 1997. Higher Education Act, 1997* as amended.

———. 1997b. *Education White Paper 3: A programme for the transformation of higher education.* Pretoria: Department of Education. July 1997. Notice 1196 of 1997.

———. 2000. *Act 46 of 2000. The Engineering Profession Act, 2000.*

———. 2003. *Act 61 of 2003:* National Health Act, 2003.

———. 2008a. *Act 67 of 2008: National Qualifications Framework Act, 2008*, as amended.

———. 2008b. *South African classification of educational subject matter.* Department of Education.

———. 2020. *Report of the Presidential Commission on the Fourth Industrial Revolution.* The Presidency: Republic of South Africa.

Rycroft-Smith, L. 2021. *Mathematics as an ethical practice.* 8 Apr 2021. https://www.cambridgemaths.org/blogs/mathematics-as-an-ethical-practice. Accessed 22 Jul 2022.

Smidt, H.J., and O. Jokonya. 2022. Towards a framework to implement a digital agriculture value chain in South Africa for small-scale farmers. *Journal of Transport and Supply Chain Management* 16: a746. https://doi.org/10.4102/jtscm.v16i0.746. Accessed 3 Nov 2022.

South African Qualifications Authority (SAQA). 2012. *Level descriptors for the South African national qualifications framework.* Pretoria.

South African Cities Report. 2015. *Agro-food value chain.* Johannesburg.

Steele, A., and B.R. Brew. 2012. The tower builders: A consideration of STEM, STSE and ethics in science education. *Australian Journal of Higher Education.* 37 (10): 118–133. https://doi.org/10.14221/ajte.2012v37n10.2.

Stellenbosch University. 2022. *Engineering: Academic programmes and faculty information. Calendar. Part 2.* www.sun.ac.za. Accessed 22 Jul 2022.

Sustainable Development Goals (SDG). 2015. *Sustainable Development Goal 2.* https://www.un.org/sustainabledevelopment/hunger/. Accessed 2 Aug 2022.

Swanepoel, F., A. Ekwama, F. Dakora, A. Modi, V. Muchenje, B. Njobe, A. Pell, A. Stroebel, V. Wedekind, J. Kirsten, and F. Terblanche. 2017. *Revitalising agricultural education and training in South Africa*. Pretoria: ASSAf. https://doi.org/10.17159/assaf.2016/0016.

Taylor, N. 2021. The dream of Sisyphus: Mathematics education in South Africa. *South African Journal of Childhood Education* 11 (1): a911. https://doi.org/10.4102/sajce.v11i1.911.

University of Pretoria. 2022. *Centre for Ethics and Philosophy of Health Sciences*. https://www.up.ac.za/centre-for-ethics-and-philosophy-of-health-sciences-/article/2169075/preview?module=cmsandslug=content-pagesandid=2169075. Accessed 22 Jul 2022.

Varma, R. 2000. Technology and ethics for engineering students. *Bulletin of Science, Technology and Society* 20: 217–224. https://doi.org/10.1177/027046760002000309.

Vithal, R., and J. Volmink. 2005. Mathematics curriculum research: Roots, reforms, reconciliation, and relevance. In *Researching mathematical education in South Africa: Perspectives, practices, and possibilities*, ed. R. Vithal, J. Adler, and C. Keitel, 3–27. Cape Town: HSRC.

Walsh, T. 2020. *Machines behaving badly: The morality of AI*. Collingwood: La Trobe University Press.

Wilkinson, A., C. Slack, C. Crews, N. Singh, J. Salzwedel, and D. Wassenaar. 2021. How can research ethics committees help to strengthen stakeholder engagement in health research in South Africa? An evaluation of REC documents. *South African Journal of Bioethics and Law* 14 (1): 6–10. https://doi.org/10.7196/SAJBL.2021.v14i1.698.

World Health Organization (WHO). 2022. *Food safety*, 19 May 2022. https://www.who.int/news-room/fact-sheets/detail/food-safety. Accessed 29 June 2022.

World Medical Association. 2013. *World Medical Association declaration of Helsinki ethical principles for medical research involving human subjects*. Published Online: October 19, 2013. https://doi.org/10.1001/jama.2013.281053. Accessed 28 Jul 2022.

Chapter 7
Ethics Education in Science, Technology, Engineering and Mathematics (STEM) in Africa: A Reflection on the Successes, Failures and the Way Forward in the Era of a Global Pandemic

M. A. Akudugu and F. K. Abagale

Abstract The COVID-19 global pandemic has proven how vulnerable human populations are and how that vulnerability could be ethically reduced using science and technology. The COVID-19 outbreak led to a complete change in the work and human relations world that had to be sustained through Science, Technology, Engineering, and Mathematics (STEM) education. Border closures and physical distancing protocols meant that virtual and related technologies became the only lifelines for keeping in touch and supplying essential goods and services to people most in need. In some countries, the use of robots through Artificial Intelligence (AI) became critical in the delivery of health services for patients in isolation centers. Countries with strong STEM education combined with robust professional codes of ethics are better able to respond to the challenges imposed by the global pandemic – ranging from vaccine development to efficient delivery of essential services such as medicines and foods during lockdowns. The economies of most nation-states in Africa suffered greatly during the pandemic despite the low infection rates, largely due to the limited access to STEM education and ethics therein. Even though STEM education in Africa has over the past two or more decades received coordinated attention from national governments and their development partners consistent with the African Union's Agenda 2063 that aspires to be a continent where "well educated and skilled citizens, underpinned by science, technology, and innovation for a knowledgeable society is the norm, and no child misses school due to poverty…", less than 25% of African higher education students are in STEM fields. And for those in STEM fields, there is little emphasis on ethics education. The pandemic also threatens the success or possible achievement of the Sustainable Development Goal 4, focusing on ensuring inclusive and equitable quality education and promoting lifelong learning opportunities for all, especially

M. A. Akudugu (✉) · F. K. Abagale
University for Development Studies, Tamale, Ghana
e-mail: makudugu@uds.edu.gh; fabagale@uds.edu.gh

E. Hildt et al. (eds.), *Building Inclusive Ethical Cultures in STEM*, The International Library of Ethics, Law and Technology 42,
https://doi.org/10.1007/978-3-031-51560-6_7

in Africa. In this Chapter, we reflect on the need for ethics education in STEM in Africa in the era of a global pandemic. We advocate for a paradigm shift from the over-focus on access to STEM education to better include ethics education in STEM to ensure that it is not used to widen inequality gaps but rather narrow them.

Keywords Africa · Ethics · Engineering · Mathematics · Science · Technology

7.1 Introduction

Ethics in education is very significant in transforming and modeling behaviours that promote safety and social cohesion in society. There are two (2) motives of ethics education, which are not directly acquired from simply learning ethics. The first motive is to grow intellectual dimensions that will authorize people to recognize ethical dimensions of issues and address ethical issues in the fields of medicine, economics, law, politics, psychology, and policy, among others. The second motive of ethics education focuses on developing critical thinking skills intentional to one's purposes, particularly the ability to reflect, and the theoretical and practical effect of personal and collective human actions.

Ethics play a critical and productive role in creating better citizens, thus providing a reason for its integration into the educational system, which is fundamental for citizen productivity. Integrating ethics in educational curricula provides opportunities for students to have a learning experience that helps them grow ethically through the creation of ethical awareness, understanding, and motivation to act ethically, especially in the world of work. This according to Ecole Globale (2020) is achieved through the four (4) principles of ethics – honesty, confidentiality, conflict of interest, and responsibility. With regards to the first principle, ethics in education inculcates honesty in students, meaning training students to be loyal, truthful, trustworthy, sincere, and fair. This is based on the fact that to be successful in today's world; students must have the morality that complements their academic qualifications and knowledge. By integrating ethics into the educational curricula, students are taught to be honest persons with the aim of making them give their honest best in whatever they do.

Confidentiality is another critical principle of ethics that must be embedded in education. With this, students are taught to uphold the commitment not to disclose or transmit information to unauthorized people. This includes information about people or organizations, which is critical for promoting ethical conduct and professionalism in the world of work. The third principle of ethics in education is conflict of interest, which can be displayed in various contexts and in different ways. Generally, conflicts of interest emerge when the best interest of one person is not in the best interest of another individual or organization to which that person incurs loyalty. The different ways that conflicts of interest can manifest include mistakenly permitting another priority to affect one's judgment or to deliberate an infraction of rules for personal benefit. Responsibility is the fourth principle of ethics in education, and it takes place when individuals acknowledge that they are responsible for the consequences of their actions and/or inactions. As a principle of ethics in the educational system, individuals are taught the responsibility to communicate respectfully and carefully with others.

In this chapter, we provide a reflection on the need for ethics education in STEM in Africa in the era of a global pandemic. We advocate for a paradigm shift from the over focus on access to STEM education to better include ethics education in STEM so as to ensure that it is not used to widen inequality gaps but rather narrow them. We discuss the concept of STEM education within the context of Africa, the state of STEM education, and how to integrate STEM into the educational system of Africa. Ethics in STEM education in Africa and the commitment of Higher Educational Institutions (HEIs) to ethics and STEM education are captured in this Chapter. We conclude by arguing that there is a need for a paradigm shift in the education curricula in Africa by placing ethics in STEM at the center of higher education in the continent. From the foregoing, there is a clear need for a new paradigm that demands higher education institutions in Africa to change their policies and practices in STEM education centered on ethics. Efforts to promote the new paradigm shift towards the incorporation of morals, values and ethics in STEM education must be situated within the context that Africa produces only 1.1% of global scientific knowledge (World Economic Forum, n.d) and has just 79 scientists per million of inhabitants compared to countries like Brazil and United States where the ratio stands at 656 and 4500 respectively.

7.2 The Concept of STEM Education

STEM education is the integration of other disciplinary knowledge into a new composite discipline. It is an interdisciplinary approach to learning by integrating the four (4) disciplines (i.e., Science, Technology, Engineering, and Mathematics) into one cohesive teaching and learning paradigm (Morrison 2006; Tsupros et al. 2009). This integration aims at the removal of the traditional barriers erected between the four (4) disciplines now branded as STEM (Morrison 2006). According to Tsupros et al. (2009), STEM education is an interdisciplinary approach to learning where rigorous academic concepts are blended with real-world lessons as students apply science, technology, engineering, and mathematics in contexts that make connections between school, community, work, and the global enterprises enabling the development of STEM literacy and with it the ability to compete in the new economy (Tsupros et al. 2009). STEM education has been defined as a standards-based, meta-discipline residing at the school level where all teachers, especially science, technology, engineering, and mathematics (STEM) teachers, adopt an integrated approach to teaching and learning, where discipline-specific content is not divided, but addressed and treated as one (Brown et al. 2011a, b).

Education has historically been focused on building character and academic proficiency (Williams 2000). Modern day education however, goes beyond only character and academic proficiency, to the need to include more skills (Ozsoylu 2017).

The need to incorporate the acquisition of life skills required in the digital age has become imperative (Resnick 2002). This means that students are not only supposed to understand the nature of science but also make scientific inquiries and applications of skills for the resolution of problems (Schwartz et al. 2007). This has led to a paradigm shift in pedagogy that promotes the teaching based on research and inquiry, including design-based engineering practices (Kelly et al. 2008). The concept of STEM, therefore, emerged as a result of the shift in approach (NAE & NRC 2009).

The development agenda of Africa must be hinged on STEM education, meaning the need for formulation of favorable policies as well as policy frameworks that promote STEM education in higher institutions in Africa. African governments need to put in place policies that encourage students to embrace and explore STEM subjects so they can potentially fill the skills gap in the science, engineering, and technology industries, apply their skills to develop scalable solutions to local and global challenges, and ensure improved quality of life and socio-economic development. Governments of nation-states in Africa and private sector players need to provide accessible resources and opportunities to tertiary institutions to carry out STEM-related research that will enable students acquire practical experiences to help proffer solutions to real-world problems. This must be accompanied by the maintenance of a robust STEM curriculum, which must be regularly updated to reflect current trends in innovation. The relevant resources must be put in place to accommodate the rapid technological changes experienced by providing lifelong learning where essential skills are taught. The introduction of student-centered innovative approaches to teaching and learning is thus critical.

7.3 The Concept of Morals, Values and Ethics

Morals, values and ethics have become more important than ever given the increased complexities of society motivated by rapid advances in science and technology. Whereas morality focuses on the universal values and standards of conduct that every rational person wants every other to follow, ethics describes the theoretical, systematic, and rational reflection upon that human behavior (Churchill 1982). Values are linked to beliefs and attitudes in a manner that shapes and directs human behavior. UNESCO (1991) contends that morals, values, and ethics are fundamentally ascribable to society, spirituality and culture. In many instances and sociocultural settings, ethics is treated as a synonym to morality and a well-established branch of philosophy that studies the sources of human values and standards, and struggle to locate them within theories of human individual and social condition (Chowdhury 2014). This gives rise to the concept of professional ethics; special codes of conduct adhered to by those who are engaged in a common pursuit.

Morality, values and ethics are part of a way of life and cannot be separated from all other aspects of life experiences (Kang and Glassman 2010). This is why STEM education in Africa in particular must emphasize character formation to promote morality, values and ethics in individuals and societies. This is especially, critical within the African context where technological advances seem to have significantly reduced the uptake of indigenous morals, values and ethics among the younger generations. It is important to ground STEM on the theoretical framework of moral education, which is supported by moral philosophy, moral psychology and moral educational practices (Han 2014). It must go beyond the scope of promoting rational pro-social skills or virtues to cover real human values based on the meaningful and personally formative knowledge that significantly transcend natural and/or social scientific understanding and explanation (Carr 2014). The focus must be on the inner change through the internalization of universal values (Halstead 2007).

Ethics is the branch of philosophy which tries to probe the reasoning behind our moral life (Chowdhury 2014). The critical examination and analysis through the concepts and principles of ethics help to justify our moral choices and actions (Reiss 1999). There have been no consensus in drawing a distinction between ethics and morals (McGavin 2013). It has been reported that there is a decreasing motivation and interest in the study of sciences (Kiemer et al. 2015), particularly STEM (Kiemer et al. 2015). This is affecting the values that emanate from science – the values associated with education, values of science and values of science education. The values in science education include values associated with teaching science in schools, epistemic values of science, societal values and the personal values of scientists. Western science values may be different from other indigenous science values (Corrigan et al. 2010). As noted by UNESCO (1991) and Witz (1996), morality, values and ethics are always connected and interrelated to society, ascribable to societal culture constantly influenced by politics.

Morals, values and ethics are particularly important in the twenty-first century where job markets require a new set of requirements with more emphasis on STEM skills (Voogt and Roblin 2012). The incorporation of morals, values and ethics in STEM education accompanied by the critical investments must be a priority (Burnett 2013), especially in high, middle, and low-income countries (Burnett 2013). The performance of African students in mathematics and science is persistently lower than international averages (Hooker 2017). Thus, investment in STEM with morals, values and ethics as the foundation blocks is critical for improved economic outcomes in Africa (Williams 2000; African American Institute 2015) through economic diversification (Tikly et al. 2018) and improved competitiveness (African American Institute 2015). Unemployment rates in Africa could be attributed to workers who do not have the skills required by employers as skills taught in Africa mostly do not match with the requirements of employers (Burnett and Jayaram 2012). STEM stimulates innovative approaches for education based on morals, values and ethics and aligning the demand and supply of skills (Burnett and Jayaram 2012; Hooker 2017). Lwakabamba and Lujara (2003) reported that engineering education in Rwanda has provided skilled personnel for industry and solutions for local development problems by boosting research capacity (Tvedten et al. 2018).

STEM education in Africa must focus on producing morally aware, value conscious and ethically driven workforce with the ability to apply critical thinking, creativity, and innovation to create applications that can be commercialized and create jobs in the continent (Kariuki 2015). This is particularly important given the shortage of such a workforce, which has been linked to poor classroom teaching and learning practices and infrastructure in Africa that are predominantly geared towards passing examinations and not towards applying knowledge acquired to solve real-life problems affecting societies.

7.4 The Status of STEM Education in Africa

Africa was projected to record enormous growth in electrification, telecommunications, healthcare, and the internet of 'everywhere' in 2020 (Campos 2019). In his opening address at the Pan African Conference on Education, the Assistant Director-General for Education at the United Nations Educational, Scientific and Cultural Organisation (UNESCO), Firmin Matoko, said: *"We need to help Africa tap into scientific inventions and discoveries that are happening around the world. ... step up investments in scientific research to enable Africans to be producers of knowledge rather than consumers by embracing the advancement in technology and equipping the youth with relevant knowledge and skills the 21st century demands"* (African Union 2018, p. 2). African countries must make it possible by allocating substantial funding to STEM education that can be translated into economic development. Besides, there must be conscious efforts to reduce the "brain drain" phenomenon that makes skilled people emigrate from Africa to other continents in search of "greener pastures", especially those with STEM skills.

It has been reported that each year 70,000 skilled professionals make this exodus, thereby diminishing the development of industries that are grounded in STEM that need highly educated nationals. Campos (2019) reports that due to the lack of a domestic STEM workforce, most STEM jobs in Africa are outsourced to other countries, including the US, China, and India. As noted by Business Tech (2018), *the biggest challenge entrepreneurs face is a lack of job seekers with critical STEM skills.* According to the Trends in International Mathematics and Science (TIMS) study and the UNESCO Global Monitoring Reports, the performance of African students in mathematics and science is persistently lower than international averages (Hooker et al. 2017). Hence, the quality of STEM education is poor in Africa.

With about 90 million people aged 20–24 on the continent, a figure projected to double over the next 30 years, it is important to focus STEM education on this age cohort. This is particularly important given that an increasing number and share of young people in tertiary education in sub-Saharan Africa are studying non-STEM programs. African countries must invest in STEM in a way that will meet the aspirations of educated young people. Both the private and public sectors need to work in tandem in order to create education and employment opportunities for young people, particularly at this critical point in time when the public sector is struggling to create spaces and to offer a high-quality education to the youth. The private sector is thus critical in filling the gap in terms of providing opportunities in education and employment. According to Clay (2016), Africa's challenge is how to create a strong higher education sector to prepare a new generation of leaders in the professions, business, and civil society. The continent has a critical shortage of highly trained scientists, technologists, engineers, and innovators, with enrolments in STEM programs below global averages. The African Capacity Building Foundation estimates that the continent suffers from a shortage of over eight million medical doctors and specialists, over one million researchers, over four million engineers, and nearly 70,000 agricultural scientists and researchers (Mukeredzi 2019). There are fewer than 2000 colleges and universities on a continent with a billion people in 55 countries. Less than 7% percent of Africans have college degrees compared to 30% of North Americans and Europeans and more than half of Africa's population is under the age of 30. An estimated 4–12% of students graduate with STEM-related degrees (https://ja-africa.org/our-work-in-stem/) with less than 25% of African students in secondary and tertiary institutions pursuing STEM-related career fields. There are more students pursuing social sciences and humanities than STEM (Ayeni 2022) and Africa lags behind other regions of the world in scientific productivity and knowledge systems.

The shortages in STEM professionals in the continent are exacerbated by infrastructural deficits in secondary and higher education institutions in Africa. There is inadequate electricity and water supply despite the ever-increasing enrolments due to the fast-growing youth population in the continent. This is made worse by poor research infrastructure, weak governance, and quality assurance. Sadly, the few STEM graduates face unemployment rates of up to 90% thus resulting in a lack of competent domestic workforce, which means African countries cannot maintain their own infrastructures or develop new industries unless they rely on expatriates at more cost. Nigeria, for example, requires 51,000 more engineers in its electric power infrastructure alone than the country currently produces.

Despite the fact that the era of digital revolution offers opportunities that remove barriers of distance and physical isolation, Africa is still isolated from the talent and knowledge generation market which impedes its ability to move into the global economy. It will be important that Africa owns, develops, manages, and deploys its talent both to advance its own development and to be among the global players in the science and technology domains as diverse as agriculture and medicine on the one hand and climate change and business management, on the other. Increased investments in STEM is necessary to halt and reverse the overreliance on international expertise and funds that make the STEM agenda in Africa mostly influenced and dictated by international donors and bilateral and multilateral initiatives such as the European Union–Africa Joint Strategy, the India–Africa Science and Technology Initiatives, US-based Foundations, and China–Africa Science and Technology Partnership (Khumbah 2018). While these initiatives provide the resources required to support STEM in Africa, they are not structured to promote African ownership, accountability, and sustainability.

The heavy influence of international agencies on STEM activity in Africa also tends to fragmentize inter-African research communities, with each of the subregions collaborating more with international partners and less with one another, as measured through their publications output. Inter-African collaborations comprise just 2% of East African, 0.9% West and Central African, and 2.9% Southern African total research output. If Africa is to realize its aspirations contained in the African Union's continental agenda 2063 and Science, Technology, Innovation Strategy for Africa (STISA), its governments have to forcefully revitalize their higher education systems towards STEM Education, as the next – even pivotal – frontier in the continent's historical evolution. It is the caliber of its university graduates in STEM fields that will produce and manage the knowledge that will give relevance to its other institutions – governance, trade, defense, agriculture, health, finance, energy, and diplomacy. It is through a vitalized STEM Education that Africa may turn its increasing demographics into a dividend to enhance its democracies (Fomunyam 2020). The task of developing African science and its future scientists is daunting with those scientists and engineers who are trained in Africa emigrating out of the continent due to limited opportunities (Kariuki 2015). The state of STEM in Africa is compounded by poor education infrastructure. For example, according to UNESCO (2017), only 22% of African primary schools in sub-Saharan Africa have access to electricity.

7.5 The Integration of STEM in the African Education System

The long-term economic prospects of Africa are being constrained by severe skills shortages in many vital sectors and the key immediate solution is STEM. According to Yusuf (2018), the progress toward achieving sustainable and inclusive development will be undermined if Africa does not start building capacity in STEM. Despite the fact that some of the world's fastest-growing economies are in Africa, it can only compete with the rest of the world if it invests in STEM education for young people.

Many African countries face significant challenges in their quest to improve STEM education due to inadequate facilities and basic amenities like electricity, restricted access to education, poor learning outcomes, low salaries of faculty, lack of research funding and equipment, as well as limited autonomy and these disincentives professors to stay in African universities. The situation is compounded by limited educational infrastructure, staff shortages, lack of electricity and water supply, low student attendance in school, and weak governance that hinder efforts to improve STEM education across the continent. There is also the problem of limited research collaboration across STEM fields (Fomunyam 2020).

Many STEM educators have failed in their efforts to collaborate with other STEM educators that teach other STEM disciplines resulting in poor skill development in giving learners an adequate sense of direction and purpose for effective learning and choice of career in STEM-related fields. Since STEM education is an integration of many disciplines with their differences and similarities, a normal approach to teaching and learning should be devised through the collaboration of the educators involved (Fomunyam 2020). Research collaborations through cluster concepts across STEM fields for integrated curriculum will enhance connectivity and information sharing among the stakeholders. Therefore, all efforts should be made to foster an increase in research collaboration among educators and partnership with industry personnel.

In order to attract and retain a new generation of learners, engineering and technology curricula need to be restructured to optimize the acquisition of relevant skills. All new teaching materials should provide clear guidelines for all anticipated teachers and students workload and classroom activities. STEM educators and students will benefit from explicit outcomes for courses, assignments, and projects. When specific and clear outcomes are identified, not only can the instructors focus their instruction on specific knowledge, but they can also link their knowledge assessment directly to the outcomes.

The poor condition of laboratory facilities and instructional media negatively affects the path of better STEM education in African higher institutions. Many schools are not equipped with the needed facility structure, tools, equipment, and required instructional media. When teaching materials are insufficient, teachers should learn to improvise (Nwanekezi and Nzokurum 2010). If changes are implemented as needed in African schools, this will enhance teachers' ability to facilitate learning activities to students, improve academic achievement and increase in state and national test scores (Ejiwale 2013). There is also inadequate practical training for students which inhibits the ability to achieving high quality STEM education in African tertiary institutions. Providing hands-on training will be an opportunity for engineering students to take practical action for the future. Practical training and internships will make students understand how STEM offers career opportunities by employing the machines used in the laboratories that are similar to the ones they would use on the job. It will help students to be able to use technology in the way it enhances their competences in the world of work in the STEM profession (Glasgow et al. 1997).

There is a lack of investment in the professional development of teachers for strong knowledge base resulting in poor student performance in STEM. Teachers need professional internships following the completion of a degree. There is poor preparation and a shortage in the supply of qualified STEM teachers. The quality of teacher preparation is crucial to helping students reach higher academic standards because under-prepared individuals or poor-quality training results in incompetent teachers producing poor students, and the cycle continuous (Sawchuk 2011).

It has also been documented that poor content delivery and method of assessment hinder the growth of STEM education in Africa. The method of teaching determines the amount of knowledge that learners acquire. The STEM educator, as a facilitator, should not only be knowledgeable in the subject but should also possess the basic and necessary skills with which to impact the knowledge of the subjects to the students and learners at all levels of learning (Nwanekezi and Nzokurum 2010). When teaching is not effective, the learners grasp little or nothing, and this reflects in their future choice of career. This implies that STEM educators should endeavor to understand the available methods and teaching strategies and select from them according to the demand of the lesson at hand, with attention to the diverse nature of the students in the classroom, their learning styles, and abilities.

There is a mismatch between the demand and supply of qualified teachers and the gap between science education research findings and what happens in the classroom. There often is an insufficient understanding of the breadth of competencies required of teachers and teacher educators for enhancing personal and collaborative achievement, innovation, and cultural and economic sustainability. Insufficient investments in strategic co-operation and development of ecosystems that would foster effective adoption of the latest research findings and emerging technologies in industry and enterprise, particularly small and mid-size enterprises (SMEs). There is also inadequate public knowledge about and understanding of the complexities of the scientific and social challenges facing humanity across Africa and globally. Limited involvement of stakeholders in science education policy, research, development, and innovation, particularly between students, families, teachers, employers and civil society in the formal education system all affect STEM education in Africa.

In a report by the African Engineering Deans Council (AEDC 2017) at the African Engineering Deans Summit held at Ota, Nigeria, AEDC noted its roles with respect to engineering education in Africa as:

(a) To achieve the aspiration as expressed in the African Union Agenda 2063, Africa must use African resources and Africa driving its own development. For a high standard of living, sound health and well-being, well-educated citizens with skills underpinned by science, technology and innovation with Africans having access to basic necessities of life including shelter, water, sanitation, energy, public transport and information and communications technology (ICT) as well as modern agriculture for increased productivity, African Engineers must be fully involved. Africa must solve Africa problems. There does not seem to be a specific policy framework to encourage African engineers collaborate to solve African infrastructure and developmental problems.
(b) Africa has largely been successful in solving its political conflicts using African peace keepers; however, Africa has not been able to use her engineers in solving African infrastructure challenges and developmental problems. In Africa, less than 33% of active African Higher Educational Institutions run engineering programmes and they hardly work together on African infrastructure and developmental issues. It is time for **Africa to solve Africa infrastructure and developmental problems by African engineers working together.** Present generations are confident that the destiny of Africa is in their hands, and that Africa must act now to shape the future she wants.
(c) As AEDC is an umbrella body of chief executives of Higher Engineering Education Institutions across the continent of Africa with the vision of creating innovative and creative engineers that solve infrastructure and developmental challenges in Africa, African leaders must chart a necessary course of utilizing African Engineers in solving Africa infrastructure and development problems thereby enhancing the realization of Africa Union 2063 Agenda.

AEDC also noted in 2017 that for industry-academic partnership the following factors were important:

(a) Engineering graduates from our African Universities are not adequately trained to fit into the industry system.
(b) Industry and the Academia need each other to adequately train the engineering students that will fit into the industry system and therefore the need for a structured partnership between Industry, the academia and government to bridge this gap must be pursued.
(c) With the aspiration of the African Union (AU) in their Africa 2063 Agenda to be a prosperous Africa based on inclusive growth and sustainable development, there is the need for engineering students and faculty to collaborate across African countries (or where applicable across the globe) in identifying and solving global challenges in Africa, the multiplicity of languages notwithstanding.
(d) Collaboration between stakeholders on regional level should be executed on regional or global associations or councils such as Western, Eastern, Northern, Central and Southern African Manufacturers Associations, African Engineering Deans Councils, African Association Universities, etc.
(e) Engineering training seeking to meet resource-based industrialization should be outcome-based and entrepreneurial in orientation. This will indirectly reduce the restiveness of our citizenry as a happy and engaged youth is a great asset to his country.
(f) To get started, implementation of Industry-Academia Partnership in Nigeria between COREN and MAN should be used as a case and modified appropriate versions adopted by other African countries. The experience of African Countries will form the basis for an Industry-Academia to be proposed for adoption by the African Union as strategy for actualizing the African Union Agenda 2063. Industry players under this arrangement will introduce and inculcate in students the ethical principles of engineering practice by transforming the courses taught mostly in theory to practice.

7.6 Commitment of African Higher Education Institutions to STEM

For Africa to fully benefit from, and cope with technological transformation, universities must adopt research-based education to improve the quality and quantity of STEM education, systematically scale-up support to STEM disciplines, and maximize the use of digital technology. African universities must undertake a review of their STEM-related curricula extensively every three (3) to five (5) years to respond to the ever-changing needs of the society – and to prepare a critical mass of youth with the skills needed to promote research and innovation for the continent's industrialization. The stand was one of several resolutions adopted at the Seventh African

Regional Conference of Vice-Chancellors and Deans of Science Education, Engineering and Technology held in Zimbabwe where nearly 60 vice-chancellors and deans of science, engineering and technology, people responsible for teaching policy in universities across Africa, attended the conference (Mukeredzi 2019). Professor Eddie Mwenje, Vice-chancellor of Bindura University of Science Education, said *"universities have resolved to improve the integration of information communication technology in university teaching, have blended learning; that means we have to do the traditional face-to-face teaching but we ought to ensure that universities also include and adopt e-learning in their teaching"* (Mukeredzi 2019).

Other resolutions included amplifying universities' linkage with industry to strengthen curricula and research programs, thereby making them more relevant to the needs of society. Bridging Africa's technological gap will require concerted efforts to reform Africa's innovation system and STEM education by making it more entrepreneurial and market oriented. The government of Nigeria has already forged partnerships and is drawing upon the technological experience of other countries to build new STEM learning opportunities for Nigerian students. The Global Partnership for Education (GPE) has allocated more than US$100 million in grants to Nigeria in an effort to improve the quality of education.

Educationists, governments, and corporations seem increasingly focused on encouraging students' interest through innovative and relatable STEM initiatives, including experimentation, robotics, coding, or even low-tech group activities that model the experience of solving engineering problems in the real world as well. Classrooms need to offer teaching and learning processes that have real impact on these disciplines in daily life and specifically in Africa. Upgrading of STEM curriculum will transform the way learners understand the world environment, giving them a more sensible outlook of everything and a practical approach to real-life experiences as well as problems and solutions. This approach of teaching STEM subjects is realizable if teachers are well trained and supported to achieve behavior changes. Curriculum reform needs to go hand-in-hand with continuous teacher professional development. These kinds of curriculum reforms require support from all stakeholders, including parents. Steps towards improving STEM education include:

(a) Creation of centers of excellence: resources in many of Africa's universities are already stretched and covering a wide range of subjects does little to help the situation. Centers of excellence can provide leadership, best practices, and research, among others, in a specific field. This allows institutions to focus their resources on a handful of key areas, and the pooled funding results in better resources and improved facilities. By addressing regional challenges, these centers can establish a sustainable business model, and their graduates can have a direct impact on improving their local community.

According to Hambly (2016), governments need to invest in education and create a legislative framework, which allows tertiary education, particularly in STEM subjects, to prosper. The Institute of Mathematics and Physics at the

University of Abomey-Calavi in Benin is an example in this regard. The institution's mission is to equip Africa with young scientists who can become future teachers in the field, to promote cooperation and partnerships in research and training within the continent, to prevent scientists from leaving the continent and contributing to so-called 'brain drain' (Hambly 2016).

(b) Continued improvement in digital technology: technological advances have had a huge impact on learning across Africa. Online learning platforms, including mass open online courses, or MOOCs, have the potential to revolutionize education, with students able to access high-quality learning materials regardless of their geographic location if they can connect to the internet.

(c) Improve links with technological hubs: technological hubs in Africa have been popping up at a considerable rate, and in all corners of the continent. Aside affording users including startup owners and entrepreneurs a unique opportunity to grow and develop their business and ideas, it also exposes them to the latest tool in the industry at a pocket-friendly cost.

(d) Measure performance and labor markets: for education to be effective, institutions should regularly monitor and evaluate their programs. Information about the local labor market should be used to determine the needs of local regions and the relevance of the institution's curriculum in being able to meet these demands. Regular inspection of other standards can ensure universities are maintaining a high level of quality and efficiency. There are a number of organizations in Africa that gather data which can be used by universities, enabling them to effectively evaluate their programs without putting a huge burden on their already limited resources. The African Capacity Building Foundation is one such organization, which works with a number of multilateral partners, African and non-regional partners to provide capacity building data. This *'knowledge hub'* publishes and disseminates regular reports on the state of play across the continent. Resources such as this can help universities and other educational institutions to tailor their programs to the needs of specific regions and sectors, without having to do the groundwork themselves (Hambly 2016).

(e) Increase links to the private sector: given that the private sector is the continent's main source of job creation, it could help universities to establish how best to equip Africa's youth with the skills needed to enter the work. An example of a partnership in action is that of Hecate Energy Africa and two (2) universities in Tanzania. The company has joined with the University of Dodoma and the Nelson Mandela African Institute of Science and Technology to implement undergraduate and graduate programs in the field of renewable energy. The aim of this collaboration is to increase access to clean and reliable power in Tanzania, and the wider region by harnessing the talents of their students in this subject area. For the continent to prosper, it is essential that 'Generation Science' - the generation empowered by an understanding and appreciation of science and technology - succeeds so young people can harness STEM skills to overcome the many challenges facing the region. (Hambly 2016).

7.7 Ethics in STEM Education

It is a common practice in non-STEM programs to incorporate components of ethics in the curricula. For example, students studying theology always have courses in ethics, and those studying business go through ethics in business. Engineering students are sometimes required to learn the legal standards of technology and the design process. Science students are however, mostly not required to take ethics courses. A mathematics student, for instance, may finish a study program without taking a course in ethics. This makes STEM students mostly unable to engage in deep ethical discussion and are thus woefully unprepared to further society's understanding of ethics. They graduate and enter the labor market as professionals without the idea that their involvement in broad ethical discussion is important. This affects society's understanding of science and technology in an attempt to address ethical issues such as abortion, stem cells, genetic engineering, cloning, superweapons, and online privacy, among others. Simply, the lack of ethics in STEM education in general and African STEM ethics in particular, has become apparent given the rising number of ethical dilemmas facing society today, which are rooted in technology and engineering.

Ethics education in STEM will enable scientists as the foremost authorities on some of the most important questions facing humanity to lend their wisdom and understanding to the ethical deliberation process. Integrating ethics in STEM education for students will enable the next generation of scientist-ethicists to influence society's discussion of ethics. This will make the new generation of scientist-ethicists to explain away the fear of technology, bringing firsthand knowledge to the table, eliminating stereotypes, and establishing a deeper way of thinking. This is particularly so given the recent advances in science that create new and ever-changing dilemmas for society to deal with. Through STEM there have been tremendous medical advances that make it possible to keep humans alive for a very long-time using machines. This raises the question of whether an obligation exists to maintain life even when a subjectively measured "quality of life" is lost.

The controversial stem cell research that presents the mind-boggling possibilities that it offers is fraught with reservations regarding their acquisition and use (Lachmann 2001). The brain-computer interface is fast making science fiction movies into realities in the day-to-day life of humans. These advances raise ethical dilemmas and the question of what it means to be human given the new frontiers in autonomous robotics that perform tasks with precision and more effectively than humans. The accuracy of satellite imaging and tracking raises questions of the right to privacy and in some cases threatens the right to life through use of satellite guided weaponry. Whereas there are positive aspects of the advances in satellite technologies, which make it possible to locate terrorists or to visualize areas hurt by a natural disaster, there are obvious potentials for misuse. The emergence of weaponry with pin-tip precision poses real danger to society, as a result of possible disastrous consequences of these powerful weapons being in the wrong hands. These are the reasons why ethics must become an integral part of STEM education, not only in Africa but across the world.

7.8 Conclusion and Implications for Policy and Practice

We conclude that there is the need for a paradigm shift in the education curricula in Africa by placing STEM at the center of higher education in the continent. The implications of this are that higher education institutions in Africa must change their policies and practices that make them take the lead in ensuring that this new paradigm shift, which is centered on four (4) main pillars – partnerships, mentorship, research, and innovation succeeds. The COVID-19 pandemic taught us how partnerships are critical in promoting collaborative global learning leveraging on improved access to STEM inventions and innovations via the internet in partner institutions, industries and research facilities. There were partnerships across the world to remove the physical barriers to access to education and learning through STEM-based inventions such as Zoom, MS Teams, Google Meet, etc. that allowed learning to proceed virtually. The learning experience offered during this period made it possible for students to build up global competence and ensured only relevant knowledge is assimilated and improved upon. Mentorship in STEM is also very critical in this paradigm shift. It provides opportunities for STEM experts to guide students until graduation. This must be accompanied by new students' assessment systems that focus on innovation level and research skills to encourage students to be innovators and inventors, with the view of grooming them to be problem solvers and researchers.

Declaration The authors declare that there is no conflict of interest and no specific grant was received from any funding agency in the public, commercial, or not-for-profit sectors for the purpose of writing this Chapter.

References

African American Institute. 2015. *State of education in Africa report 2015*. www.aaionline.org/wp-content/uploads/2015/09/AAI-SOE-report-2015-final.pdf

African Engineering Deans Council. 2017. *African Engineering Deans Council 3rd Summit*. Main Report and Draft Communique: African Engineering Deans Council Summit.

African Union (AU). 2018. *The Pan African Conference on Education (PACE 2018) calls on National Education Policies to integrate Continental Education Strategy for Africa (CESA) and SDG 4 Commitments*. https://au.int/sites/default/files/pressreleases/34226-pr-pr_063-_pace_technical_meeting_3.pdf. 15 Nov 2022.

Ayeni, O. 2022. *The reality of STEM education in Africa (present, future and challenges)*. https://getbundi.com/the-reality-of-stem-education-in-africa-present-future-and-challenges/. 2 Oct 2022.

Brown, J., R. Brown, and C. Merrill. 2011a. Science and technology educators' enacted curriculum: Areas of possible collaboration for an integrative STEM approach in public schools. *Technology & Engineering Teacher* 71 (4): 30–34.

Brown, R., J. Brown, K. Reardon, and C. Merrill. 2011b. Understanding stem: Current perceptions. *Technology & Engineering Teacher* 70 (6): 5–9.

Burnett, N. 2013. *Pathways to employability: Lessons and case studies for closing the youth skills*. Washington, DC: Results for Development Institute. http://www.aaionline.org/wp-content/uploads/2014/09/R4D-Pathways-to-Employability.pdf.

Burnett, N., and S. Jayaram. 2012. *Innovative secondary education for skills enhancement: Skills for employability in Africa and Asia*. Washington, DC: Results for Development Institute. https://www.r4d.org/.../InnovativeSecondaryEducationSkillsEnhancement-PhaseI-Syn.

Carr, D. 2014. Metaphysics and methods in moral enquiry and education: Some old philosophical wine for new theoretical bottles. *Journal of Moral Education* 43 (4): 500–515. https://doi.org/10.1080/03057240.2014.943167.

Campos, Z. 2019. How focusing on education in Africa can help raise young, successful entrepreneurs. *Forbes*. https://www.forbes.com/sites/theyec/2019/10/15/how-focusing-on-education-in-africa-can-help-raise-young-successful-entrepreneurs/?sh=2d98a7593942. 2 Oct 2022.

Chowdhury, M.A. 2014. The necessity to incorporate TQM and QA study into the undergraduate chemistry/science/engineering curriculum. *The TQM Journal* 26 (1): 2–13. https://doi.org/10.1108/TQM-06-2012-0043.

Churchill, L.R. 1982. The teaching of ethics and moral values in teaching: Some contemporary confusions. *The Journal of Higher Education* 53 (3): 296–306. https://doi.org/10.2307/1981749.

Clay, P. 2016. *It's Africa's time to rise: Let's end our academic isolation*. https://www.theguardian.com/science/blog/2016/mar/07/its-africas-time-to-rise-lets-end-our-academic-isolation. 2 Oct 2022.

Corrigan, D., R. Cooper, S. Keast, and D.T. King. 2010. *Expert science teacher's notion of scientific literacy*. Paper presented at the first international conference of STEM in education. Queensland University of Technology, Brisbane, Australia.

Ecole Globale International Girls' School. 2020. *Ethics in education*. https://www.ecoleglobale.com/blog/ethics-in-education/. 3 Oct 2022.

Ejiwale, J. 2013. Barriers to successful implementation of STEM education. *Journal of Education and Learning*. 2: 63–74.

Fomunyam, K.G. 2020. Towards enhancing science, technology, engineering, and mathematics (STEM) education; a case for higher education in Africa. *International Journal of Engineering Research and Technology* 13 (7): 1516–1524.

Glasgow, K.L., S.M. Dornbusch, L. Troyer, L. Steinberg, and P.L. Ritter. 1997. Parenting styles, adolescents' attributions, and educational outcomes in nine heterogeneous high schools. *Child Development* 68: 507–529. pmid:9249963.

Halstead, J.M. 2007. Islamic values: a distinctive framework for moral education? *Journal of Moral Education* 36 (3): 283–296. https://doi.org/10.1080/03057240701643056.

Hambly, S. 2016. *Seven ways to improve Africa's STEM education*. https://www.scidev.net/sub-saharan-africa/opinions/ways-to-improve-africa-s-stem-education/. 2 Oct 2022.

Han, H. 2014. Analysing theoretical frameworks of moral education through Lakatos's philosophy of science. *Journal of Moral Education* 43: 32–53. https://doi.org/10.1080/03057240.2014.893422.

Hooker, C., A. Capon, and J. Leask. 2017. Communicating about risk: Strategies for situations where public concern is high but the risk is low. *Public Health Research Practice* 27 (1): e2711709. https://doi.org/10.17061/phrp2711709.

Hooker, M. 2017. *A study on the implementation of the strengthening innovation and practice in Secondary Education Initiative for the preparation of Science, Technology, English and Mathematics (STEM) teachers in Kenya to integrate Information and Communication Technology (ICT) in Teaching and Learning*. PhD thesis. Queen's University Belfast. https://www.thebusinessyear.com/nigeria-2018/stem-to-roots/focus

Kang, M.J., and M. Glassman. 2010. Moral action as social capital, moral thought as cultural capital. *Journal of Moral Education* 39 (1): 21–36. https://doi.org/10.1080/03057240903528592.

Kariuki, Tom. 2015. *Africa produces just 1.1% of global scientific knowledge – But change is coming*. https://www.theguardian.com/global-development-professionals-network/2015/oct/26/africa-produces-just-11-of-global-scientific-knowledge

Kelly, A.E., R.A. Lesh, and J.Y. Baek. 2008. *Handbook of innovative design research in science, technology, engineering, mathematics (STEM) education*. New York: Taylor & Francis.

Khumbah, N. 2018. *Stem education and African development*. http://africapolicyreview.com/2018/03/07/stem-education-and-african-development/

Kiemer, K., A. Gröschner, A.K. Pehmer, and T. Seidel. 2015. Effects of a classroom discourse intervention on teachers' practice and students' motivation to learn mathematics and science. *Learning and Instruction* 35: 94–103. https://doi.org/10.1016/j.learninstruc.2014.10.003.

Lachmann, P. 2001. Stem cell research – Why is it regarded as a threat? *EMBO Report* 2: 165–168. http://embor.embopress.org/content/2/3/165?ijkey=11fe746e06da2bd47ddd69d85a4857fef88adf8b&keytype2=tf_ipsecsha.

Lwakabamba, S., and N.K. Lujara. 2003. Effective engineering training: The case of Kigali Institute of Science, Technology and Management. *Global Journal of Engineering Education* 7 (1): 71–76.

McGavin, P.A. 2013. Conversing on ethics, morality and education. *Journal of Moral Education* 42 (4): 494–511. https://doi.org/10.1080/03057240.2013.817330.

Morrison, J. (2006). *TIES STEM education monograph series, attributes of STEM education.*

Mukeredzi, T. 2019. *African vice-chancellors agree to review STEM curricula*. https://www.universityworldnews.com/post.php?story=20191125124727355. 28 Sept 2022.

National Academy of Engineering [NAE] & National Research Council [NRC]. 2009. *Engineering in K-12 education understanding the status and improving the prospects*. Eds: Katehi, L., Pearson, G. & Feder, M. Washington, DC: National Academies.

Nwanekezi, A.U., and J.C. Nzokurum. 2010. Science teaching in Nigerian primary schools: The way forward. *African Journal of Education and Developmental Studies* 7 (1): 68–73.

Ozsoylu, A.F. 2017. Industry 4.0 (Endustri 4.0). *Journal of Cukurova University Faculty of Economics and Administrative Sciences* 21 (1): 41–64.

Reiss, M.J. 1999. Teaching ethics in science. *Studies in Science Education* 34 (1): 115–140. https://doi.org/10.1080/03057269908560151.

Resnick, M. 2002. Rethinking learning in the digital age. In *The global information technology report: Readiness for he networked world*, ed. G. Kirkman, 32–37. Oxford: Oxford University.

Sawchuck, S. 2011. *EWA research brief studies say about teacher effectiveness*. London: Sage.

Schwartz, R.S., G. Akom, B. Skjold, H.H. Hong, R. Kagumba, and F. Huang. 2007, April. *A change in perspective: Science education graduate students' reflections on learning about NOS*. In International meeting of the National Association for Research in Science Teaching, New Orleans, LA, vol. 18.

Tikly, L., M. Joubert, A. Mbogo Barrett, D. Bainton, L. Cameron, and H. Doyle. 2018. *Supporting Secondary School STEM Education for Sustainable Development in Africa*. University of Bristol. https://www.bristol.ac.uk/media-library/sites/education/documents/Supporting%20Secondary%20School%20STEM%20Education%20for%20Sustainabale%20Development%20in%20Africa.pdf

Tsupros, N., Kohler, R., and Hallinen, J. (2009). *STEM education: A project to identify the missing components*. Intermediate Unit 1 and Carnegie Mellon, Pennsylvania.

Tvedten, I., A. Byabagambi, J. Lindström, and M. Tedre. 2018. Evaluation of the Sida supported research capacity and higher education development programme in Rwanda, 2013–2017. *Sida Decentralised Evaluation*. https://www.sida.se/contentassets/.../de2018_3_62126en.pdf

UNESCO. 2017. http://www.unesco.org/new/en/dakar/education/higher-education/UNESCO. Retrieved from: http://uis.unesco.org/en/topic/education-africa

United Nations Educational Scientific and Cultural Organization. 1991. *Values and ethics and the science and technology curriculum*. Bangkok: Asia and the Pacific Programme of Educational Innovation for Development.

Voogt, J., and N.P. Roblin. 2012. A comparative analysis of international frameworks for 21st century competences: Implications for national curriculum policies. *Journal of Curriculum Studies* 44 (3): 299–321.

Williams, M.M. 2000. Models of character education: Perspectives and developmental issues. *Journal of Humanistic Counseling, Education and Development* 39 (1): 32–40. https://doi.org/10.1002/j.2164-490X.2000.tb00091.x.

Witz, K.G. 1996. Science with values and values for science education. *Journal of Curriculum Studies* 28 (5): 597–612. https://doi.org/10.1080/0022027980280504.

Yusuf, K. 2018. Lack of STEM education is holding back growth and opportunities in Africa. First published in: *Accounting Business Magazine*. https://www.accaglobal.com/africa/en/member/member/accounting-business/2018/06/insights/stem-education.html

Chapter 8
Ethics Education in STEM in Eastern Europe, Moral Development or Professional Education?

Aive Pevkur

Abstract In the Soviet Union, STEM education was high-quality. On the one hand, the ambition to have technological predominance and to be in a world-leading position guaranteed relevant financing; on the other hand, less ideological influence allowed the field to develop. At the same time, ethics was highly ideologized and linked to the image of the 'Soviet people' rather than to professional behavior, and STEM ethics was seen more as a matter of quality and safety.

The changes that have occurred over the past 30 years have influenced higher education in general and STEM ethics education in particular. While in the Soviet Union, there were unified curricula for the teaching of all subjects, including philosophy, ethics, and STEM, the destruction of the Soviet hemisphere cultivated different approaches to STEM ethics in former Soviet Union republics, now independent countries.

This chapter presents preliminary evidence of the ethics culture in STEM education in the former Soviet countries.

Keywords STEM ethics · STEM education · Values · Former Soviet countries

8.1 Introduction

Hans Jonas starts the preface to the English edition of his book *The Imperative of Responsibility. In Search of an Ethics for the Technological Age* with the statement, "*Modern technology, informed by an ever-deeper penetration of nature and propelled by the forces of market and politics, has enchanted human power beyond anything known or ever dreamed of before*" (Jonas 1984, ix). In this sentence, he envisioned interrelations between technology, environment, and human society,

A. Pevkur (✉)
Department of Business Administration, Tallinn University of Technology, Tallinn, Estonia
e-mail: aive.pevkur@taltech.ee

E. Hildt et al. (eds.), *Building Inclusive Ethical Cultures in STEM*, The International Library of Ethics, Law and Technology 42,
https://doi.org/10.1007/978-3-031-51560-6_8

three parts that are today recognized as equal goals for global sustainable development.

Science and technology play an enormous role in transforming our lives, as well as the planet itself. The key question is, are we aware of the consequences of our actions, and can we foresee or predict the future? STEM students are taught to be knowledgeable in their field, but also more and more universities recognize their responsibility in shaping student attitudes and designing the ideals of the future by introducing ethics or ethics-related courses for STEM students.

The beginning of the 20 s in the twenty-first century will be described in history as years heralding new types of crises. The Covid-pandemic started at the beginning of 2020, and the war in Europe in 2022 influenced not only people's mindsets but also the understanding of how interconnected the world and the people are. The globalization and internationalization of the labor market for science, technology, and engineering professionals raise the question of if the core knowledge in the STEM field is the same all over the world, can we claim the same for ethical attitudes and culture?

In this chapter, I analyze two questions to shed light on the problem of global understanding of ethics and values for STEM students and professionals. The first question concerns the global recognition and acknowledgment of STEM ethics, especially in the former Soviet region. This is partly a question of STEM education integration in countries separated for half a century from the international STEM community. The second question is about differences in understanding of core concepts in STEM ethics in the Western world compared to the former Soviet region. To answer those questions, we analyze shreds of evidence of engineering and science ethics in publications on the pedagogical approaches in engineering education in the former Soviet hemisphere.

8.2 A Quest for Unified Features of Ethical Culture in STEM

STEM ethics is broader than separate discussions of ethics in science, technology, and engineering. Despite differences in natural and technical sciences (Andrianov et al. 2015), the question is, what is the content of discussions about STEM ethics and how widespread are the discussions about ethics in STEM disciplines globally.

Those discussions have taken place in the United States for many decades (National Academy of Engineering 2009). In the Western part of Europe, Japan, India, and China, there are continuous discussions about the role of ethics in engineering (Ladikas et al. 2015a; Brom et al. 2015; Downey et al. 2015; Mitcham and Nan 2015). Under the European Commission, the European Group on Ethics in Science and New Technologies (EGE) is established to advise policy making and ensure an independent, interdisciplinary perspective on the ethical questions posed by scientific and technological innovation.[1] At the same time, there is not much evidence of STEM ethics debates in the eastern part of Europe, especially in countries not part of the European Research Area (ERA)[2] and, therefore, not integrated into European collaborative research projects under the European Research Council.[3] The reasons for this might be historical as well as contemporary. In the Soviet Union, there were unified curricula for the teaching of all subjects, including philosophy, ethics, and STEM. Education on ethics-related issues was ideologized (de George 1967) and was a moral education (Zajda 1988; Dunstan 1981) rather than professional, ethical training. In addition, as technical experts experienced low individual involvement in making ethically sensitive decisions, they considered humanities-related debates about society and culture irrelevant for their activity due to their ideologization (Khasanova, et al. 2013).

The destruction of the Soviet hemisphere cultivated different approaches to STEM ethics in former Soviet Union republics, now independent countries. Many former Soviet countries are still relatively separated from the international scientific and engineering communities. According to empirical evidence (Khasanova, et al. 2013), engineering ethics has not been incorporated until recently into STEM education in Russia.

Nowadays, a strong culture of teaching and developing STEM disciplines in former Soviet countries continue. The tradition of high-level engineering education also transferred into information technology education. Geopolitical changes in Europe caused a wave of immigration also among technology and science intelligence from former Soviet countries, especially from Russia and Ukraine. Due to the lack of engineering professionals, many western countries, the United States, Germany, the United Kingdom, or the Netherlands, ease immigration for science and engineering experts. To have a good working environment, people moving from one cultural and educational tradition to another should adapt not only to work tasks but also to the work and ethics culture of the recipient country.

There is a fundamental issue, should we appreciate differences in STEM ethics education in different countries or regions, or should the international engineering and science community embrace certain fundamental STEM ethics values and their implementation in ethics education? In this chapter, I consider STEM ethics education as a preparation for future professional activity and follow Mike Martin's approach to professional morality as a set of binding moral obligations to which professionals ought to be committed because of their unique skills, functions, working milieu, etc. and which derives from the special roles of professionals (Martin 1981, 631). That means all engineering or science professionals should have certain common principles due to their professional role duties and responsibilities, and

[1] European Group on Ethics in Science and New Technologies (EGE) | European Commission (europa.eu), Home | National Academies

[2] Members of the ERA are members of the European Union, form the group of former Soviet countries namely Estonia, Latvia and Lithuania, see: European research area (ERA) | European Commission (europa.eu)

[3] ERC: European Research Council | (europa.eu)

ethics education should address them. The more the STEM ethics education embeds principles of professional ethics and resembles between countries, the fewer misunderstandings, conflicts, and problems it causes in work life.

8.3 Ethics Topics in STEM Education

In discussions about the aim and content of ethics education in STEM, there is a question of topics that should be included or covered during the studies. As the professional activity in the field of STEM influences almost all aspects of human life – ecological and built environment, technological enablers of human interaction in society, and the future of life on the planet, a selection of questions need to be covered in forming of ethics culture for STEM professionals.

A commonly accepted view in the western world is that every professional education has to contain aspects of ethical culture characteristic of that discipline (National Academy of Engineering 2009), but also to develop a sense of professional responsibility to society (Vesilind 2004). Despite recognizing that engineering (and STEM) ethics follow diverse local trajectories (Downey et al. 2015), some general discussions about ethics in science and engineering could be brought out.

The recognition that globalization influences the developments and raises issues in the field of science and technology ethics (Ladikas, et al. 2015a, b; Downey et al. 2015; National Academy of Engineering 2009; Brom et al. 2015) is widespread. Discussions about the "ethicization" of the public discourse on science, technology, and innovation, the need for the plurality of voices, including women, people with different cultural backgrounds, or informal laypersons' viewpoints in discussions on technology and innovation developments, questions about global corporate responsibility and negative consequences of technology and innovation are ongoing.

In the aforementioned discussions, the STEM professionals as experts have a leading role in raising and deliberating the problems and finding appropriate solutions. To address the problems having in mind societal, environmental, and ethical considerations expects a strong ethical culture of those professionals. Developing an ethical culture can be done by developing professional attitudes. Ethical culture and attitudes should be considered integral parts of understanding professionalism. In general, "*Engineering ethics is professional ethics, as opposed to personal morality. It sets the standards for professional practice and is only learned in a professional school or professional practice. It is an essential part of professional education because it helps students deal with issues they will face in professional practice*" (Harris et al. 1996, 93). The development of professionalism should encompass the development of ethical professional attitudes. Professional ethics is different from personal morality. Professional ethics standards are expressed in professional codes of ethics. Personal morality is not the concern of professional education.

Professionals in the STEM field, as in any other field, serve society. Serving societal goals makes concepts like public interest, trust, and responsibility relevant in engineering and science ethics discussions (National Academy of Engineering 2009; Harris et al. 2009). Critical attitudes towards technology or authorities and critical thinking (Harris et al. 2009) help students acquire skills and knowledge to make ethical decisions and achieve societal ideals. The opposite situation is described as occurring in the Soviet Union where "*an individual was separated from decision-making and simply implemented state-developed policy, technical activity was depersonalized, and ethical dilemmas were resolved at a state-run level*" (Khasanova et al. 2013, 340). Failure to value the critical approach might lead to a deviation from the professional ideal of serving the good. In the empirical part of this study, I take a closer look, how the widespread understanding of ethics in STEM during the soviet times have evolved and what kind of changes are made.

The power of technology and the risks of technological development have added a burden to the ecological environment. In STEM education, the Sustainable Development Goals of the United Nations influence curricula, especially in technical universities. Education should promote competencies helping to design a sustainable future (Vintere et al. 2021). For example, despite ecological awareness not being seen as an essential part of engineering competencies (Didier and Talin 2015), globalization and disasters caused by nuclear (Andrianov, et al. 2015), chemical or other technologies force scientists and engineers to be aware of ecological and environmental issues.

When talking about ethics in an academic and educational environment, questions about research misconduct and cheating are raised. Researchers in STEM are expected to follow good academic norms, such as publication and openness, credit allocation, authorship practices, confidentiality, and avoiding fabrication, falsification, and plagiarism (National Academy of Engineering 2009).

Thus, the following five STEM ethics questions can be brought forth as parts of forming the ethical culture: professionalism and professional ethics, public interest and trust, critical attitude and critical thinking, sustainable development and environmental issues, and academic misconduct and cheating.

8.4 Evidence of STEM Ethics Culture in Former Soviet Union Countries: An Empirical Study

When talking about the aims of the educational process, there is an expectation that every educational project in the STEM field has to (at least) a certain extent reflect the general understanding of ethics and/or values. To understand the values and principles of ethical culture in STEM curricula in the former Soviet hemisphere, the most accessible method is to look at how researchers describe their activities, projects, and approaches. The common understanding is that ethics education should be embedded in the educational process (National Academy of Engineering 2009).

After the fall of the Soviet Union in 1991, there are almost no comparative analyzes of STEM ethics education developments in former Soviet republics, now independent countries. Looking at articles describing STEM education-related projects in technology universities allows us to draw some conclusions about understanding the importance of ethics issues and cultivating an ethics culture in those universities. That's why an analysis was conducted based on articles, representing the studies and activities in technology universities in former Soviet republics. Describing the research projects or extracurricular activities, the researchers reflect the basic beliefs on ethics, its role in education and teaching methods.

8.4.1 Sources of the Study

For the analysis, articles presented in the thematic conference of engineering and STEM education at the 23rd International Conference on Interactive Collaborative Learning, Educating Engineers for Future Industrial Revolutions (ICL2020) (Auer and Rüütmann 2021a, b) were chosen. Altogether there were 162 published articles in two volumes. All papers were thematic and related to STEM education on various topics and approaches in science and humanities. This method allowed for a cross-section during a certain time point of the understanding and approaches to STEM ethics. As the conference took place virtually in September 2020, it allowed researchers to participate and present the research despite the Covid pandemic in their home country. In addition, as the Tallinn University of Technology organized the conference, Estonia, there was an expectation that researchers from the former Soviet countries would actively participate. About half of the articles presented (77 articles, 47.5%) were from researchers from former Soviet countries.

Seventy-seven papers by authors from former Soviet countries were selected for the analysis. Most of the papers were presented by more than one author. When the paper was written by authors of more than one country, it was assigned as an example of each country presented in the paper. I also included in the analysis collaborative papers where at least one of the researchers was from former Soviet countries. Only ten papers (13%) were collaborative, but the variety of countries was quite wide: Belgium, Finland, Germany (four collaborative articles), Iran, Italy, Norway (two articles), Peru, Poland, and the United States (two articles).

All former Soviet countries have universities delivering STEM and/ or engineering education. The expectation was that most of the fifteen former Soviet countries would be represented. In reality, only six countries presented their research projects. In many countries, there were only a few papers that did not allow one to draw any valid conclusions but only point to certain tendencies (Fig. 8.1).

The largest number of representatives was from the Russian Federation, and 37 (72.5%) papers out of 51 were presented by researchers from the Kazan National Research Technological University. The large number of papers from Russia is not surprising, as Russia is still the most significant country among former republics, but the substantial role of one university is notable. The active work on engineering and STEM education in the Kazan National Research Technological University is a long-lasting tradition (Khasanova et al. 2013).

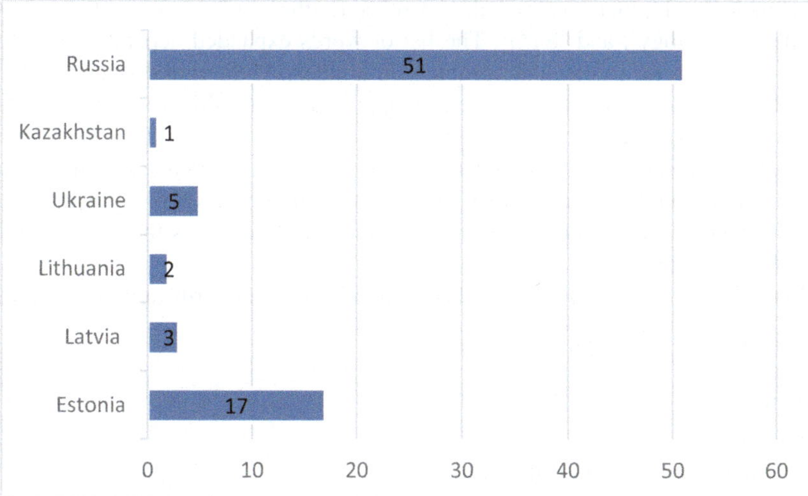

Fig. 8.1 Papers represented by researchers by countries, n = 79

Research Questions

For the analysis, two research questions were posed. The first question concerns the recognition and acknowledgment of STEM ethics and its integration into education in the former Soviet region. The second question focused on the particular topics, amount, and extent of ethics-related debates in educational projects. In articles, I was searching for evidence of five STEM ethics questions as parts of forming the ethical culture: professionalism and professional ethics, public interest and trust, critical attitude and critical thinking, sustainable development and environmental issues, and academic misconduct and cheating.

8.4.2 Research method

I used the quantitative standardized content analysis method to find an answer to the first research question, which meant searching for indicator words referring to ethical thinking and ethical culture. To emphasize only clearly reflected signs of ethical culture and how the researchers understand and reflect values and principles, I read the whole text and searched for words referring to five aspects of ethical culture (professionalism and professional ethics, public interest and trust, critical attitude and critical thinking, sustainable development and environmental issues, and academic ethics and cheating). In addition, I was searching for shreds of evidence about awareness of ethics challenges due to globalization.

I did not limit the search to predefined words but rather tried to grasp a wider spectrum of any evidence of ethical culture. Limiting the search to only keywords would not give an adequate picture. For example, when searching for indications of sustainable development issues, I also searched for the words ‘sustainability’, ‘environment’, ‘ecology’, and ‘STG’. The list of words expanded depending on other words used to refer to the same problem in articles. At the initial stage, the table with the 19 keywords (ethics, values, principles, culture (multi), vocational/ professional (development, education), environment, sustainability/ sustainable development, STG, ecological intelligence, social justice, discrimination, innovation, responsibility, public interest, trust, critical thinking, cheating, plagiarism, global mobility) and 77 article entries with the indication of countries the authors represented was created.

In the later stage of the analysis, I compounded the keywords into nine groups,

1. Professional (development, education)
2. Ethics
3. Values
4. Culture (multi)
5. Public interest, responsibility, trust
6. Sustainability & sustainable development, STG, environment, ecology
7. Critical thinking
8. Plagiarism, cheating
9. Global mobility.

When performing the content analysis, a specific context was considered. For example, the word ‘value’ might also appear in the numeric or data value sense. In these cases, appearance was not taken into account. This method is not very accurate, but it gives a glimpse of some general trends.

The second question had a qualitative character; what kind of ethics-related issues were discussed, and how was the qualitative content analysis made? It shows what questions are on the agenda of the researchers. The related question was, how are ethics and value-related issues discussed among the representatives of different countries? It is clear that behind every kind of theoretical generalization of practices or activities, there is a specific frame of values and beliefs which might not distinctively be brought out but form the general expectations towards the right and wrong, good and bad. For example, if the authors mention self-dependent learning, then the idea of autonomy lies behind this. If the authors did not refer to the general principle, this was indirect evidence of missing awareness.

8.4.3 Results

In analyzing the results, I discuss what topics got more attention in articles and how the researchers discussed these topics. There was no expectation that all the educational projects described in the papers should touch on the issues of ethics culture. The general assumption was that the higher the rate of research projects that reflect the issues of values and STEM ethics in a particular country, the greater the awareness of STEM ethics and culture. As the text corpus is relatively large, I only reference those papers that illustrate or provide an example of the situation. All other articles are analyzed generally by bringing out the main characteristics, see Fig. 8.2.

In STEM ethics debates, globalization and ethics problems are important topics. In 77 articles I analyzed for this study, globalization or global mobility is not the issue. Many projects, described in articles were about implementing specific teaching methods or tools into the teaching or learning process in a particular university, and the influence of ethics questions due to globalization of technology or innovation developments to the education process was modest or not mentioned. Many Russian projects were targeting language and cultural differences, noting that international contacts are not very widespread. Issues of globalization were discussed in two papers from Estonia, one from Ukraine, and six from Russia.

Most of the papers discussed professional development and/or professional education, all together in 55 (64%) of papers acknowledging that teaching STEM students means educating future professionals to have a broad range of skills and knowledge. Including professional ethics in these necessary competencies was mentioned only in one paper: "*As ethical behavior has become a part of the professional identity and practice of engineers, an important goal of teaching ethics to engineering students is to enhance their ability to make well-reasoned ethical decisions in their engineering practice: a goal in line with the stated ethical codes of professional engineering organizations*" (Durst et al. 2021, 781). Considering the importance of resolving ethical problems and the complicated value choices intrinsic to science and technology, the papers' extremely low representation of science and engineering ethics issues was surprising. In total, only eight articles mentioned the word 'ethics': four articles by Estonian researchers (23.5% of all Estonian

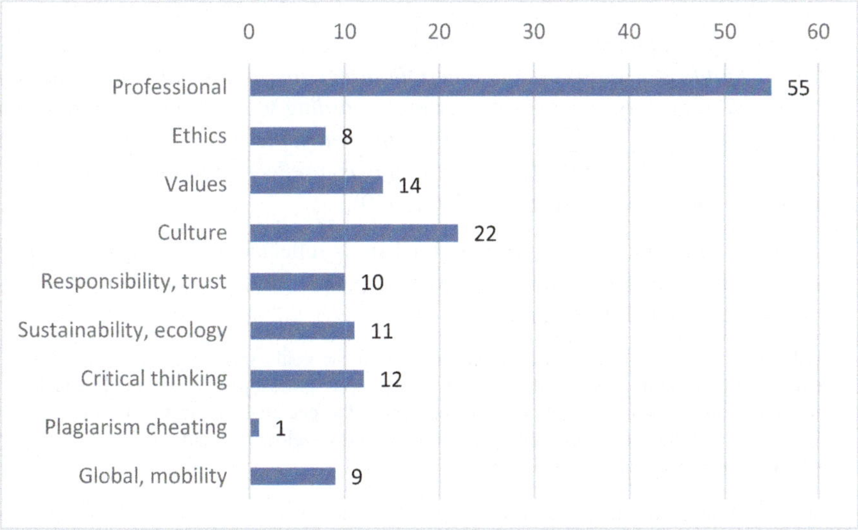

Fig. 8.2 Categories and entries mentioned in articles (n = 79)

papers) and four Russian articles (7.8%). An interesting detail is that three papers from Estonia and one paper from Russia, where ethics was mentioned, were collaborative papers with researchers from western countries. The Estonian paper mentioned the concept of 'ethics' in connection with promoting trust, transparency, student control over data, the right to access, and accountability (Maennel et al. 2021). The mention of values was not in connection with ethics but rather an essential part of planning and designing educational tools and methods.

Despite the concept of professionalism being mentioned in most articles, there are only a few further elaborations on the concept. Looking at the papers from Russian researchers (professional is mentioned in 70.5% of papers), professional education is seen as a molding or shaping instrument to develop a personal character that would fit future professional roles. It is not so much about the specific knowledge, skills, and attitudes necessary to fulfill professional tasks but rather something more resembling personal attitudes or worldviews. Below are some examples to illustrate that point:

1. *"Understanding professional competency implies selecting the principles essential for acquiring the necessary knowledge. And the first principle we put forward is the principle of convergence. Studying the phenomenon of convergence, A.S. Arsenyev comes to the conclusion that "convergence consists in the fact that if you seriously delve into the philosophical, religious, mystical, mythological teachings of both the West and the East, you begin to notice a certain alignment of all these teachings and ideas of a man about the world and about himself. In all these teachings and reflections there is the attempt of a person to define himself in the world and in his relationship with the World as a whole"* (Lopukhova and Makeeva 2021, 842) or
2. *"In V.N. Obnosov's opinion, professional perceptions should be considered as an individual and peculiar system of knowledge, beliefs, and experiences of a person which are associated with a certain profession, i.e., "a totality of information available to a subject about a certain profession, his or her competence on the world of professions, their estimation according to the scale of prestige and attractiveness; this is a dynamic information entity, the structure, and content of which depends on its intended purpose; this is a reflection of the human "Self" through a profession"* (Gulk et al. 2021, 440).

An interesting understanding of professionalism is reflected as training of the elite, who has higher moral demands than average people, and this higher morality is cultivated via personal development.

> Today it is necessary: a systematic representation of the goals and values of engineering activities in the future; taking into account the emerging philosophy of vocational education; taking into account the personal characteristics of a specialist engineer in his own way of entering the engineering culture; installation on self-development and professional creativity (Khatsrinova et al. 2021, 13).

In other articles, all these quotations directly and indirectly continuity of understanding of ethics in STEM education in Russia. Ethics is not seen as a professional or quality matter but rather as a matter of personal aspiration, and if the person is "moral", his or her professional activity will benefit. This refers to a still-existing conception of moral education as something separate from professional preparation but instead having civic or personal character. In articles, the concept of culture is mentioned as a specific culture of teaching or learning. No ethical culture or codes of ethics were mentioned in any article.

In the articles analyzed, there are no elaborations on the topic of public interest, responsibility, or trust. These categories are mentioned as necessary qualities or skills:

> The modernization of higher and postgraduate education is carried out in the context of global and interdisciplinary competencies of the 21st century (volunteering, civic and social responsibility, leadership, communication, research skills, entrepreneurship, etc.). (Karstina 2021, 69).

Critical thinking was mentioned in 11 papers (14%) as a necessary skill to cope with future challenges, and its rising importance was emphasized.

An interesting tendency emerges when looking at the categories of sustainability, sustainable development, and ecology. Only in three papers from Russia are those topics touched upon. At the same time, sustainability issues are raised in four papers from Estonia, two from Latvia, and one from Lithuania and Ukraine. After the general categories 'professional' or 'values', sustainability issues received the third greatest attention in papers from countries belonging to the European Research Area. The researchers from Baltic states emphasized that: "*the study process should be focused on education promoting the development of socially active, responsible, leading and professionally competent specialists protecting positions of sustainable future*" (Vintere et al. 2021, 427) (Table 8.1).

Despite all the articles being about education in science and engineering and many papers touching on the issue of a fair assessment of students, only one article from Estonia discussed the problem of cheating and plagiarism.

8.5 General Observations on STEM Ethics in Former Soviet Countries

The results of both quantitative and qualitative methods were used to analyze the situation and draw some preliminary conclusions. In the quantitative approach, a simple calculus was used to calculate how many articles specific categories are mentioned. In a deeper, qualitative reading of the articles, I analyzed the type of STEM ethics culture described in a particular project. I did not pay attention to concrete results of projects, for example, if the project compared the sustainable attitudes of students in the three Baltic States. For this analysis, it was not important in which country the students are the most sustainable consumers.

An interesting observation from reading the descriptions of the research projects is that even though the project titles indicated professional development, research,

Table 8.1 Categories mentioned in articles per country

	Papers per country	Professional	Ethics	Values	Culture	Responsibility, trust	Sustainability, ecology	Critical thinking	Plagiarism, cheating	Global, mobility
Estonia	17	12	4	3	4	3	4	2	1	2
Latvia	3	2	0	2	0	0	2	1	0	0
Lithuania	2	1	0	2	0	0	1	0	0	0
Ukraine	5	3	0	1	2	0	1	0	0	1
Kazakhstan	1	1	0	1	0	1	0	0	0	0
Russia	51	36	4	5	16	6	3	9	0	6
SUM	79	55	8	14	22	10	11	12	1	9

and globalization, there is little or no indication of the ethics, values, or culture in the description of the actual research project. For example, suppose the research project is about awareness of sustainable development or dilemma situations. In that case, the authors do not draw the broader frame, why sustainability issues should be included in the education of STEM students. Projects are seen and described as practical issues supporting forming a professional, as developing particular skills necessary for the engineering activities of students but not as a good professional whose obligation is to perform according to ethical norms. Professional competency is seen as a set of knowledge and skills, necessary competencies, and not as servants of people and society.

The papers presented by Russian researchers often sought more general concepts such as intellectual skills and competence, general cultural competencies, cross-cultural communicative competencies, digital competencies, metalinguistic awareness, soft skills, personal traits, and self-determination, and their research interest was focused on students' personal development. The researchers from other countries had more concrete research questions concerning the educational process, for example, the evaluation of the specific teaching or assessment methodology or learning model. Suppose researchers from European Research Area (ERA) countries see soft skills, including ethics, as one of the traits among other knowledge and skills that need to be taught during the academic preparation of future professionals. In that case, non-ERA researchers see ethics as a part of character education and more as a matter of personal virtue than a topic for formal professional training.

In papers presenting the research projects conducted in Russian universities, there was much emphasis on the need for integration into the global educational area. Notably, definitions of many concepts necessary for describing professional education and forming the professional, ethical culture agreed upon among Western researchers are on the stage of formalization in articles from non-ERA researchers, for example, competencies in general and pedagogical competencies in particular. Looking at scientific cooperation, in Russian papers, there are few references to non-Russian authors and sources, and vice versa.

8.6 STEM Ethics, Moral Development, or Professional Education?

During the Soviet times, ethics education in universities was a matter of politically informed philosophy. The aim of ethics education was linked to political objectives (de George 1967) and to raise a good, "moral" citizen carrying soviet ideals (Zajda 1988; Dunstan 1981). The current study shows that the ideology of "moral education" is still prevalent in Russian technical universities. The main difference is that moral education is not aimed to develop a citizen with certain political ideals but rather to encourage students to self-fulfillment.

An alternative approach to ethics education is to see it in the frame of professional education. STEM professionals, engineers, or scientists must acquire standards of professional practice, and that can be done only via professional education by learning norms and standards recognized to be praiseworthy in that field of activity. STEM ethics can be seen as the study of principles and standards underlying the professions' responsibilities and conduct. It examines ethical dilemmas and challenges met by the practitioners of professions, the way in which professionals organize and develop ethical standards for members of their profession, and how these standards are applied in everyday practice. Without having ideals, principles, and standards, it is complicated to lead students to make ethically sound choices.

The question is not choosing between moral development and professional ethics education but between local and global approaches. The complexity of ethics challenges students and tomorrow's experts face the need for ethically competent and morally sensitive people. Rather the question is, if we are not providing specific, STEM-related ethics education and limiting it only to the encouragement of moral self-fulfillment, some important professional principles and standards might be overlooked.

8.7 Concluding Remarks

A close reading of the papers presented by researchers from the former Soviet countries gave evidence of the dividedness of the STEM research communities. While belonging to the European Research Area has fostered the integration of three Baltic states into the international research community, historical, cultural, and language-specific reasons have kept Russian researchers from close cooperation with western countries. The war between Ukraine and Russia widens the gap even more.

Although physics and mathematics are universal and technology works or not, the mentality of scientists or engineers seems to be the key to the global future. The principal question is, should it be the process of orienting non-western countries towards the Western ideas of professional, ethical education, or should it be created a commensurable understanding of global ethical culture in STEM education which would serve the global ideals of professional activity but be sensitive to local differences.

References

Andrianov, A., V. Kanke, I. Kuptsov, and V. Murogov. 2015. Reexamining the ethics of nuclear technology. *Science and Engineering Ethics* 21: 999–1018. https://doi.org/10.1007/s11948-014-9578-0.

Auer, M.E., and T. Rüütmann. 2021a. Educating engineers for future industrial revolutions. In *ICL 2020. Advances in intelligent systems and computing*, vol. 1329. Cham. Vol. 1: Springer. https://doi.org/10.1007/978-3-030-68198-2.

——— (Eds.). 2021b. Educating engineers for future industrial revolutions. In *ICL 2020. Advances in intelligent systems and computing*, Vol. 1329. Cham: Springer. Vol. 2 https://doi.org/10.1007/978-3-030-68201-9.

Brom, F.W.A., S. Chaturvedi, M. Ladikas, and W. Zhang. 2015. Institutionalizing ethical debates in science, technology and innovation policy: A comparison of Europe, India, and China. In *Science and technology governance and ethics*, ed. M. Ladikas, S. Chaturvedi, Y. Zhao, and D. Stemerding. Cham: Springer. https://doi.org/10.1007/978-3-319-14693-5_2.

de George, R.T. 1967. The foundations of Marxist-Leninist ethics. In *Philosophy in the Soviet Union*, ed. E. Sovietica Laszlo, vol. 25. Dordrecht: Springer. https://doi.org/10.1007/978-94-011-7539-5_4.

Didier, C., and K. Talin. 2015. Engineer's ecoskepticism as an ethical problem. In *Engineering identities, epistemologies and values*, Philosophy of engineering and technology, ed. S. Christensen, C. Didier, A. Jamison, M. Meganck, C. Mitcham, and B. Newberry, vol. 21. Cham: Springer. https://doi.org/10.1007/978-3-319-16172-3_13.

Downey, G.L., J. Lucena, and C. Mitcham. 2015. Engineering ethics and engineering identities: Crossing national borders. In *Engineering identities, epistemologies and values*, Philosophy of engineering and technology, ed. S. Christensen, C. Didier, A. Jamison, M. Meganck, C. Mitcham, and B. Newberry, vol. 21. Cham: Springer. https://doi.org/10.1007/978-3-319-16172-3_5.

Dunstan, J. 1981. Soviet moral education in theory and practice. *Journal of Moral Education* 10 (3): 192202. https://doi.org/10.1080/0305724810100305.

Durst, S., A. Pevkur, and V. Parts. 2021. Ethical attitudes among engineering students: Some preliminary insights. In *Educating engineers for future industrial revolutions*, ICL 2020. Advances in intelligent systems and computing, ed. M.E. Auer and T. Rüütmann, vol. 1329. Cham: Springer. https://doi.org/10.1007/978-3-030-68201-9_77.

Gulk, E.B., M.V. Olennikova, T.A. Baranova, and V.V. Khoroshikh. 2021. Professional perceptions of students of the polytechnic university. In *Educating engineers for future industrial revolutions*, ICL 2020. Advances in intelligent systems and computing, ed. M.E. Auer and T. Rüütmann, vol. 1329. Cham: Springer. https://doi.org/10.1007/978-3-030-68201-9_43.

Harris, C.E., Jr., M. Davis, M.S. Pritchard, and M.J. Rabins. 1996. Engineering ethics: What? Why? How? And When? *Journal of Engineering Education* 85: 93–96. https://doi.org/10.1002/j.2168-9830.1996.tb00216.x.

Harris, C.E., M.S. Pritchard, and M.J. Rabins. 2009. *Engineering ethics: Concepts and cases*. Belmont: Wadsworth.

Jonas, H. 1984. *The imperative of responsibility. In search of an ethics for the technological age*. Chicago: University of Chicago Press.

Karstina, S.G. 2021. Educators training in the context of socio-economic and technological trends of Kazakhstan. In *Educating engineers for future industrial revolutions*, ICL 2020. Advances in intelligent systems and computing, ed. M.E. Auer and T. Rüütmann, vol. 1329. Cham: Springer. https://doi.org/10.1007/978-3-030-68201-9_7.

Khasanova, G., V. Ivanov, and Y. Sokolova. 2013. Ethics in training of engineers. In 2013 International Conference on Interactive Collaborative Learning (ICL), 339–341. https://doi.org/10.1109/ICL.2013.6644593.

Khatsrinova, O., J. Khatsrinova, and V. Bronskaya. 2021. Realization of elite education at engineering higher education institution. In *Educating engineers for future industrial revolutions*, ICL 2020. Advances in intelligent systems and computing, ed. M.E. Auer and T. Rüütmann, vol. 1328. Cham: Springer. https://doi.org/10.1007/978-3-030-68198-2_2.

Ladikas, M., S. Chaturvedi, Y. Zhao, and D. Stemerding. 2015a. Introduction: Embedding ethics in science and technology policy – A global perspective. In *Science and technology governance and ethics*, ed. M. Ladikas, S. Chaturvedi, Y. Zhao, and D. Stemerding. Cham: Springer. https://doi.org/10.1007/978-3-319-14693-5_1.

———. 2015b. Science and technology governance and ethics: A global perspective from Europe, India and China. In *Science and technology governance and ethics*, ed. M. Ladikas, S. Chaturvedi, Y. Zhao, and D. Stemerding. Cham: Springer. https://doi.org/10.1007/978-3-319-14693-5_1.

Lopukhova, J., and E. Makeeva. 2021. Developing teacher professional competency under the integration of Russian Universities into the world academic area. In *Educating engineers for future industrial revolutions*, ICL 2020. Advances in intelligent systems and computing, ed. M.E. Auer and T. Rüütmann, vol. 1329. Cham: Springer. https://doi.org/10.1007/978-3-030-68201-9_83.

Maennel, K., O. Kivimägi, S. Maennel, S. Sütterlin, and M. Ernits. 2021. Remote technical labs: An innovative and scalable component for university cybersecurity program admission. In *Educating engineers for future industrial revolutions*, ICL 2020. Advances in intelligent systems and computing, ed. M.E. Auer and T. Rüütmann, vol. 1329. Cham: Springer. https://doi.org/10.1007/978-3-030-68201-9_51.

Martin, M.W. 1981. Professional and ordinary morality: A reply to freedman. *Ethics*: 631–633. https://doi.org/10.1086/292276.

Mitcham, C., and W. Nan. 2015. From engineering ethics to engineering politics. In *Engineering identities, epistemologies and values*, Philosophy of engineering and technology, ed. S. Christensen, C. Didier, A. Jamison, M. Meganck, C. Mitcham, and B. Newberry, vol. 21. Cham: Springer. https://doi.org/10.1007/978-3-319-16172-3_17.

National Academy of Engineering. 2009. *Ethics education and scientific and engineering research: What's Been Learned? What should be done? Summary of a Workshop*. Washington, DC: The National Academies Press. https://doi.org/10.17226/12695.

Vesilind, P.A. 2004. *The right thing to do. An ethics guide for engineering students*. Woodvile: Lakeshore Press.

Vintere, A., E. Aruvee, and D. Rimkuviene. 2021. A comparative study on the environmental behavior of engineering students. In *Educating engineers for future industrial revolutions*, ICL 2020. Advances in intelligent systems and computing, ed. M.E. Auer and T. Rüütmann, vol. 1329. Cham: Springer. https://doi.org/10.1007/978-3-030-68201-9_42.

Zajda, J. 1988. The moral curriculum in the Soviet School. *Comparative Education* 24: 389–404. https://doi.org/10.1080/0305006880240310.

Chapter 9
Engineering Ethics Education in China: Development, Promoters, and Challenges for the Future

Lina Wei and Jian Yuan

Abstract Engineering ethics education has recently been essential in China's engineering education. This chapter describes the development, promotors, and challenges of engineering ethics education in China. First comes an overview of the development of engineering ethics education in China. The brief history of such topics within engineering ethics education is described. The second part of the chapter concentrates on the promoters of engineering ethics education in China. It closes with the challenges for the future. Some suggestions for such topics for research in engineering ethics education are given.

Keywords Engineering ethics · Engineering education · Chinese characteristics

9.1 Introduction

This chapter describes the development, promotors, and challenges of engineering ethics education in China. The growing attention to such topics is due to the constant and conscious efforts of government agencies, enterprises, universities, and educators in China in recent years to reform and improve the teaching effectiveness of engineering ethics to meet the needs of the times. The national call for promoting engineering ethics instruction has become more imperative than ever due to several factors:

Firstly, the governance of science, technology, and engineering puts forward practical requirements for developing engineering ethics education in China. Science, technology, and engineering are the engine to promote human progress and a powerful lever for the industrial revolution, economic development, and social

L. Wei
China Jiliang University, Hangzhou, China

J. Yuan (✉)
Zhejiang Shuren University, Hangzhou, China

E. Hildt et al. (eds.), *Building Inclusive Ethical Cultures in STEM*, The International Library of Ethics, Law and Technology 42,
https://doi.org/10.1007/978-3-031-51560-6_9

progress. In the speech at the joint session of the 19th Meeting of the Members of the Chinese Academy of Sciences and the 14th Meeting of the Members of the Chinese Academy of Engineering, Xi Jinping, general secretary of the Communist Party of China (CPC) Central Committee, pointed out that "with the advent of the 21st century, an unprecedented level of scientific and technological innovation has intensified around the globe. A new round of scientific and technological revolution and industrial transformation is redrawing the map of world innovation and reshaping the global economic structure. Never before have we seen science and technology exert such a profound influence on our country's future and our people's lives." (Xi Jinping 2020, pp. 287–288) The influence of contemporary science and technology on society is mainly through engineering activities.

With the universal application of modern science and technology represented by information science and technology and biological science and technology, the impact of modern engineering on society and nature (including benefits and risks) is becoming more prominent, deeper, and more complex, which not only involves the economic activities of applying science and technology to production. The project's design, decision-making, implementation, and operation management will influence many aspects, such as social politics, law, culture, and the environment. In addition, the cross-application of multi-disciplinary knowledge, the combination of different technologies, the participation of various social groups, and the distribution of benefits, costs, and risks permeate engineering activities with human values. Engineers and technicians play an essential role in engineering. Therefore, if engineers and technicians only master professional knowledge but lack the understanding of ethical norms and moral judgments related to professional activities, they cannot view their professional responsibilities from different perspectives, cannot sufficiently understand the interests of varying service objects, and cannot satisfactorily deal with the conflicts among various responsibilities (Cao Nanyan 2004, p. 37).

Secondly, the goal of becoming a country with a robust engineering sector puts forward practical requirements for developing engineering ethics education in China. Since the reform and opening up of the Chinese economy, engineering in China has grown on an unprecedented scale and speed. For instance, the construction of infrastructures such as high-speed rail, roads, bridges, ports, and airports has been rapidly promoted. While fulfilling its political and economic missions in a specific historical period, Chinese engineering also faces many practical problems, such as insufficient original innovation, lack of normative consciousness, apparent desire for quick success and instant benefit, serious ecological damage, and uneven engineering quality, which resulted in many ethical problems (serious quality accidents, resource waste, and environmental pollution, etc.) in engineering activities. In the "*Jiu Cheng Gai Zao Gong Cheng*" (an old city rebuilding project), the business interests are given priority, and the protection of cultural heritage, humanistic care for disadvantaged groups, and the protection of the ecological environment are ignored (He Yongqiang 2004). These problems seem purely technical or economic but are permeated with ethical problems of value hierarchies and coordination. As China has ushered in a new historical period of economic development transformation, sound engineering has become increasingly concerning. Whether a project is "good" is fundamentally a question of engineering ethics. Good engineering is not simply technically adeptness (though it is that) but also morally good (in the sense of serving the moral ideal of improving the material condition of humanity or the like without violating any justified moral standard) (Lina Wei and Davis 2020). The projects described above require engineers and technicians to take reasonable care of ethical issues such as resource conservation, environmental protection, fairness, justice, and sustainable development while pursuing technological excellence of original innovation.

In a word, Chinese society urgently needs talents who can solve the ethical problems of Chinese engineering, guide Chinese engineering to thoroughly pursue with resolve the concept of innovative, coordinated, green, open, and shared development, give more prominence to the concepts of people-oriented and ecological harmony, build a "good project" with economic efficiency, technological excellence, and ethical excellence, strongly support the implementation of the "Made in China 2025", the strategic document issued by the State Council of the People's Republic of China in May 2015, which aims to comprehensively solve the outstanding contradictions and problems faced by China's manufacturing industry, accelerate the establishment of a modern industrial system, and provide guidance for the development of the manufacturing industry in the next 10–30 years (Miao 2015), strategy and the long-term goal of 2035, according to the strategic arrangement of building a socialist modern country in an all-round way, China will basically achieve socialist modernization by 2035 (The State Council of the People's Republic of China 2021), and strive to build a beautiful, harmonious, safe, and sustainable China. Science and engineering students are crucial sources of future engineers and scientific and technical personnel. Suppose they lack the necessary engineering ethics quality and are not good at dealing with moral problems in engineering activities. In that case, engineering projects risk hurting the harmonious development of the economy, society, and nature (Liu 2018).

Thirdly, global engineering development puts forward practical needs for researching engineering ethics education in China. Engineering has increasingly become a global profession in the era of the global economy. As more and more Chinese engineering projects are going globally, primarily driven by the "Belt and Road Initiative ",[1] Chinese engineers have more opportunities to participate in the design, decision-making, construction, operation, and maintenance of engineering projects in other countries around the world. In this process, Chinese engineers will inevitably be impacted by the local culture of the project site and face global engineering ethical issues such as cross-cultural communication, fairness, justice, and

[1]The Belt and Road Initiative is officially known as the "Silk Road Economic Belt" and "the 21st Century Maritime Silk Road," which were first implemented by Chinese President Xi Jinping in 2013. Inspired by the ancient Silk Road, the initiative aims to establish a new cooperation platform to actively develop economic networks with countries along the Belt and Road and jointly build a community with a future shared by mankind. The information is available (in Chinese) at https://www.yidaiyilu.gov.cn/info/iList.jsp?tm_id=540 (accessed 29 Nov 2022).

bribery. How should Chinese engineers make independent choices in the face of differences in technical standards, codes of conduct, customs, and other aspects? Such issues have become a practical starting point for developing engineering ethics education in China.

Education should meet the needs of the times. China strives to move from a large engineering country to a robust one. "Powerful" refers to enhancing quality, level, and innovation ability and includes improving values and ideas. The training of engineering talents should also be improved and transferred from emphasizing instrumental rationality to highlighting value rationality. This change has begun in the education sector and has expanded to the whole country and society. Drawing on the engineering development experience of developed countries and regions, people of insight in China-proposed that relevant research on engineering ethics should be started as soon as possible. Engineering ethics courses aimed at cultivating engineers' ethical awareness, moral imagination, and ability to solve ethical problems in engineering practice creatively should be set up and actively. practiced so that engineering projects can better benefit human society (Li Zhengfeng et al. 2016, Preface 2).

Since the late 1990s, after more than two decades of development, and especially in the last 5 years, China's engineering ethics education has made significant progress. However, the quality, scale, and speed of development still need to be improved, and there are still some problems and difficulties in implementing engineering ethics education (Xia Song et al. 2020). In short, engineering ethics education started in China but has a long way to go. It requires the joint efforts of the government, academia, industry, and society (Yang Bin et al. 2017). In this chapter, we focus on the development process and current situation of engineering ethics education in China, its driving factors, and the challenges to be faced in its future development from the perspective of the combination of macro background and micro details. We start with a brief discussion of China's situation concerning the teaching of engineering ethics.

9.2 An Overview of Engineering Ethics Education in China

In this section, we attempt to sketch the current state of engineering ethics education in China. Hopefully, this will create an intelligible picture of engineering ethics education in China. We start by giving a brief history of teaching this kind of topic. Then, we elaborate on the general outlook and contents of the teaching. We also describe what sources the educators use for their course material and cases.

9.2.1 Brief History and State of Affairs

Compared with the development speed of Chinese engineering, the development of engineering ethics education in China is relatively slow. The lagging development of engineering ethics education is not only related to the stage of China's economic and social development but also to the general tendency of higher education to emphasize science and engineering and neglect liberal arts (In Chinese, it is called "*Zhong Li Qing Wen*"). Under the long-term guidance of this concept, on the one hand, engineering students think that their major is to apply scientific and technological knowledge to engineering, and another ability related to humanities and society is insignificant. On the other hand, students have many vague and wrong views on science, technology, and engineering (e.g., some engineering students think that the consequences of engineering on society are not the problem that engineers need or can consider) because they undervalue the research and teaching of humanities and social sciences (Cao Nanyan 2004).

As a late-developing country of engineering ethics education, China's engineering ethics education is also transforming from localization to Sinicization. The "localization" of engineering ethics education here refers to importing the research results of engineering ethics education from other countries to China, especially the western developed countries of engineering ethics education, to interpret the phenomenon or discover and solve the problems of engineering ethics education in China. The "Sinicization" of engineering ethics education here refers to combining the basic principles of engineering ethics education with the specific reality of Chinese engineering talents, engineering practice, and Chinese cultural context, and constantly producing the research results of engineering ethics education with Chinese characteristics.

The development of engineering ethics education in China is based on engineering ethics research. In the late 1990s, Chinese scholars began to pay attention to engineering ethics education. This does not mean that in the past, in the engineering curricula in China, no attention has been paid to the ethical and social aspects of engineering or the responsibility of engineers. They were usually taught in "two courses" (Marxist theory courses and ideological and political education courses offered in ordinary colleges and universities) and Dialectics of Nature. Many colleges and universities are still teaching in this way. However, these courses did not focus on ethical aspects of the engineering profession or practice. Xiao Ping, Professor at Southwest Jiaotong University, has been pioneering engineering ethics research and developing engineering ethics courses in China Mainland. In 1998, Xiao Ping took charge of the first research project on "engineering ethics" in the Chinese Mainland funded by the National Social Science Fund of China. Then in 1999, she published her monograph "*Gong Cheng Lun Li Xue*" (Engineering Ethics), which served as the first textbook of engineering ethics education in the Chinese Mainland. In 2002, Li Bocong's book "*Gong Cheng Zhe Xue Yin Lun*" (An Introduction to the Philosophy of Engineering) was published. Since then, "engineering" has become a philosophical concept distinct from science and technology (Liu and Zhu 2015), and more Chinese philosophers are studying engineering ethics.

With China's progress in politics, economy, culture, science, technology, and ecology, Chinese scholars started to introduce more research results of engineering ethics from western countries on a large scale, especially engineering ethics in America, in the early twenty-first century. In 2005, Li Shixin published his monograph "*Gong Cheng Lun Li Xue Ji Qi Ruo Gan Zhu Yao Wen Ti Yan Jiu*" (Study on

Engineering Ethics and Its Several Important Questions) that systematically introduced related theories of western engineering professional ethics. In 2006, Cong Hangqing et al. translated the classical American textbook "*Engineering Ethics: Concepts and Cases (third edition)*" edited by Harris, C. E. et al., which has a broad impact on engineering ethics in China and still is used as an essential reference for engineering ethics education in China. In 2007, Tang Li's monograph "*Mei Guo Gong Cheng Lun Li Yan Jiu*" (Research on American Engineering Ethics) systematically introduced the formation and development, theoretical core, methods, and practical strategies of American engineering ethics. In addition to the general introduction of western engineering ethics theory, Chinese engineering ethics scholars have also made further theoretical research on significant issues such as engineers' ethical awareness, ethical issues, ethical choices, and ethical responsibilities, and tried to view Chinese engineering from the perspective of western engineering ethics theories. In 2007, Zhejiang University and the *Journal of Philosophical Research*, the Engineering and Social Research Center of the Graduate School of the Chinese Academy of Sciences, Tsinghua University, and other more than ten institutions jointly held the Academic Conference on Engineering Ethics, which the first academic seminar on engineering ethics in China. Developing engineering ethics education in colleges and universities was a hot issue of this seminar. The scholars attending the meeting generally believed that developing engineering ethics education suitable for China's national conditions was necessary. They specifically discussed teaching issues, including teaching materials and methods.

The top-down engineering education reform promoted research on Chinese engineering ethics education from different perspectives, including engineering, education, philosophy of science and technology, and ethics. In 2010, the Ministry of Education formulated the "Outstanding Engineer Education and Training Plan" (Ministry of Education of the People's Republic of China 2013). As a primary measure to cultivate a large number of high-quality engineering and technical personnel, the plan is a critical demonstration and guidance to comprehensively improve engineering education's quality. Article 1 of the general standard for outstanding engineer education and training plan is "having good engineering professional ethics, the attitude of pursuing excellence, the spirit of patriotism, dedication, and hard work, a strong sense of social responsibility and a sound humanistic quality" (Ministry of Education of the People's Republic of China 2013)[2] In the same year, the Ministry of Education also issued "Measures for the Implementation of National Certification of Engineering Education" (Ministry of Education of the People's Republic of China 2010), specifying requirements for certification of engineering education in 14 professional fields (Wang and Yan 2019).

In 2014, the engineering ethics education forum "Engineering Calls for Ethics: Dialogue between Academia and Business" was held at Tsinghua University. Representatives participating in the conference proposed that engineering education should strengthen the aspect of "ethics" and take value-shaping as one of the core

[2] This direct quotation is originally in Chinese. The author of the chapter translated it into English.

objectives of engineering education. In 2015, the National Engineering Education Steering Committee launched engineering ethics courses (e.g., organized experts to jointly write and edit a textbook, make an online course, and train engineering ethics teachers) and decided to carry out engineering ethics education in the cultivation of engineering masters. In 2016, China became a formal signatory of the "Washington Accord",[3] and China's engineering education began to integrate with that of the world. China Engineering Education Accreditation Association (CEEAA) regulates that graduates to be trained must "have humanities and social science accomplishments and a sense of social responsibility, be able to understand and abide by engineering professional ethics and norms in engineering practice and fulfill their responsibilities."[4] The professional certification of engineering education promotes the research and development of engineering ethics education in China.

At the Ideological and Political Work of National Colleges and Universities Conference in 2016, President Xi Jinping pointed out that all courses in colleges and universities should shoulder the responsibility of ideological and political education (Ministry of Education of the People's Republic of China 2016). The combination of ideological and political education and engineering ethics education is a realistic path that science and engineering colleges and universities can operate. Integrating connotation of ideological and political education into engineering ethics education can, on the one hand, give full play to the guidance and infiltration of ideological and political education in the colleges and universities, on the other hand, guide engineering students to uphold the scientific concept of development and adhere to ecological civilization with Chinese characteristics, hold paramount the safety and well-being of the public (Mao Tianhong 2018). On August 25th, 2016, the National Engineering Education Steering Committee released the proposal on strengthening the construction of engineering ethics courses, promoting the teaching of engineering ethics, and cultivating postgraduates with professional competence and political integrity in the Engineering Ethics Education Forum and Press Conference (National Graduate Education Steering Committee for Professional Engineering Degree 2016). In 2018, the Academic Degrees Committee of the State Council issued the guiding opinions on formulating the training program for masters of engineering, which formally included the engineering ethics course into the compulsory courses of an Engineering Master's degree, further highlighting the necessity of strengthening the research and practice of engineering ethics education (Ministry of Education of the People's Republic of China 2018, p.4). In 2020, the Ministry of Education issued the guiding outline for the construction of ideological and political theories

[3]This multinational agreement set the progression toward the mutual recognition of engineering accreditation in motion. Initiated in the UK by the Engineering Council, the Washington Accord, signed in 1989, is an international agreement among bodies responsible for accrediting engineering degree programs. It recognizes the substantial equivalency of programs accredited by those bodies. It recommends that graduates of programs accredited by any of the signatory bodies be recognized by the other bodies as having met the academic requirements for entry to the practice of engineering (Washington Accord (accreditation.org)).

[4]This direct quotation is originally in Chinese. The author of the chapter translated it into English.

taught in all courses of colleges and universities, which placed a particular emphasis on strengthening students' engineering ethics education in engineering majors to cultivate students' craftsman spirit of excellence and stimulate students' sense of mission to serve the country by science and technology (Ministry of Education of the People's Republic of China 2020a, b). In 2022, as the first group standard in the field of talent training quality assessment of higher education in China, the certification standard of engineering education was included in the framework of the national standard system (Table 9.1).

Regulations at the macro level, the construction of engineering ethics teaching resources, and the training for teachers at the micro level jointly promote the development of China's engineering ethics. A consensus seems to have developed that teaching engineering ethics is very significant. In recent years, Tsinghua University, the University of the Chinese Academy of Sciences, Zhejiang University, Dalian University of Technology, Southwest Jiaotong University, and many other colleges and universities have attached varying degrees of importance to engineering ethics education and have successively developed their engineering ethics courses. For instance, Tsinghua University has set up engineering ethics courses in different fields, such as landscape architecture, civil engineering, water conservancy, environment, industrial engineering, electricity, chemical industry, materials, nuclear

Table 9.1 Crucial years for the development of engineering ethics education in China

Year	Key events
1998	The first research project on "Engineering Ethics" in the Chinese Mainland was funded by the National Social Science Fund of China
1999	The first monograph on engineering ethics "Gong Cheng Lun Li Xue" in Chinese Mainland was published
2002	The book "*Gong Cheng Zhe Xue Yin Lun*" (An introduction to the philosophy of engineering) was published
2006	The "Engineering Ethics: Concepts and Cases (third edition)," edited by Harris et al., was translated into Chinese
2007	The first academic conference on engineering ethics was held
2010	The "Outstanding Engineer Education and Training Plan" was formulated. "Measures for the Implementation of National Certification of Engineering Education" was issued
2014	The engineering ethics education forum "Engineering Calls for Ethics: Dialogue between Academia and Business" was held
2016	China's Engineering Education officially joined the "Washington Accord". The proposal on strengthening the construction of engineering ethics courses, promoting the teaching of engineering ethics, and cultivating postgraduates with professional competence and political integrity was issued. The national planning textbook of engineering ethics was published
2018	Engineering ethics has become compulsory for a master's degree in engineering
2019	The second edition of the national planning textbook of engineering ethics was published
2020	The Ministry of Education issued the guiding outline for the construction of ideological and political theories teaching in all courses of colleges and universities
2022	The certification standard of engineering education is included in the framework of the national standard system

energy, information, etc. According to incomplete statistics, as of April 2019, 80.6% of the training institutions for engineering degrees have engineering ethics courses (Zhang and Wang 2020).

9.2.2 Goals of Engineering Ethics Education

The goals of engineering ethics education in China mainly include three aspects: knowledge, skill, and attitudes. In terms of knowledge and skill, Chinese engineering ethics education shares similar goals to those elaborated by the Hastings Center: (1) stimulating the ethical imagination of students; (2) helping students recognize ethical issues; (3) helping students analyze key ethical concepts and principles that are relevant to the particular profession or practice; (4) helping students deal with ethical disagreement, ambiguity, and vagueness; (5) encouraging students to take ethical responsibility seriously; (6) increasing student knowledge of relevant standards; (7) improving ethical judgment; (8) increasing a student's ethical will-power (Harris et al. 1996). The knowledge and skills will enable engineering students to recognize and analyze the ethical aspects and problems of their future professional practice and will allow them to conduct a solution-oriented debate about such problems.

Regarding attitudes, Chinese engineering ethics education has unique characteristics, emphasizing the ideological and political correctness of engineering talents (engineers and technicians, etc.). Many colleges and universities in China still regard engineering ethics as combining professional knowledge teaching and ideological and political education. As a part of moral education, engineering ethics courses are usually classified as Marxist theory courses and ideological and political education courses offered in ordinary colleges and universities. A critical goal of engineering ethics education is cultivating the national spirit with patriotism as the core and encouraging engineering talents to take the country's prosperity, national rejuvenation, and people's happiness as the purpose in engineering practice and promoting self-fulfillment while realizing the national and collective interests. For instance, the guiding outline for the construction of ideological and political theories taught in all courses of colleges and universities in 2020 stipulates that "Engineering courses should pay attention to strengthening students' engineering ethics education, cultivating students' craftsman spirit of excellence, and stimulating students' sense of mission to serve the country by science and technology" (Ministry of Education of the People's Republic of China 2020a, b).[5] This feature is the inheritance and development of Chinese traditional mutual interaction of ethical politics and political ethics. For example, Guihong Cao observed that:

[5] This direct quotation is originally in Chinese. The author of the chapter translated it into English.

> *With over 5,000 years of civilized history, China follows Taoism, Confucianism, and Buddhism. Thus, morally, Chinese science, technology, and engineering educations tend toward civilization and public benefit, which are embodied implicitly in gong cheng [engineering].... China has established a happy medium and collectivism-centered mode, focusing on the big picture and small sacrifices for the sake of the big in order to achieve harmony*. (Cao 2015, p. 1621)

Zhu and Jesiek also pointed out:

> *As the most influential school of thought in Chinese culture and philosophy, Confucianism originated as an "ethical-sociopolitical teaching". Confucianism is a sociopolitical philosophy that examines engineering and technology through ethical and political lenses. An overarching Confucian philosophy of technical projects embraces a "sociopolitical practicality", which posits that technical projects should contribute to the state's and its people's social welfare. Late in the Ming dynasty (1368-1644), this idea was systematically developed as a national philosophy or Jingshi zhiyong*. (Zhu and Jesiek 2015, pp. 155–156).

9.2.3 Teaching Mode and Content

Engineering ethics education in China is still in the exploratory stage. Colleges and universities offering engineering ethics courses actively explore the mode and content of teaching according to the characteristics of their own schools. Although engineering ethics teaching varies with the institutions, there are similar trends:

Firstly, there is widespread agreement that using cases is a critical way to teach engineering ethics (Chen Wen 2018, p. 183). The approach of case studies is commonly used in an engineering ethics course. For example, the School of Automation Science and Electrical Engineering of Beihang University, in teaching courses with aerospace characteristics, trains students' ability to analyze and solve ethical problems through case studies of ethical problems (Liu Jinkun 2018). The engineering ethics teaching team of Kunming University of Science and Technology has collected and sorted out 25 typical landslide accidents in China in recent years and preliminarily established the engineering ethics education case library for the "slope engineering" course. In the "slope engineering" teaching process, the established engineering ethics education case library was organically combined with the course knowledge instruction (Wang Chao et al. 2019).

It is usual to adopt cases as a central starting point to discuss the ethical aspects of engineering. Moreover, role-playing and group discussion are commonly used in case analysis. Students are organized to play different roles (representing various stakeholders) and carry out group discussions. Through the discussion, students can deeply understand the difficulties of dealing with ethical problems in engineering practice.

Wu Linlin et al., teachers from the College of Quality & Safety Engineering of China Jiliang University, researched the application of the case discussion teaching method in engineering ethics teaching. Taking students as the main body, they guided students to participate in the discussion actively. They helped students learn and apply the knowledge of ethics and engineering technology from different perspectives to enhance students' ethical awareness and ability to identify and respond to ethical problems.

- They adopted corresponding teaching design for different students, cases, and teaching chapters. Taking the case of the Wenzhou multiple unit train accident,[6] corresponding to the source and prevention of engineering risks in Section 2.1 of the textbook "*Engineering Ethics*" (Li Zhengfeng et al. 2019) as an example, they briefly introduced the application of case discussion method in the engineering ethics course. The teacher of the engineering ethics course discussed and guided the same accident case three times at different levels and with different concerns and completed the integration teaching of ethical knowledge and engineering knowledge.
- They introduced the case to discuss the causes of the accident and learn the primary sources of engineering risks. Through watching relevant report videos and animations of the Wenzhou multiple unit train accident in class, students were guided to think about and discuss the causes of the Wenzhou multiple unit train accident and summarize the technical, environmental, and human factors of engineering risk sources. By exposing the causes of the accident, such as inadequate design consideration, insufficient emergency awareness, and ability, the teacher strengthened students' ethical awareness and sense of professional mission of respecting nature and valuing life.
- The case was introduced for the second time to discuss the technical factors and preventive measures in the cause of the accident. Based on the first discussion, the teacher guided students to use their professional knowledge to deeply discuss the technical problems and preventive measures exposed by the accident, which helped to cultivate students' professional pride and sense of mission. At the same time, students' engineering knowledge can be strengthened by analyzing accident causes in professional qualification examinations, such as registered safety engineer, to improve students' initiative and depth in learning professional courses.
- The case was introduced for the third time to expand the discussion on improper information disclosure after the accident by watching the video of the Wenzhou multiple unit train accident conference and other ways, such as role-playing. Based on the first two discussions, students were guided to discuss how enterprises and government authorities involved in the matter should better release information to improve students' people-oriented social care and ethical awareness. The teacher introduced relevant knowledge, such as public crisis management.

[6] At 20:30:05 on July 23, 2011, in Wenzhou City, Zhejiang Province, a multiple-unit train crash occurred between the D301 train from Beijing South Railway Station to Fuzhou Station and the D3115 train from Hangzhou Station to Fuzhou South Railway Station resulting in 40 deaths, 172 injuries, 32 h and 35 min of suspension, and a direct economic loss of 193,716,500 yuan. After investigation, it was determined that the traffic accident was caused by severe design defects of the train control center equipment, lax use inspection, and ineffective emergency response after equipment failure caused by lightning stroke (Li Zhengfeng et al. 2019, p. 36).

- Summarized the cases. After the discussion, the teacher and students summarized the main factors of the accident sources and the lessons learned from the accident and emphasized safety awareness. Based on the analysis of accident causes and risk sources, the main principles of risk prevention in the next chapter were introduced.

The teaching method of case discussion was carried out in the first semester of "Engineering Ethics" teaching for master's degree students in safety engineering in 2017. The satisfaction of the comprehensive teaching effect evaluation reached 99% (Wu Linlin et al. 2019, pp. 182–184).

Some classical case materials, the Challenger disaster (Wang Pugong et al. 2022a, b, p. 89), Silent Spring (Li Zhengfeng et al. 2016, p. 84), Hydrolevel (Sheng and Tingting 2007, pp. 46–50), etc. from American textbooks were, of course, used. However, this is less desirable for a couple of reasons. On the one hand, by and large, case materials are related to the role played by codes of conduct in the American context or at least a dominant role is played by the codes. However, the situation is very different in China. On the other hand, the distance between China's engineering practice and that described in American case materials is quite considerable for students. Therefore, collecting and analyzing Chinese cases is preferable, especially living ethics cases practicing engineers encounter in their everyday experience. In other words, Chinese engineering ethics teachers focus more on the relevance of ethics pedagogies to the technical practice (e.g., learning practical wisdom for tackling ethical problems from frontline engineers) and social practice (e.g., teaching ethics through going into the field and collecting firsthand data) of engineering (Zhang and Zhu 2021). In addition, scholars in related fields believe that China's engineering ethics education should be committed to localization. Scholars and teachers tend to collect and analyze case materials from the following perspectives: the thought on socialism with Chinese characteristics, the four histories (the history of the Communist Party of China, the history of New China, the history of reform and opening-up, history of socialist development), socialist core values, national development strategy, Chinese excellent traditional culture (e.g., the theory that man is an integral part of nature, Self-discipline, and Social Commitment), the development context of national industry, and Outstanding figures (i.e., ethical engineers) working on Chinese engineering projects (Shen Yan 2022).

In summary, a case library of engineering ethics education with Chinese characteristics can present the features of China's engineering and ethical engineering problems in the context of the new era better, such as disputes over the development of Nujiang Hydropower Station, the light rail project in Mecca, Saudi Arabia undertaken by China Railway Construction Corporation Limited. Nujiang River is a river reach with rich hydropower resources and excellent development conditions in southwest China, but it is also one of the poorest areas in China. The relevant departments of Yunnan Province proposed the Nujiang River hydropower development plan as an important way for the region to become rich. However, the Nujiang River hydropower development has caused many disputes. The main reasons of the opponents include the following aspects: the construction of hydropower stations may affect the tourism of Nujiang River; hydropower stations will change the hydrology, geomorphology, and ecological integrity of natural rivers and reduce their biological and aesthetic value; hydropower stations will destroy the unique local, the national culture of multi-ethnic settlement in Nujiang region; from the long-term goal of national ecology, it should be preserved as an ecological river; the immigration problem is not easy to solve.

Meanwhile, some experts believe that resource development and environmental protection can be realized simultaneously as long as attention is paid to environmental protection in development and scientific development mode is adhered to. The dispute over the Nujiang River hydropower development has lasted for more than 10 years and has become a landmark event in the debate over environmental protection and development (Li Zhengfeng et al. 2019, pp. 5–6). The Mecca Light Rail Project in Saudi Arabia is China's first light rail project in the overseas market. During the implementation of the project, China Railway Construction Corporation Limited guaranteed the construction period in the face of a substantial increase in project workload and cost input and completed the project in only 16 months. The project was economically unprofitable but obtained good social benefits, established a brand for Chinese projects, and strengthened the relationship between China and Saudi Arabia (Zhou Xiaodong 2016, pp. 345–368).

Secondly, most engineering ethics courses are jointly taught by staff members of the philosophy department and the relevant engineering faculty. The complementary advantages of the co-teaching arrangements prove rather evident in this respect. Humanities and social sciences faculty are skilled in conceptualizing ethical issues in engineering practice. In contrast, engineering faculty are generally more aware of what is happening in their discipline than philosophers. In general, there are two modes of co-teaching: one is that teachers of philosophy are responsible for teaching the general theory, that is, the relevant classical theories of engineering ethics (e.g., Kantian moral theories, Utilitarian, virtue theories), and the engineering teachers are responsible for teaching engineering ethics in specific engineering fields. Another method is where philosophy teachers are responsible for training engineering teachers, and then philosophy teachers and engineering teachers independently complete their engineering ethics teaching. Since each method has its advantages and disadvantages, each teaching team is still exploring its own way.

The third method focuses on inviting industry experts to give lectures. Engineering ethics issues are often multidimensional, complicated issues that involve more than just engineering and ethical perspectives. Teaching practical engineering competencies requires diverse perspectives from engineering and philosophy and legal studies, policy, management, and public administration (Zhang and Zhu 2021). Integrating knowledge in different fields into engineering ethics can broaden students' horizons and encourage them to discuss actively to cultivate their ability of ethical innovation. In addition, industry experts generally possess more resources and information than staff members in colleges and universities. Usually, guest lecturers focus on a specific case in which the guest speaker has been involved. Guest lectures make it possible to present real-life cases, which makes ethical problems, it seems, more pressing and more a part of real life for students (van de Poel et al. 2001).

Moreover, the involvement of practicing engineers in engineering ethics education allows students to learn the tools, tactics, and practical wisdom that engineers often use to tackle ethical problems or other ethical considerations which may not be included in most existing textbooks (Zhang and Zhu 2021). Engineering activities are an essential way for science and technology to affect social development, and their results are closely related to the social economy, culture, law, and ecological environment. To enrich the classroom teaching content and more genuinely reflect the ethical scene of the front-line engineering practice, some engineering ethics courses invited several outside experts to teach students about cases in the process of environmental law legislation and law enforcement, ethical cases in safety evaluation, and the practice of responsibility care, to give students practical ethical guidance. For example, a chemical engineering ethics course offered by the department of chemical engineering of Tsinghua University invited experts from foreign enterprises, the government, and the legal profession to participate in the teaching. The enterprise representatives shared the excellent cases, experiences, and practices of foreign enterprises from the social responsibility perspective. The Ministry of Environmental Protection officials analyzed engineering ethics' critical role in society and the public from the government's perspective. Legal experts taught the legislation and sentencing of relevant laws from the legal level (Qin and Jinsong 2020).

The fourth method is to take the national planning engineering ethics textbook and its corresponding MOOC (Massive Open Online Courses) version as authority teaching materials (Li Zhengfeng et al. 2016, 2019). In the spring of 2015, the National Engineering Education Steering Committee organized more than a dozen experts from Tsinghua University, Beijing University of Technology, Peking Union Medical College, Dalian University of Technology, Zhejiang University, and other institutions to jointly compile the national planning engineering ethics textbook for the graduate education of engineering degree (Li Zhengfeng et al. 2016, 2019). Compiling this textbook has gathered the scientific research and teaching teams of China's authoritative experts in engineering ethics. They included scholars engaged in theoretical research in science and technology and society and engineering ethics and experts with long-term scientific research and teaching experience who have set up courses related to engineering ethics in specialized fields. The textbook, published in 2016 and revised in 2019 by Tsinghua University Press, analyzes the common ethical problems faced by engineering practice (general topics) and considers the characteristics and requirements of different engineering fields (subtopics). The textbook compilation fully considered the characteristics of engineering ethics as a kind of "practical wisdom" and highlighted the requirements of case teaching and interactive teaching. Table 9.2 gives an overview of the subjects covered in these texts.

To relieve the pressure brought by a limited number of teachers and the great difficulty in teaching engineering ethics courses, the MOOC of engineering ethics has also been launched on the online platform with the strong support of the National Engineering Education Steering Committee. These MOOCs align with the national planning textbook chapters and are jointly taught by teachers from Tsinghua

Table 9.2 Subjects covered in the national planning textbook of engineering ethics (2019)

General topics	
Chapter 1	Engineering and Ethics (Gong Cheng yu Lun Li)
Chapter 2	Risks, Safety and Responsibilities in Engineering (Gong Cheng Zhong de Feng Xian, An Quan yu Ze Ren)
Chapter 3	Value, Benefit and Justice in Engineering (Gong Cheng Zhong de Jia Zhi, Li Yi yu Gong Zheng)
Chapter 4	Environmental Ethics in Engineering Activities (Gong Cheng Huo Dong Zhong de Huan Jing Lun Li)
Chapter 5	Professional Ethics of Engineers (Gong Cheng Shi de Zhi Ye Lun Li)
Sub topics	
Chapter 6	Ethical Issues in Civil Engineering (Tu Mu Gong Cheng de Lun Li Wen Ti)
Chapter 7	Ethical Issues in Hydraulic Engineering (Shui Li Gong Cheng de Lun Li Wen Ti)
Chapter 8	Ethical Issues in Chemical Engineering (Hua Xue Gong Cheng de Lun Li Wen Ti)
Chapter 9	Ethical Issues in Nuclear Engineering (He Gong Cheng de Lun Li Wen Ti)
Chapter 10	Ethical Issues in Information and Big Data (Xin Xi yu Da Shu Ju de Lun Li Wen Ti)
Chapter 11	Ethical Issues in Environmental Engineering (Huan Jing Gong Cheng de Lun Li Wen Ti)
Chapter 12	Ethical Issues in Biomedical Engineering (Sheng Wu Yi Yao Gong Cheng de Lun Li Wen Ti)
Chapter 13	Engineering Ethics in the Global Perspective (Quan Qiu Hua Shi Ye Zhong de Gong Cheng Lun Li)

These subjects are originally in Chinese. The author of the chapter translated them into English

University, Zhejiang University, Beijing University of Technology, Dalian University of Technology, Nanjing Forestry University, and other universities.

Fifthly, pay attention to "*Xue Yi Zhi Yong*" (apply what one has learned). This implies a focus on a combination of theory and practice. Zhao Jinsong et al. believed that the separation of theory and engineering practice is the fundamental to engineering ethical problems (Zhao Jinsong et al. 2021, p. 117). At the end of some engineering ethics courses, the students are usually arranged to write an essay on a topic of their choice relating to ethics and engineering. Substantial elements of the course contents should be used or discussed.

In some cases, the integration of empirical description and ethical and theoretical reflection is demanded. The essay is written either individually or in groups. In their essays, students can show their grasp of the issues and apply the learning materials to real-life cases. For example, in the course of chemical engineering ethics at Tsinghua University, students were required to use the Internet to search the directory of crucial pollutant discharge enterprises in the city in which they live, find out an enterprise that does not disclose environmental information according to law, explain the necessity of timely disclosing environmental information according to legal provisions and ethical basis, and take actions to urge the enterprise to disclose environmental information. Students should complete their practical assignments within 5 weeks and submit a written report showing the evidence and results of the report. Ninety-four students (taking the engineering ethics course) actively participated in the practical investigation. The practice areas were distributed in 30 provinces, autonomous regions directly under the Central Government in 30 provinces, autonomous regions, and municipalities directly under the Central Government. In total, 106 units were reported. Eighty-two students received feedback from government departments or companies on rectification. The average period for students from conducting research to obtaining rectification feedback was about 24 days. Through practice, students have contributed to protecting their hometown's ecological environment with practical actions, which realizes the joint promotion of professional teaching and ideological and political guidance of engineering ethics education (Qin and Jinsong 2020).

In 2018, in the "Integrated Design" course for sophomores of civil engineering majors at Southwest Jiaotong University, students were guided to design a bridge between the library and the gymnasium on campus. Students were divided into groups of 4–5 people in each group. Each group was required to complete the following work in the form of a team within 4 weeks: ① investigate and get the environmental information near the proposed bridge site, including (but not limited to) surrounding buildings, traffic conditions, geological conditions, people flow, vegetation information, environmental changes, etc. ② select the bridge location and type, determine the basic parameters of the bridge, and draw the outline sketch of the bridge. ③ complete the written report and make slides to introduce their designs in class. In the design process, teachers guided students to comprehensively consider various technical and non-technical influencing factors (social, economic, environmental, and other influencing factors in the project) and put forward precise requirements for engineering ethics to students. The purpose was to improve students' comprehensive ability to solve problems and their sense of professional ethics and to obtain an optimal plan that balanced the interests of most people and social factors. Although these students had just started their sophomore studies, and their professional knowledge and engineering ethics knowledge were still limited, they comprehensively considered many factors (social environment, economy, structure, etc.) and made the best plan they carefully considered. This method cultivated students' ability to solve specific problems in the real world with engineering ethics awareness, evaluate and make choices with a responsible attitude, and realize the value of engineering. This method aims at cultivating students' ability to solve specific problems in the real world (Xia Song et al. 2020).

Sixthly, highlight the resources and tools of engineering ethics with Chinese characteristics. American courses on engineering ethics have been an important example and point of reference. However, engineering ethics aspects and problems cannot be adequately analyzed and reflected upon without serious attention to ethical reflection in the relevant context (van de Poel et al. 2001). Chinese engineering ethics instructors suggest that more emphasis should be placed on the virtue ethics of Confucian and Taoist ethics, which are the most influential schools of thought in the Chinese tradition. For instance, when dealing with the relationship between "*Li*" (interest) and "*Yi*" (righteousness) in engineering practice, they tend to follow the Confucian principle of "*Zheng Yi Mou Li* (seek interests in a moral way)". When handling the relationship between man and nature in engineering practice, they prefer to absorb the traditional wisdom of "*Dao Fa Zi Ran*" (following the laws of nature) of Taoism and "*He He Gong Sheng*" (harmonious coexistence between man and nature) of Confucianism. Some Chinese engineering ethics teachers strive to integrate Chinese engineering culture into engineering ethics education. For example, the teachers from Shanghai Polytechnic University pay attention to applying "*Lao Mo Wen Hua*" (model worker culture) in engineering ethics teaching. *Lao Mo Wen Hua* is one of the advanced socialist cultures with Chinese characteristics, which is formed in socialist practice under the guidance of Marxism and has distinctive characteristics of the times and the nation. It condenses the moral concepts and value orientation of an era for life and occupation. It has become a significant spiritual and cultural force to promote the development of socialist productive forces (Li and Shiliang 2019).

9.3 Promotors of the Development of Engineering Ethics Education

Some economically developed countries (the United States, the Netherlands, Britain, Germany, Japan, etc.) in the world attach great importance to engineering ethics education. Strengthening engineering ethics education has become a universal consensus of international higher engineering education. As a big country of engineering education, from the perspective of overall scale, China ranked first in engineering education in the world before 2000. From the personnel training level perspective, all engineering education levels in China reached the largest scale in 2003 (Lin and Zheng 2018, p. 15). To enhance the global competence and competitiveness of engineering talents and meet the requirements of the high-quality development of engineering, engineering ethics education must become an important part of engineering education in China. Therefore, the government, academia, industry, and society jointly promote the development of engineering ethics education (Fig. 9.1 shows the main promotors of the development of engineering ethics education).

9.3.1 The Academic Community's Vigorous Promotion

The academic community has conducted much research to provide teaching resources for engineering ethics education in recent years, aiming to promote the development of engineering ethics education better. On the one hand, engineering

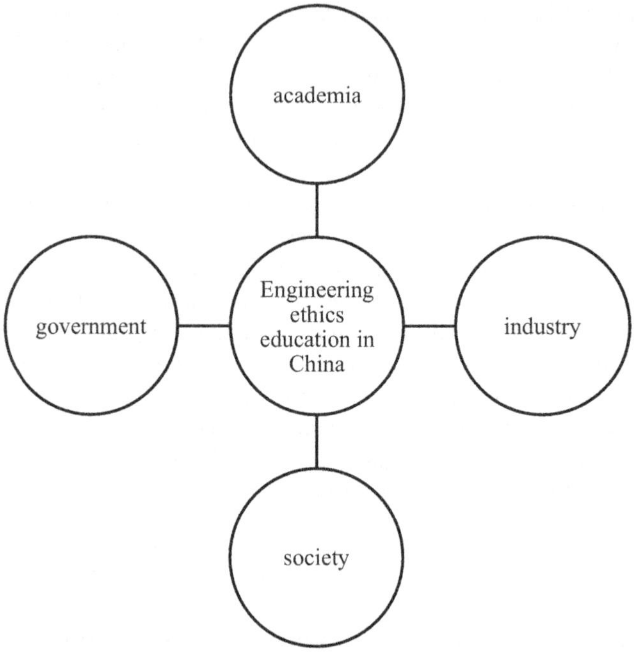

Fig. 9.1 Promotors of the development of engineering ethics education

ethicists are constantly introducing foreign advanced theoretical resources and tools of engineering ethics. On the other hand, they are consciously learning from the history and reality of Chinese engineering practice and putting forward theoretical resources and tools of engineering ethics with Chinese characteristics. They provide teaching references and materials for engineering ethics educators in monographs, textbooks, papers, case databases, etc., and cultivate engineering ethics educators through workshops, seminars, lectures, and MOOCs. Engineering teachers who teach engineering ethics try to view their professional knowledge from an ethical perspective and develop effective teaching approaches according to their professional characteristics.

9.3.2 The Voice and Action of the Industry

As the most authoritative institution in China's engineering field, the Chinese Academy of Engineering attaches great importance to engineering ethics and its education. In the book "*Gong Cheng Zhe Xue*" (Philosophy of Engineering), academician Yin Ruiyu strongly called on engineers to conduct an ethical evaluation of engineering activities in addition to scientific, technical, and economic assessment, consciously assume the responsibility for human health, safety, and welfare, and hold paramount the safety, health, and welfare of the public (Yin Ruiyu 2013).

More and more enterprises in China gradually recognize the importance of a sense of social responsibility. To pursue the sustainable development of the enterprise, they must bear responsibility for the direct stakeholders, such as customers, shareholders, and employees, and shoulder the social responsibility for the direct and indirect impact of enterprise activities on society (Yang Bin et al. 2017). Therefore, engineering enterprises pay increasing attention to employing staff members with high-level professional ethics. To achieve the win-win effect of industry and education alliances, many engineering enterprises actively cooperate with colleges, universities, and scientific research institutes to build platforms for engineering students to apply what they have learned in class.

Zhejiang University has improved the teaching quality of engineering ethics courses for graduate students by integrating industry and education. The university and enterprise jointly formulated the postgraduate training program to ensure that the proportion of practical teaching courses exceeds 50% and together built 15 brand courses with well-known enterprises at home and abroad. For example, the chief designer of the Long March 8 launch vehicle and other significant projects participated in collaborative education, which considerably deepened the students' love for the motherland and inspired the students' enthusiasm to strive for the prosperity of the motherland (Ministry of Education of the People's Republic of China 2020a, b). Meanwhile, the enterprise can cultivate the talents they need.

9.3.3 *Overall Promotion of the National Engineering Education Steering Committee*

"Postgraduate education of engineering professional's degree is a meaningful way to cultivate high-level, applied, and compound engineering and technological talents in China" (Yang Bin et al. 2017, p. 4). The National Engineering Education Steering Committee, established by the Academic Degrees Committee of the State Council, the Ministry of Education, and the Ministry of Human Resources and Social Security, is responsible for guiding the graduate education of engineering degrees in more than 400 colleges and universities. The National Engineering Education Steering Committee attaches great importance to the overall development of graduate students majoring in engineering and believes that engineering ethics is related to the sustainable development of engineering activities and even human society. Handling engineering ethics problems should become necessary for future engineering and technological talents. Therefore, it actively set the cultivating goal "*De Cai Jian Bei*" (have both political integrity and professional competence) for the graduate education of engineering degree, launched the construction of engineering ethics courses (e.g., organized the compilation of engineering ethics textbooks and engineering ethics case base, held national training courses for critical teachers of engineering ethics) and incorporated engineering ethics courses into the compulsory system in the training program of masters of engineering.

9.3.4 Promotion of Social Level

In 2015, China put forward the "Made in China 2025" strategy, hoping to achieve the upgrading and leapfrog development of the manufacturing industry. The then premier Li Keqiang pointed out: "the quality revolution of 'Made in China" needs to rely on the craftsman spirit and technological innovation." The "craftsman spirit" of Chinese engineers should not only be reflected in the high level of technology but also take customers as the center, designing and manufacturing more products and services with high quality and creativity, which can better meet the needs of consumers. Since then, the research, publicity, and reporting on the craftsman spirit have increased. The national broadcast of documentaries on engineering achievements and engineering models, such as "*Da Guo Gong Jiang*" (Craftsman of Great Nation) and "*Chao Ji Gong Cheng*" (Great Engineering), makes the image of the engineering profession clearer and more prominent. The public, on the one hand, has a better understanding of the contribution of the engineering profession to national prosperity, national rejuvenation, and people's happiness. On the other hand, the public expects the engineering profession to do better projects to meet their yearning for a better life. All the above described require engineers to hold paramount the safety, health, and well-being of the public in engineering practice.

9.4 Challenges for the Future

As mentioned above, we can find that China's engineering ethics education has begun and has made specific achievements. The call to popularize engineering ethics education in China has become increasingly strong. Nevertheless, we also need to recognize that the future development of engineering ethics education still faces many challenges, mainly reflected in the following aspects: (1) Compared with the development scale and speed of China's engineering and the number of engineering students, engineering ethics education in China has not received due attention. (2) The number of teachers competent in engineering ethics teaching is very limited. (3) The resources for engineering ethics teaching are not enough. (4) The support from engineering organizations for engineering ethics education is minimal. Most engineering societies in China have not made codes of professional ethics for engineers. The articles of engineering societies' code of ethics generally involve ethical requirements for the members from the scientific research perspective and rarely put forward ethical requirements for engineering practice activities (Yang Bin et al. 2017). Cao and Su concluded: "Although the ethical obligations of engineers have become an important part of the qualification standard for Chinese registered engineers, engineering societies of mainland China still lack clear and comprehensive recognition about the ethical responsibilities of engineers, and lack a moral ideal, beyond civic morals, exclusively belonging to engineers" (Cao Nanyan et al. 2013, p. 211). In addition, the existing professional qualification examination for engineers in China rarely involves the ethical knowledge of the engineering profession except for the specialized knowledge and relevant industry regulations.

The reasons for the above problems are complex, and the methods to solve them are also diverse. The construction of engineering ethics education in the future is a long-term, systematic, comprehensive, and strategic task. It requires the government, universities, industries, and other relevant parties to manage and promote it step by step jointly. This chapter does not put forward suggestions in all aspects. We only put forward the two main suggestions:

Provide implementation guarantee for developing engineering ethics education at the college and university level. The college and university's status and role are crucial to promoting and implementing ethics education. Firstly, taking effective measures to arouse the colleges and universities to pay attention to engineering ethics education. For example, bringing engineering ethics courses into the requirements of compulsory courses in the cultivating program and bringing the knowledge of engineering ethics into the teaching evaluation, etc. Secondly, strengthening the training of teachers. Colleges, universities, and enterprises need to work together to create conditions and build platforms to enhance the communication of teaching and scientific research cooperation between engineering ethics teachers and enterprise experts through seminars, workshops, forums, and other ways in a planned and organized manner. Thirdly, exploring and innovating teaching resources and methods.

On the one hand, engineering ethics scholars should continue introducing international advanced research results of engineering ethics comprehensively and deeply. On the other hand, actively extract engineering ethics theoretical resources and methods from Chinese engineering culture and engineering practice. For instance, He Jing proposed applying the "*Chang Jing Xu Shi*" (scene narrative) to engineering ethics teaching. In the case of teaching, the "narrative" method is adopted in combination with the "scene" of engineering practice. By comprehensively considering engineering professionals' specific moral psychology, living situation, and social roles, the students can better understand the interaction among engineering, human, nature, and society in the actual engineering activities (He and Hangging 2017). Fourthly, paying attention to "*Zhi Xing He Yi*" (the unity of knowing and doing). "'*Zhi Xing He Yi*' is the fundamental requirement of engineering ethics education" (Liu and Qian 2022, p. 61). It is far from enough for the ethics education of engineering and scientific talents to only rely on classroom training. Practice is a meaningful way to carry out engineering ethics education for engineering students. In the process of teaching engineering ethics, colleges, universities, and industrial enterprises should jointly establish an engineering ethics education base so that students can have the opportunity to go deep into the reality of enterprises, study on the spot, perceive the corporate culture and the actual engineering ethics problems they may face, explore the ways to solve problems and improve their cognition and judgment of engineering ethics problems.

Provide institutional support for the development of engineering ethics education. Engineering ethics education is an important content of the national strategy "*Yi De Zhi Guo*" (govern the country by virtue). Therefore, it is necessary to give play to the social linkage effect and promote the national top-level design. The government should promulgate relevant policies, make long-term plans for goals, contents, and development of engineering ethics education, and guide all relative parties in society to attach importance to engineering ethics education. For example, engineering ethics education should be regarded as an important content of engineering discipline construction and a necessary condition for engineering education evaluation and application for engineering education certification. In addition, the government should further encourage the construction of an "engineering spirit" based on the "craftsman spirit," further strengthening and enriching the content of engineering culture. China has a long and brilliant engineering history. Since the founding of the people's Republic of China, China's engineering has created several wonderful engineering stories, such as the construction of the Chengdu-Kunming Railway and Red Flag Canal, and cultivated outstanding engineers, such as Lin Ming (a Chief engineer of Hong Kong-Zhuhai-Macao Bridge Island tunnel project), and Qiao Sukai (a senior engineer of China Nuclear Power Operations Co., Ltd.). The above described are good resources and materials for studying the Chinese engineering spirit.

9.5 Conclusion

The development of China's engineering ethics education is deeply influenced by China's politics, economy, science and technology, culture, ecology, and society. In the development of more than 20 years, China's engineering ethics education is undergoing a process from mainly relying on the introduction of international advanced research results of engineering ethics, especially the introduction of American-style engineering ethics, to focusing on the development of engineering ethics research results with Chinese characteristics. Therefore, there are some similarities between Chinese engineering ethics education and American-style engineering ethics education, but Chinese engineering ethics education also has its own characteristics. For instance, China's engineering ethics education emphasizes the ideological and political education function of engineering ethics, highlighting community ethics, such as the values that national and collective interests are higher than individual interests and that people and nature should coexist harmoniously. In addition, compared with theoretical research, Chinese engineering ethics educators emphasize the practical research of engineering ethics, focusing more on collecting and analyzing the living cases of practicing engineers rather than fictitious ones.

With the joint promotion of the government, universities, industry, and society, China's engineering ethics education has made significant progress in setting teaching goals, constructing courses, accumulating teaching resources, and training teachers. Through empirical research such as questionnaires and interviews with engineering students who have received engineering ethics education, we can find that most of the students participating in the survey believe that their awareness and understanding of engineering ethics issues, such as the moral ideal of engineering (e.g., improving the material condition of humanity) and the professional responsibility of engineers (e.g., the safety, health, and well-being of the public shall be seen as paramount) have been improved (Lina Wei and Davis 2020). The case analysis report submitted by students shows that the students can examine the engineering practice from the ethics perspective and propose methods to solve engineering ethics problems. Therefore, it can be said that China's engineering ethics education has achieved some positive teaching results. However, compared with the development speed of Chinese engineering and the training scale of engineering students, the development of Chinese engineering ethics education is still relatively slow. It is still necessary to improve engineering ethics education in China from the following aspects: Firstly, it still needs the joint efforts of the government, academia, industry, and society to reach a common consensus on fundamental issues such as what should be taught in an engineering ethics course? Who should teach engineering ethics? How should engineering ethics be taught?

Moreover, when should engineering ethics be taught? Aiming at further promoting the systematic and long-term development of China's engineering ethics education. Secondly, while constructing Chinese engineering ethics education teaching resources with Chinese characteristics, engineering ethics scholars must consider what resources can be shared with other countries. It is hoped that the globalization of Chinese engineering ethics education can contribute Chinese wisdom to developing global engineering ethics education.

References

Cao, Nanyan. 2004. Reflections on Engineering Ethics Education in Chinese Colleges and Universities. *Research in Higher Education of Engineering* 48 (5): 37–39.

Cao, G.H. 2015. Comparison of China-US Engineering Ethics Educations in Sino-Western Philosophies of Technology. *Science and Engineering Ethics* 21: 1609–1635.

Cao, Nanyan, Su Junbin, and Hu Mingyan. 2013. Ethical Awareness in Chinese Professional Engineering Societies: Textual Research on Constitutions of Chinese Engineering Organizations. *Philosophy and Engineering: Reflections on Practice, Principles, and Process*: 203–214.

Chen, Wen. 2018. Study on the Necessity and Reform of Case Teaching in Engineering Ethics Education. *Journal of Fujian University of Technology* 16: 183–187.

Harris, Charles Edwin, Michael Davis Jr, Michael S. Pritchard, and Michael J. Rabins. 1996. Engineering Ethics: What? Why? How? And When? *Journal of Engineering Education*: 93–96.

He, Yongqiang. 2004. Ethical Thinking on Urban House Demolition. *The Journal of Yunnan Administration* 4: 120–122.

He, Jing, and Cong Hangqing. 2017. Research on Practice Innovation of Engineering Ethics Education in China. *Jiangsu Higher Education* 6: 29–33.

Li, Dan, and Zheng Shiliang. 2019. Innovative Practice of Integrating Model Worker Culture into Engineering Ethics Education of Graduate Students. *Academic Degrees & Graduate Education* 4: 40–45.

Li, Zhengfeng, Cong Hangqing, Wang Qian, et al. 2016. *Engineering Ethics*. Beijing: Tsinghua University Press.

———. 2019. *Engineering Ethics*. 2nd ed. Beijing: Tsinghua University Press.

Lin, Jian, and Lina Zheng. 2018. From the Big Country to the Strong One: China's Engineering Education in the 40 years of Reform and Opening-Up. *Tsinghua Journal of Education* 39: 1–17.

Liu, Jinkun. 2018. Study on Engineering Ethics Cases with Aerospace Characteristics. *University Education* 7: 39–41.

Liu, Hongzuo, and Wang Qian. 2022. On the Approach and Mode of Engineering Ethics Consciousness Education of New Engineering Talents. *Journal of Engineering Studies* 14: 56–64.

Liu, Z., and Q. Zhu. 2015. Engineering Ethics: Chinese Perspectives. In *Ethics, Science, Technology, and Engineering: A Global Resource*, ed. J.B. Holbrook and C. Mitcham, 2nd ed., 84–87. New York: MacMillan Library Reference.

Mao, Tianhong. 2018. Research on the Theory and Practice of Multi-dimensional Integration of Engineering Ethics Education into Ideological and Political Education in Colleges and Universities. *Journal of Higher Education* 23: 71–73.

Miao, Wei. 2015. *Made in China 2025: Towards a Manufacturing Powerful Country* (in Chinese). http://finance.people.com.cn/n/2015/0526/c1004-27054769.html. Accessed 29 Nov 2022.

Ministry of Education of the People's Republic of China. 2010. *Guanyu yinfa < quanguo gongcheng jiaoyu zhuanye renzhen shidian gongzuozu he fenweiyuanhui guanli banfa (shixing) > de tongzhi* (jiao gao si han [2010]189 hao) (Notice on Printing and Distributing the Administrative Measures of the National Engineering Education Certification Pilot Working Group and Sub-committee (Trial) (Ministry of Education, Higher Education Department, Letter [2010] no.189)). http://old.moe.gov.cn/publicfiles/business/htmlfiles/moe/s3861/201110/125418.html. Accessed 12 Sept 2019.

———. 2013. *Notice of the Ministry of Education and the Chinese Academy of Engineering on Printing and Distributing the General Standards for the Education and Training Plan of Outstanding Engineers* (in Chinese). http://www.moe.gov.cn/srcsite/A08/moe_742/s3860/201312/t20131205_160923.html. Accessed 29 Nov 2022.

———. 2016. *The Ideological and Political Work of National Colleges and Universities Conference* (in Chinese). http://www.moe.gov.cn/jyb_xwfb/s6319/zb_2016n/2016_zb08/201612/t20161208_291276.html. Accessed 28 Nov 2022.

———. 2018. *Notice on Forwarding the Guidance on Formulating the Training Program for Postgraduates of Engineering Master's Degree* (in Chinese). http://www.moe.gov.cn/s78/A22/tongzhi/201805/t20180511_335692.html. Accessed 29 Nov 2022.

———. 2020a. *Notion of the Ministry of Education on Printing and Distributing the Guiding Outline for Ideological and Political Construction of Colleges and Universities* (in Chinese). http://www.gov.cn/zhengce/zhengceku/2020-06/06/content_5517606.htm. Accessed 29 Nov 2022.

———. 2020b. *Zhejiang University Creates Outstanding Postgraduate education in the high-level integration of industry and education* (in Chinese). http://www.moe.gov.cn/jyb_sjzl/s3165/202002/t20200225_423797.html. Accessed 29 Nov 2022.

National Graduate Education Steering Committee for Professional Engineering Degree. 2016. *Engineering Ethics Education Forum and Press Conference Held at Tsinghua University* (in Chinese). http://meng.tsinghua.edu.cn/xxfb/kyjy/2335.htm. Accessed 29 Nov 2022.

Qin, Ruoshi, and Zhao Jinsong. 2020. Research on Teaching Methods and Extracurricular Practice of Engineering Ethics Education Under the Background of COVID-19 Epidemic. *Chemical Engineering Higher Education* 6: 1–6.

Shen, Yan. 2022. Exploring and Practicing on Engineering Ethics Education in the New Era. *The Theory and Practice of Innovation and Entrepreneurship* 2: 150–152.

Sheng, Xiaoming, and Ma Tingting. 2007. Watergate in Engineering History-The Analysis of the Ethical Significance in Hydrolevel Case. *Studies in Dialectics of Nature* 8: 46–50.

The State Council of the People's Republic of China. 2021. *The 14th Five-Year Plan and the Outline of the Long-term Goals for 2035 for National Economic and Social Development of the People's Republic of China* (in Chinese). Accessed 29 Nov 2022.

van de Poel, I.R., H. Zandvoort, and M. Brumsen. 2001. Ethics and Engineering Courses at Delft University of Technology: Contents, Educational Setup, and Experiences. *Science and Engineering Ethics* 7: 267–282.

Wang, Q., and P. Yan. 2019. Development of Ethics Education in Science and Technology in Technical Universities in China: Commentary on "Ethics 'upfront'": Generating an Organizational Framework for a New University of Technology. *Science and Engineering Ethics* 25: 1721–1733.

Wang, Chao, Zhang Chengliang, Liu Lei, and Zhang Xiaoqiang. 2019. Construction of Engineering Ethics Teaching Case Base of "Slope Engineering" for Mining Engineering Postgraduates. *Education Modernization* 6: 126–128.

Wang, Pusheng, Li Ping, and Liu Lidong. 2022a. *A Collection of Engineering Ethics Cases*. Beijing: Tsinghua University Press.

Wang, Pusheng, et al. 2022b. *A Collection of Engineering Ethics Cases*. Beijing: Tsinghua University Press.

Wei, Lina, and Michael Davis. 2020. China's Unwritten Code of Engineering Ethics. *Business & Professional Ethics Journal* 39: 169–206.

Wu, Linlin, Chen Yongliang, Xu Wang Qiang, Liu Hui Zhiling, An Yan, and Qian Xiaoyao. 2019. The Application of Case Discussion Method in Teaching of Engineering Ethics. *Education Modernization* 6: 182–184.

Xi, Jinping. 2020. *The Governance of China III*. Beijing: Foreign Languages Press Co. Ltd.

Xia, Song, Wang Yilin, Fu Xiao Ping, and Haiying. 2020. Integration of Engineering Ethics in Civil Engineering Education. *Research in Higher Education of Engineering* 1: 172–176.

Yang, Bin, Zhang Man, and Shen Yan. 2017. Promoting the Future-Oriented Education on Engineering Ethics in China. *Tsinghua Journal of Education* 38: 1–8.

Yin, Ruiyu. 2013. *Gong Cheng Zhe Xue*, 251–257. Beijing: Higher Education Press.

Zhang, Man, and Sun-yu Wang. 2020. The Practice and Exploration of Engineering Ethics Education at Universities. *Journal of Shanxi Normal University (Social Science Edition)* 47: 103–107.

Zhang, Hengli, and Qin Zhu. 2021. Instructor Perceptions of Engineering Ethics Education at Chinese Engineering Universities: A Cross-Cultural Approach. *Technology in Society* 65: 1–11.

Zhao, Jinsong, Qiu Tong, and Chen Bingzhen. 2021. The Role and Practices of Engineering Ethics Education in the Liberal Arts Education for Engineering Students. *Journal of Dialectics of Nature* 43: 115–120.

Zhou, Xiaodong. 2016. *The Belt and Road Practices*. Beijing: China Machine Press.

Zhu, Q., and B.K. Jesiek. 2015. Confucianism, Marxism, and Pragmatism: The Intellectual Contexts of Engineering Education in China. In *International Perspectives on Engineering Education*, Philosophy of Engineering and Technology, Vol. 20, ed. S.H. Christensen, C. Didier, A. Jamison, M. Meganck, C. Mitcham, and B. Newberry, 151–170. Cham: Springer.

Chapter 10
Building Ethical Awareness Using Culturally Relevant Practices in STEM Departments

Karina Vielma

Abstract Many minoritized students (traditionally underrepresented and socially marginalized) in science, technology, engineering, and mathematics (STEM) fields often feel the need to assimilate or repress their identities in order to succeed in their college majors. Others who successfully complete STEM degrees leave the workforce for an occupation in a field that is more welcoming of their identities and cultures. College STEM departments can make a lasting impact toward ethical awareness on all students by using culturally relevant practices in their courses, mentoring, research, leadership, and professional and personal development. This chapter presents ways to incorporate culturally pertinent practices of STEM educational spaces, including research and courses. It provides examples of what these practices may look like in a STEM department.

Keywords Culturally relevant pedagogy · Broadening participation · Ethical teaching

10.1 Introduction

Benitez (2010) used the term "minoritized" to "refer to the process [action vs. noun] of student minoritization…[that] assumes that there is a history of structural and institutional actions that have over time-limited access to, and led to a lack of presence among students of color in higher education labeled as racially and ethnically different from the norm" (p. 131). Minoritized students include those from underrepresented groups in science, technology, engineering, and mathematics (STEM) fields; for example, students from Hispanic/Latinx, American Indian, or Black

K. Vielma (✉)
Department of Biomedical Engineering and Chemical Engineering,
University of Texas at San Antonio, San Antonio, TX, USA
e-mail: karina.vielma@utsa.edu

E. Hildt et al. (eds.), *Building Inclusive Ethical Cultures in STEM*, The International Library of Ethics, Law and Technology 42,
https://doi.org/10.1007/978-3-031-51560-6_10

backgrounds, females, and veterans. Students from low-income backgrounds and rural geographic areas are also often underrepresented in STEM. This is primarily due to fewer school resources, including less qualified STEM teachers, and less exposure to role models and career opportunities in science, technology, engineering, and mathematics disciplines.

The National Science Foundation and other federal organizations (NCSES 2023) acknowledge the challenge of underrepresentation in STEM fields. Along with the potential to forfeit global competitiveness, problems most challenging our nation's stability and sustainability require diverse perspectives to solve effectively. At the systemic level, companies and organizations that understand the research behind diverse teams look for diverse individuals to join their groups (Stahl et al. 2010; Wang et al. 2019). However, ethical considerations require us to think differently about who benefits from the technologies developed and which communities are harmed by the creation of the technologies or the exclusion of their availability for certain portions of society.

Individually, those underrepresented and minoritized in STEM fields often feel like outsiders, opting to assimilate rather than fight to be included in work and school spaces (McGee 2020). Many underrepresented STEM majors who fight to be seen and heard in engineering feel they must constantly prove their abilities to be respected (Alexander and Hermann 2016). This often leads to battle fatigue, which can affect different minoritized groups differently. Racial battle fatigue is "the physical and psychological toll of constant and unceasing discrimination, microaggressions, and stereotype threat" (Fasching-Varner et al. 2014, p. xvii; Smith 2004). Microaggression is the "brief and commonplace daily verbal, behavioral, or environmental indignities, whether intentional or unintentional, that communicate hostile, derogatory, or negative racial slights and insults toward people of color" (Sue et al. 2007, p. 271). Aside from battle fatigue, impostor syndrome often plagues the thoughts of minoritized students in the STEM spaces by causing them to feel like they don't belong, and even though they are well-prepared to do the work, they feel like they may be removed from their post or be "found out" that they do not belong there. This impostor phenomenon is described as "feelings of fraudulence," especially when people are unable to "attribute their success to their abilities despite many achievements and accolades" (Parkman 2016, p. 52). This, of course, is an unrealistic perception; however, because of small microaggressions that accumulate over the years through being in the STEM field, it is not uncommon for minoritized individuals to feel impostor syndrome.

Similarly, stereotype threat is another perceived reality that impacts individuals to the point of affecting their performance (Steele 1997; Steele and Aronson 1995). Stereotypes in society, i.e., women are not good at math, are perpetuated by the deficit ideologies of those in positions of power such as faculty, professors, or employers. When stereotypes are believed and communicated, the individuals will either fight to prove the stereotypes wrong and potentially get battle fatigued or can succumb to the stereotype, giving in and performing subpar. These threats cause personal effects on performance and are part of a more significant societal perception communicated in micro- or macro-aggressions.

Therefore, it's important to acknowledge that minoritized individuals who persist in STEM fields have done so with many challenges. They not only bring new, different perspectives so solve societal problems. They also contribute with resilience and pride in their work. Purpose is a large part of their motivation and includes helping society through problem-solving. Making meaningful connections between society and engineering work is critical to retain talented individuals. Minoritized students especially need to see links between their work in STEM fields and the relevance to positively serve their communities and using their unique backgrounds and talents. Culturally relevant pedagogies and practices can help students make these connections through technical courses that appear to be stripped of contextual and societal relationships.

The field of education offers culturally relevant pedagogies (Ladson-Billings 1995) and culturally responsive practices (Gay 1975) to help bridge connections between society and the work of engineering. These practices also help acknowledge the value that diverse backgrounds of students bring to engineering. Culturally relevant and responsive education can broaden the participation of minoritized students and bring diverse thought into science, technology, engineering, and math disciplines.

10.2 Culturally Relevant Education

How do educators highlight and honor learners' backgrounds? Culturally relevant pedagogies give us three key components to do this (Ladson-Billings 1995). First of all, learners are at the center of the academic spaces. Bringing in student perspectives and finding ways to include them (rather than exclude them) are at the center of this way of teaching. Activities that help students connect a home or personal experience to the material are critical to incorporate into lessons. Developing avenues to bring in content that students have learned in ways they have learned outside of academic settings is also important. This is the first component of culturally relevant educational practices. Examples include activities that include dialogue at the beginning of a course or when introducing a new topic to find out what students know about the subject and how they learned this material and help them connect what they know to what will be learned throughout the semester or year. Then, after assessing and connecting the overall ideas of the course, it is essential to incorporate ideas that give students the opportunities to use this information and methods they already know within the course to learn new concepts.

The second component of culturally relevant pedagogy recognizes the power of the educator in the learning process. With this power comes the potential to influence in both positive and negative ways the students' learning experiences. This is why this acknowledgment is so critical to culturally relevant pedagogy. The educator holds control of grading, assignments, activities, and beliefs about students that can all impact the final academic outcomes for learners. Believing that all students can succeed in the course and in the field is essential to culturally relevant pedagogy.

This means that educators must not view students and their backgrounds with a deficit perspective but rather with an asset-based lens. Students' backgrounds, beliefs, and practices cannot be seen as barriers to their success in STEM. Educators must often reflect on their beliefs about individuals or groups of students, including how these beliefs may impact class interactions or how grading is conducted. Educators are asked to find ways to include students' experiences, ways of learning, and values in the curriculum while viewing this knowledge as valuable to learning and each individual student's academic success.

Finally, the last component of culturally relevant pedagogy is developing social-political consciousness amongst the students through their academic learning. Exposing systems that seek to oppress individuals and thinking critically about ways to deconstruct these destructive forces and narratives through teaching and learning becomes essential in culturally relevant educational practices. In traditional engineering courses, this is often seen as unnecessary. Many engineering faculty members decide early in their career that their work is objective and does not have or take social or political sides. However, upon further examination, most engineering faculty members agree that there are many components to practicing engineering, and even within engineering research, that impact community, groups of people, and society in intentional and unintentional ways. Students taught that engineering is neutral often join the workforce or begin having internships in the field and quickly see and learn that engineering practice is far from neutral. Therefore, the research on culturally relevant pedagogy challenges educators to reflect on the factors that our engineering work impacts, such as who benefits from the technologies developed and who is excluded or harmed by the technologies of engineering. Engaging in this reflective practice from the early stages of the educational trajectory can help engineers-in-training understand their role in sociopolitical actions impacted by their work.

10.3 Building Connections Using Culturally Relevant Practices

So how do engineering educators build connections between students and content using culturally relevant practices? What ways can scaffold engineering educational activities to bring students' knowledge into teaching spaces? Several areas of educational research point to how engineering educators can create connections and bridge student learning. These include the use of students' funds of knowledge (Moll et al. 1992), community cultural wealth (Yosso 2005), traditional ecological knowledge (Pierotti 2011), cultural modeling (Lee 2001, 2003), and third space (Gutierrez 2008) to help make these connections.

Funds of Knowledge (Moll et al. 1992) Research on funds of knowledge recognizes that all members in the academic and cultural environment have valuable contributions to learning. This includes students, students' families, and members of the communities where students come from and communities impacted. Engineering heavily focuses on problem-solving, and the various ways students and their families solve problems at home, for example, are funds of knowledge. This may not be part of the traditional ways of solving problems at the university or academia, and it can be incorporated into lessons and valued highly. This practice requires engineering educators to learn about the cultural practices of students and respect and honor how they learn. Most importantly, engineering educators should incorporate this knowledge and methods into their instruction.

Community Cultural Wealth (Yosso 2005) Educational research on community cultural wealth seeks to bring to the forefront knowledge traditionally excluded from academic spaces of socially marginalized groups. In engineering fields, these marginalized groups include minoritized populations such as women, people living in poverty, Black, Indigenous, and Latinx racial and ethnic groups, veterans, and persons living with a disability. The use of community cultural wealth in engineering and STEM education incorporates the knowledge, skills, and abilities of these socially excluded groups and acknowledges their valuable contributions to learning.

Traditional Ecological Knowledge (Pierotti 2011) Another way to bridge connections between students and learning involves using the knowledge that generations have handed down within a culture to incorporate as the center of learning practices. Indigenous communities, for example, hold stories sacred to their culture and pass them along to new generations through traditional ceremonies and storytelling. In the same way, members of Latinx cultures hold knowledge about customs and practices passed down through various means and have special meanings in the way learning occurs. An engineering example of this knowledge was passed down to me by my father when he was looking for underground water in a plot of land he purchased. My great-grandfather showed my dad how to find water using a wire hanger; my father showed me. Upon further research, I later discovered that this is called "water locating" or "water dowsing." This knowledge can be incorporated into academic settings, and others can be welcomed to help study the various elements of STEM disciplines.

Cultural Modeling (Lee 2001) Because cultural modeling draws on cognition research, learning through cultural modeling attempts to connect previously learned concepts to new material intimately. To bridge learning, cultural modeling draws on interests, knowledge, motivation, and other socia. When designing lessons and preparing to teach new concepts to students, cultural modeling gives educators a framework for incorporating students' familiar knowledge into the learning environment. In this model, students become the experts as they attempt to solve problems. Therefore, to use cultural modeling in learning, educators must learn about the knowledge students bring to the classroom and design lessons that center students as critical knowledge holders.

Third Space (Gutierrez 2008) Another method for incorporating culturally relevant practices in STEM and engineering education spaces is using the norms within the traditional courses to think critically about what they promote and challenge the practices. Occupying a "third space" involves reflection to observe and acknowledge how we learn and the practices associated with engineering. Having this knowledge then can lead engineering educators and students to evaluate the values and outcomes of such practices and innovate ways to deconstruct them. To be prepared to present learning in this way, it's essential for the educator to survey the current landscape; then, together with the students, think critically about these spaces and how they may be promoting specific individuals and values and excluding and pushing out others. Knowing these hidden forces that are constantly at play can empower educators and learners to create space for themselves and others in teaching and learning.

10.4 A Model for Inclusion of Culturally Relevant Practices in Engineering Learning

Drawing from education best practices, engineering educators can incorporate reflection and formative evaluation with lesson planning and designing practices to include culturally relevant practices in courses, laboratories, and other formal and informal STEM learning spaces. A model for incorporating culturally pertinent techniques of engineering learning spaces follows. The steps can be observed in order, or portions can be used to integrate into various components of the educational planning process. However, regardless of how they are used, these elements will contribute significantly to bridging connections for students to the content and their cultures and identities.

Assessing Student Knowledge, Spaces, and Cultures Educators can actively evaluate the knowledge that students bring to their courses from the start of the course and throughout the duration of the course. Using surveys and questionnaires with several open-ended items can help get a general idea of the knowledge and trends present in student enrollment. However, there is added value in the social and public dialogue of these discussions. Students learn about other student's experiences, knowledge, and skills. There is a community that starts to build when students can connect in multiple ways, perhaps with a standard way of doing things or family values that they can relate to. And students can also draw from each other's experiences once they know the various expertise of their peers. Educators must also participate in these discussions as part of the community of knowledge available to draw from.

Some questions that educators can ask when assessing student knowledge include the following:

- What attributes do students possess prior to entering the class?
- What cultures are represented in student enrollment?

- How have these cultures historically, politically, or socially contributed to the content?
- Why are students interested in the course?
- What is the motivation for students to learn this material?
- How do students hope to use the content in the future?
- What knowledge and skills about the course subject do students bring to the class?
- How have students acquired this knowledge in the past? At home? In their communities? Socially?

It remains important to avoid making assumptions or generalizations about the students in the course. Instead, it is recommended that professors ask students about their cultures, knowledge, and assets to allow learners to reflect on their experiences and home lives; this will help them connect the content and their own experiences. Reflecting this way is a form of validation for the students and a way to assess prior knowledge that may be relevant to the course.

Designing Learning Spaces As the leader of the learning community, educators must create activities that connect desired to learn objectives and outcomes with students' knowledge, cultures, and learning methods. This requires reflection, creativity, and a spirit of prototyping the educational spaces. Using Bloom's Taxonomy (Bloom et al. 1956; Bloom 1969; Anderson et al. 2001), engineering educators can draw on the objectives they want to achieve with each lesson or unit. Bloom's Taxonomy is a tool used to connect desired learning goals with activities and assessments. It presents different levels required to participate in learning; the higher-order thinking skills are assigned to tasks requiring creating, evaluating, and analyzing activities while remembering, understanding, and applying activities are considered lower-order thinking skills. Educators can be creative about other objectives that can be included that may not be the traditional ABET objectives, such as analyzing engineering practices of different cultures and creating new ways to solve STEM problems that involve cultural knowledge. These objectives will help advance culturally relevant practices in the course. Then, educators can incorporate activities that would achieve these learning goals. In this step, thinking about how students will be involved is critical. Some questions to ask when designing the activities are the following:

- **Cultural Modeling**: How will students' knowledge be central to the learning?
- **Funds of Knowledge:** How will the activities involve drawing from the various ways in which students solve problems at home or in their communities?
- **Community Cultural Wealth:** How will the knowledge of marginalized groups be brought to the center of learning?
- **Third Space:** How will the engineering practices be questioned, challenged, and deconstructed for their value and power?
- **Traditional Ecological Knowledge:** What knowledge and practices passed down through generations can be incorporated into the students' activities?

If needed, working with an engineering or STEM education faculty member can faculty design activities that address components of cultural relevance into the course. Both designing and leading educational activities that integrate culturally relevant pedagogy will create a different atmosphere for student learning.

Protecting and Maintaining a Safe Space for Learning Because students may not be accustomed to engaging in dialogue in their courses or contributing in these new ways, it will require care to protect and maintain a safe space for all students. This includes not discounting students' input, acknowledging that all forms of knowing and learning are valid and valuable, addressing challenges critically and proactively, and providing opportunities for all students to participate in every class session. To take proactive steps, the instructor must collaborate with students at the beginning of the course to create rules of engagement for the students that answer the following questions:

- How will we deal with conflict? As a group? Individually?
- What steps will be taken to make difficult decisions?
- What will be the process when we disagree on a decision?
- How do you want to be shown respect?
- How will we show respect to others?
- What will happen when we feel disrespected in the learning space?
- What are the various ways in which we will communicate?
- What are the comfort levels for communicating in these various ways?
- How can we ensure that everyone is engaged in and contributing to learning?
- In what ways can we validate others' contributions?
- How do we want our contributions to be valued and/or shown value?

Educators must remain aware of the signals in the course that point to a need that must be addressed and revisit the rules of engagement often by incorporating frequent reflection and check-ins. Giving students guidance to address these conflicts will help them develop into engineers that can respectfully engage with others in their practice.

Critical Reflection Incorporated as part of maintaining and creating a safe learning environment is the essential practice of reflection. In education research, being a reflexive leader in learning settings involves rethinking what happened, how it happened, why it happened, and what can be done differently when presented with a similar situation. This can also be done for future activities by reflecting on the following questions:

- What has happened in the past? In the field of engineering? In the course? In a particular situation?
- How has it happened? What events have made it possible for these events to occur?
- Why did this happen this way? Who were the leaders in the enactment of these events?
- What can be done differently?
- Why is it essential to implement these activities differently?
- Who benefits? Who is included?
- Who is harmed? Who is excluded from the process?

Being a reflective engineering and STEM educator and engaging in reflection with students will help maintain the key purpose of the activities at the forefront of decisions and interactions. It will also help build community amongst all learners and give them a voice and contribution. No two people think perfectly alike. This is why it remains essential to think critically about the practices that we engage in. Reflecting on our own and in the social learning space fosters a more substantial meaning and connection to the work being done. Critical reflection can also allow for a safe space for students to reflect critically on their learning and think ahead of the practices they wish to engage in or avoid.

10.5 Examples of Culturally Relevant Educational Activities

There are many examples of culturally relevant educational activities, from incorporating student voices in lessons to developing user-centered design technologies to critically debating the processes of selecting where chemical plants are placed. To help illustrate the various ways in which coursework can incorporate culturally relevant practices in lessons, STEM-specific examples are presented.

Connecting Students' Culture and Interests to Biomedical Engineering In an introductory biomedical engineering course, students explore the various topics and career pathways to undertake. Early in the course, students are assigned a project with multiple deliverables to develop a positive engineering identity and create connections between their interests and career.

First, students reflect on why they seek to pursue a biomedical engineering career. Faculty leading the course prompts students to think about how that purpose connects with the student's identity, including their background, community, and culture. A rubric is distributed to include these elements as part of the assessment. Students create a visual of their purpose for seeking a biomedical engineering degree, including images, pictures, and a timeline beginning before their undergraduate journey—when they became interested in the field—and ending in the various roles they will take as leaders in their careers. This presentation is shared, and influential mentors and family is invited to participate in person or via a virtual setting.

For the next assignment, students will work in groups to meet with and interview members of the Biomedical Engineering community in the local area and learn more about the everyday work and impact. Students will analyze what they learned from the interviews and create a list of skills and knowledge they wish to gain during their undergraduate coursework. Students will think about how they will achieve these skills outside of the course and will be required to apply for an internship or research position. The rubric for this assignment will include submitting interview questions before the interview for instructor feedback, analysis of the interview (what was learned), skills and knowledge that students wish to gain, and how they will achieve these skills. Proof of application submission to an internship will help motivate students to go through the process.

As a final deliverable for the course, students interview family and community members to learn more about the biomedical engineering technologies needed in their communities. Students think critically about the needs, what caused the social and/or historical conditions, and how they, as biomedical engineers, can help be part of the solution. The rubric for this assignment includes opportunities for students to share who has benefited from biomedical engineering solutions, who has been left out, and how those left out can be included. There is also a rubric item for the authentic voices of community members, portions of society, and family most impacted by the positive advances and the missed opportunities of biomedical engineering technologies. Throughout the course, students are encouraged to share their stories. Other students also contribute to the solutions and critically dialogue to create space to conceptualize change through their current and future work.

10.6 Applying Culturally Relevant Practices to STEM Ethics Education

Applying culturally relevant practices in STEM ethics courses is essential for several reasons. First, the professor can learn about students' identities and values. Also, instruction can be tailored to address the controversies within an ethics course with the students' viewpoints. Additionally, students will get differing knowledge of the history and future of ethical issues, including emotions that support or reject specific topics; for example, an engineering design be racist? Finally, helping students learn how to deconstruct and co-construct their ideas around ethics is a leadership skill for future engineering and scientific ethical issues that students may confront as technologies and society advance.

Reflections on Positionality and Social Impact Several activities have been used to bring culturally relevant practices into STEM ethics education. One activity uses the social identity map (Jacobson and Mustafa 2019) to reflect on personal and professional identities and how society perceives these identities. This exercise helps all students (and faculty) understand how identities and perceptions are socially constructed, many times through historical, cultural, or political events that then lead to the unfair, unjust treatment of targeted individuals. Ways we can ethically respond to identities within the STEM fields are included as a concluding exercise to this activity and can be tailored to any audience. Questions that are asked at the end of this activity include: (for researchers) Which populations does your research benefit from? Are there portions of society that are harmed by the research? How can your research be more inclusive of other parts of society? (For teaching assistants) Who benefits from your instruction? How can you engage all students, especially those who are struggling?

Some notes about this activity: Often, words charged with emotions come up, such as "racism", "sexism", "privilege" and "discrimination", which can lead the discussion toward talking about sensitive experiences. Therefore, it is important to set ground rules at the beginning of this activity, including participants by asking for suggestions for the ground rules and providing a set of fundamental rules. Recommendations for ground rules include: (1) Be respectful of each other. (2) Listen with care. (3) Speak from a personal point of view. (4) Personal stories shared in this room do not leave this space. Having some ground rules is important to creating a safe space for conversation.

An extension of the social identity map can include conversations surrounding the various causes and impacts of sociocultural and sociopolitical identities of STEM disciplines, research, and specialties, including how capitalism and utilitarianism may drive engineering and scientific technologies. The intersection of social challenges and technical "needs" creates a learning space for engaging in analyzing social responsibility. Examples of these kinds of engineering ethics problems are illustrated by Jonassen and Cho (2011).

Cultural Relevance in STEM Ethics Courses
When designing a STEM ethics course, the content presented should also be critically examined. Questions faculty educators can ask are: What are the dominant forms of ethics knowledge presented? Who is the "author" of this knowledge, and why is that knowledge most important? How can other knowledge be incorporated into the course topics? How can students' ethical views be honored in an ethics course? How can personal values, STEM ethics, and social responsibilities be simultaneously learned?

As an instructional leader, learning about different ethical perspectives and including them in the syllabus topics is a form of honoring diverse cultures' ethics knowledge while challenging the norms of ethics education. A wide lens to capture multiple cultures' ideologies through ethics research and education will help students value diverse ways of thinking. Ethics in religion, cosmovision, indigenous cultures, Hispanic and Black traditions, and others can help students bring their ethical knowledge into STEM fields. Resources that provide examples of how educators incorporated non-Western ethics education can help spur ideas for innovative teaching and learning (e.g., Gustein and Peterson 2006, 2013; Leydens and Lucena 2018; Baillie et al. 2012; Mejia et al. 2022).

10.7 Engineering Education Leaders Build Connections

Incorporating culturally relevant pedagogies and engaging in reflective practices are part of becoming engineering education leaders that build connections for all students (Thevenot 2023). Engaging in these educational practices can also help students prepare for the engineering work they will conduct in the communities that will be impacted. To encourage and sustain these practices, it is essential to create a community of practice among engineering faculty that values inclusive learning systems.

Following the same model for inclusion presented, faculty and administrators can bridge connections between students, faculty, families, cultures, and communities served. In departmental meetings, community building can be initiated by assessing and dialoguing about the cultures, knowledge, and skills that each member brings to the group. Learning about our own inclusivity barriers and discussing them can deconstruct and dismantle these. It then remains important to share educational practices, professional development, or research acquisition to help overcome these obstacles. There is not one activity that will work for everyone. Faculty and educational leaders view the world differently as professionals with distinct backgrounds and perceptions. Therefore, different activities can be tried and assessed for the impact on the long-term sustainability that moves inclusivity forward. Faculty can also share effective teaching practices in their courses and serve as mentors for others attempting to incorporate these practices in their courses. Participating in research with engineering education faculty or education faculty can be a great way to learn more about faculty's effective pedagogies in various contexts. Regular workshops and discussions centered on inclusive educational practices can also promote a teaching and learning community within the engineering faculty.

Hosting professional development sessions with engineering education specialists that can help incorporate research-based, culturally relevant educational practices in courses can also serve as a tool to bridge normative practices with new practices. This entails creating space for faculty to remain comfortable with the pedagogies in the course while attempting new teaching methods with the help of education researchers. For example, professors comfortable with lecture-style teaching can begin by incorporating dialogue between students and space for students to think critically about a topic. Education faculty can observe and collect qualitative and quantitative data on the methods used. The two faculty could debrief to think of other ways to incorporate culturally relevant practices in the course, develop a lesson plan for the activity, and assess the impact on the students. This iterative process could continue as the faculty becomes more comfortable and creative in designing activities that draw on students' knowledge and experiences for course content implementation.

Sharing challenges and how they can be overcome may motivate engineering faculty to be proactive as they navigate these new experiences. There are many ways to bridge students' identities, backgrounds, and communities to the STEM content within a course, including through (A) various ways of knowing and learning, knowledge acquisition, (B) diverse forms of understanding and communicating, comprehension, (C) applying knowledge in innovatively to reach underserved individuals and portions of society, (D) analyzing scientific technologies with a critical social and cultural lens, (E) designing technologies that consider the users in all parts of the design process, and (F) evaluating to be inclusive rather than exclusive of others. Examples of these include (1) critically questioning the impact of technologies on various underserved populations, (2) including rubric items for projects that inspire students to think about the connections to their backgrounds, (3) including familial knowledge and other social networks to provide solutions to the problems presented in the course, (4) assigning groups to research the racialized or gendered differential treatments in their STEM fields and designing solutions through the technologies created. Understanding the pathways to bridge connections between students and content may be one of the most important tools for faculty to begin implementing culturally relevant educational practices.

The addition of these activities requires buy-in from administrators and faculty. For this reason, all educators and administrators must allow time and space for everyone to engage in critical reflection centered on educational pedagogies used to engage students in learning engineering concepts. The ideas resulting from these reflections should be honored and respected with an appreciation for the courage it takes to share perspectives and junctures in ideologies and with the understanding that educational change takes time and patience.

With these educational practices, engineering educators can help "students to recognize and honor their own cultural beliefs and practices while acquiring access to the wider [engineering] culture" (Ladson-Billings 2006, p. 36). Bridging these connections will ultimately help engage more students, retain them for the long-term, and foster an ethical engineering discipline and field for generations to come.

References

Alexander, Q.R., and M.A. Hermann. 2016. African-American Women's Experiences in Graduate Science, Technology, Engineering, and Mathematics Education at a Predominantly White University: A Qualitative Investigation. *Journal of Diversity in Higher Education* 9 (4): 307–322. https://doi.org/10.1037/a0039705

Anderson, L.W., Krathwohl, D.R., and B.S. Bloom. 2001. A taxonomy for learning, teaching, and assessing: A revision of Bloom's taxonomy of educational objectives. New York: Longman.

Baillie, C., A. Pawley, and D. Riley. 2012. *Engineering and Social Justice: In the University and Beyond*. Ashland: Purdue University Press.

Benitez, M., Jr. 2010. Resituating Culture Centers Within a Social Justice Framework: Is There Room for Examining Whiteness? In *Culture Centers in Higher Education: Perspectives on Identity, Theory, and Practice*, ed. L.D. Patton, 119–134. Sterling: Stylus.

Bloom, B.S. 1969. Taxonomy of educational objectives: The classification of educational goals: Handbook I, Cognitive domain. New York: McKay.

Bloom, B.S., Krathwohl, D.R., and B.B. Masia. 1956. *Taxonomy of educational objectives; the classification of educational goals*. New York: Longman Group.

Fasching-Varner, K.J., Albert, K.A., Mitchell, R.W., Allen, C., and W.A. Smith. 2014. Racial Battle Fatigue in Higher Education: Exposing the Myth of Post-Racial America.

Gay, G. 1975. Organizing and Designing Culturally Pluralistic Curriculum. *Educational Leadership* 33: 176–183.

Gutierrez, K.D. 2008. Developing a Sociocritical Literacy in the Third Space. *Reading Research Quarterly* 43 (2): 148–164.

Gutstein, E., and B. Peterson. 2006. *Rethinking Mathematics: Teaching Social Justice by the Numbers*. Milwaukee: A Rethinking Schools Publication.

———. 2013. *Rethinking Mathematics: Teaching Social Justice by the Numbers*. 2nd ed. Milwaukee: A Rethinking Schools Publication.

Jacobson, D., and N. Mustafa. 2019. Social identity map: A reflexivity tool for practicing explicit positionality in critical qualitative research. *International Journal of Qualitative Methods* 18. https://doi.org/10.1177/160940691987

Jonassen, D.H., and Y.H. Cho. 2011. Fostering Argumentation While Solving Engineering Ethics Problems. *Journal of Engineering Education* 100: 680–702.

Ladson-Billings, G. 1995. Theory of Culturally Relevant Pedagogy. *American Educational Research Journal* 32 (3): 465–491.

———. 2006. “Yes, But How Do We Do It?” Practicing Culturally Relevant Pedagogy. In *White Teachers, Diverse Classrooms: Creating Inclusive Schools, Building on Students' Diversity, and Providing True Educational Equity*, ed. J.G. Landsman and C.W. Lewis, 33–46. Sterlin: Stylus.

Lee, C.D. 2001. Is October Brown Chinese? A Cultural Modeling Activity System for Underachieving Students. *American Educational Research Journal* 38 (1): 97–142.

———. 2003. Toward a Framework for Culturally Responsive Design in Multimedia Computer Environments: Cultural Modeling as a Case. *Mind, Culture, and Activity* 10 (1): 42–61.

Leydens, J., and J. Lucena. 2018. *Engineering Justice: Transforming Engineering Education and Practice*. New York: IEEE Press/Wiley.

McGee, E.O. 2020. Interrogating Structural Racism in STEM Higher Education. *Educational Researcher* 49 (9): 633–691.

Mejia, J.A., A. Esquinca, E. Tucker-Raymond, and A. Wilson-Lopez. 2022. *Literacies of Design: Studies of Equity and Imagination in Engineering and Making*. West Lafayettee: Purdue University Press.

Moll, L.C., C. Amanti, D. Neff, and N. Gonzalez. 1992. Funds of Knowledge for Teaching: Using a Qualitative Approach to Connect Homes and Classrooms. *Theory Into Practice* 31 (2): 132–141.

National Center for Science and Engineering Statistics (NCSES). 2023. *Diversity and STEM: Women, Minorities, and Persons with Disabilities 2023*. Special Report NSF 23-315. Alexandria: National Science Foundation. Available at https://ncses.nsf.gov/wmpd.

Parkman, A. 2016. The Imposter Phenomenon in Higher Education: Incidence and Impact. *Journal of Higher Education Theory and Practice* 16 (1): 51–60.

Pierotti, R. 2011. *Indigenous Knowledge, Ecology, and Evolutionary Biology*. New York: Routledge.

Smith, W. A. (2004). Battle fatigue on the front lines of race: Teaching about race and racism at historically white institutions. In *A long way to go: Conversations about race by African American faculty and graduate students*, ed. D. Cleveland, 141–151. New York: Peter Lang.

Stahl, G.K., M.L. Maznevski, A. Voigt, and K. Jonsen. 2010. Unraveling the Effects of Cultural Diversity in Teams: A Meta-analysis of Research on Multicultural Work Groups. *Journal of International Business Studies* 41: 690–709.

Steele, C.M. 1997. The Threat in the Air: How Stereotypes Shape Intellectual Identity and Performance. *The American Psychologist* 52 (6): 613–629.

Steele, C.M., and J.K.A.W. Aronson. 1995. Stereotype threat and the intellectual test performance of African Americans. *Journal of Personality and Social Psychology* 69 (5): 797–811.

Sue, D.W., C.M. Capodilupo, G.C. Torino, J.M. Bucceri, A.M.B. Holder, K.L. Nadal, and M. Esquilin. 2007. Racial Microaggressions in Everyday Life: Implications for Clinical Practice. *The American Psychologist* 62 (4): 271–286.

Thevenot, Y. 2023. Culturally Responsive and Sustaining STEM Curriculum as a Problem-Based Science Approach to Supporting Student Achievement for Black and Latinx Students. *Voices in Urban Education* 50 (1): 60–69. https://doi.org/10.33682/bhr8-6z56

Wang, J., G.H.L. Cheng, T. Chen, and K. Leung. 2019. Team Creativity/Innovation in Culturally Diverse Teams: A Meta-analysis. *Journal of Organisational Behaviour* 40: 693–708.

Yosso, T.J. 2005. Whose Culture Has Capital? A Critical Race Theory Discussion of Community Cultural Wealth. *Race Ethnicity and Education* 8 (1): 69–91.

Part III
Introduction: Embedding Ethics Education in Practice Contexts and Labs

Part III of the volume explores novel approaches that focus on ethics in practices: out of the theoretical contexts of classrooms and directly into the laboratory or workplace. Recent years have seen an increase in STEM ethics education initiatives that aim to overcome traditional classroom education's limitations by immersing students in practice contexts that provide students with immediate learning experiences. These approaches promise to provide meaningful and relevant insights into workplace environments, laboratories, and practice contexts. In addition, embedding ethics education in practice contexts can provide a promising way to facilitate the discussion of ethical aspects and societal implications of STEM research and STEM-based developments. Students who see the practical implications of what they learned in the classroom may be more prone to reflect thoroughly on the issues involved.

In the U.S., online or traditional theoretical classroom education in Responsible Conduct of Research (RCR) has proven ineffective; there is a need to address these complex issues in the research lab itself and include all stakeholders in this solution.

As stressed by various researchers in the field of research ethics, good RCR training and ethics education in STEM extends beyond learning laws and rules but require discussion of ethical issues in their relevant context and involvement of trainees' peers, mentors, and supervisors (Holsapple et al. 2012; Kalichman 2014; Plemmons and Kalichman 2013, 2014). Recent research has revealed many factors that can affect a student researcher's individual experience in the lab. This includes the university environment, amount of funding available, individual characteristics including gender, race, ethnicity, and immigration status, their supervisor, colleagues, and the presence or lack of support structures to help guide their learning experience (Dubois and Antes 2018; Mumford et al. 2007).

The contributions of this section of the collected volume discuss promising innovations being pioneered by the chapter authors, such as focusing on workplace-related virtues, embedding moral reasoning and teamwork training, or encouraging interdisciplinary neuroethics perspectives. Others move ethics education directly into the laboratories in that they develop new approaches to safety training or ask students to build lab-specific guidelines based on a reflection on the ethical issues they encounter in their research environments.

The chapter "Character Comes from Practice: Longitudinal Practice-Based Ethics Training in Data Science" by Louise Bezuidenhout and Emanuele Ratti describes their ethics program for data scientists that focuses on microethics and uses the language of virtue ethics. The authors describe the growing need for RCR training in this discipline around data creation, analysis, and reuse and explain how traditional approaches to RCR education often fall short. They stress that as micro and incremental technical choices are always value-laden and can show up at any stage of the design process, students in this field need to be able to scrutinize their decisions throughout this process and not simply consider the downstream consequences. The authors then provide an overview of their program, which links ethical decisions to highly repeated tasks, a fact that supports the internalization of norms by means of habituation. The program seeks to foster self-reflection in students by using daily research examples. By helping students reflect on the moral and political dimensions of micro-technical choices, this approach both focuses students on the aspirational motivations of RCR education. Also, it shows ethics as integral to the daily practice of data science. The authors end with a case study exemplifying how this approach worked as part of a two-week residential course for early career data science researchers.

In Melinda Box and Maria Gallardo Williams' chapter, "Encouraging Transparency in Lab Safety via Teachable Moments and Positive Feedback," the authors look at how performance evaluations and lab safety incident reports can provide teachable moments and help build ethically responsible cultures in research labs. They begin by stating that frequently, safety training provides an overload of requirements, regulations, rules, and restrictions and happens when students and researchers are not primed to receive this critical information. Instead, providing information to recipients when they are prepared for it can result in more meaningful comprehension and bring about more long-term, fundamental improvement. The authors describe how they developed "Lessons Learned" memos that provide summaries of real safety incidents in story format with useful takeaway information. By combining this with providing positive feedback during lab safety inspections, the authors sought to increase transparency in the lab around safety, allowing researchers to admit mistakes, ask for help, and express concerns. The authors provide examples of the "Lessons Learned" memos and how positive feedback was delivered during the safety inspections. They end with a reflection on how communication, learning, mindset, and institutional structure can be changed to help increase transparency around safety in lab settings and end with some thoughts on how other institutions might adopt a similar approach.

The chapter "In Situ Ethics Education Within Research Laboratories: Insights into the Ethical Issues Important to Research Groups and Educational Approaches" describes a research ethics education project's results. It asked students to join a Student Ethics Committee and work collaboratively to develop context-specific codes-of-ethics-based guidelines for their departments. After describing the workshop series, the authors discuss how the guidelines developed by these committees reflect the current culture of the labs and highlight ethical issues the students thought were important. The guidelines also provided intriguing potential solutions to looming ethical issues. The authors discuss how participation in the workshops helped empower student participants to address the ethical problems in the lab actively. They also describe some of the challenges of this approach and suggest potential future work in the area of in situ ethical education.

In the chapter "Engineering an Ethical Ethos: Reframing Ethics Education for Engineers and Researchers", the authors describe how the Rehab Neural Engineering Labs (RNEL) at the University of Pittsburgh set out to make ethics engagement an intrinsic part of the lab culture by creating an *ethical ethos*. Over the last three years, students, post-docs, staff, and faculty participated in interactive seminars designed to encourage personal reflection through small-group discussion. After giving an overview of ethics education provided to members of the lab, the authors describe what they mean by an ethics ethos, where ethics is a deeply integrated part of the research culture, and how their approach rests on action theory, where the *agent* is the individual researcher and the *situation* is the overarching institutional, societal, and historical context in which the research takes place. In this approach, researchers are encouraged to explore their beliefs and values about a situation and begin to understand how their professional actions can reflect these beliefs. The authors describe the seminar series they developed and how their structure and open dialogue helps facilitate thoughtful reflection and learning around research ethics and help researchers take ownership of their beliefs and convictions and apply them in all stages of their research. The authors end with a discussion of how the concept of conscientious design has become increasingly popular and how their ethical ethos approach fits into this movement, as well as suggestions of how other institutions might adopt this educational approach.

References

DuBois, J.M., and A.L. Antes. 2018. Five dimensions of research ethics: A stakeholder framework for creating a climate of research integrity. *Academic Medicine* 93(4): 550–555. https://doi.org/10.1097/ACM.0000000000001966

Holsapple, M.A., D.D. Carpenter, J.A. Sutkus, C.J. Finelli, and T.S. Harding. 2012. Framing faculty and student discrepancies in engineering ethics education delivery. *Journal of Engineering Education*, 101(2): 169–186.

Kalichman, M. 2014. Rescuing responsible conduct of research (RCR) education. *Accountability in Research* 21: 68–83.

Plemmons, Dena K., and Michael W. Kalichman. 2013. Reported goals of instructors of responsible conduct of research for teaching of skills. *Journal of Empirical Research on Human Research Ethics* 8(2): 95–103.

———. 2014 *Research ethics workshop: Promoting ethics in research.* Version 10, 13 Feb 2014.

Chapter 11
Character Comes from Practice: Longitudinal Practice-Based Ethics Training in Data Science

Louise Bezuidenhout and Emanuele Ratti

Abstract In this chapter, we propose a non-traditional RCR training in data science that is grounded in a virtue theory framework. First, we delineate the approach in more theoretical detail by discussing how the goal of RCR training is to foster the cultivation of certain moral abilities. We specify the nature of these 'abilities': while the ideal is the cultivation of virtues, the limited space allowed by RCR modules can only facilitate the cultivation of superficial abilities or proto-virtues, which help students to familiarize themselves with moral and political issues in the data science environment. Third, we operationalize our approach by stressing that (proto-)virtue acquisition (like skill acquisition) occurs through the technical and social tasks of daily data science activities, where these repetitive tasks provide the opportunities to develop (proto-)virtue capacity and to support the development of ethically robust data systems. Finally, we discuss a concrete example of implementing this approach. In particular, we describe how this method is applied to teach data ethics to students participating in the CODATA-RDA Data Science Summer Schools.

Keywords Responsible conduct of research · Virtue ethics · Micro-ethics · Data science · Open science

11.1 Introduction

In recent years, data science has spread across virtually all working, managerial, and research environments. The tools provided by this new interdisciplinary endeavor have made the analysis of large data sets possible, allowing the actionable potential of Big Data to be unleashed. Despite the advantages already realized by

L. Bezuidenhout
Centre for Science and Technology Studies (CWTS), Leiden University, Leiden, Netherlands

E. Ratti (✉)
Department of Philosophy, University of Bristol, Bristol, UK

E. Hildt et al. (eds.), *Building Inclusive Ethical Cultures in STEM*, The International Library of Ethics, Law and Technology 42,
https://doi.org/10.1007/978-3-031-51560-6_11

the data revolution, advances have been accompanied by the emergence of complex ethical, political, and societal issues. Unsurprisingly, scientific research using data science tools is not immune from these issues. For this reason, there is increasing pressure on academic institutions to promote the responsible use of data science tools in research.

Many institutions formulating training in data ethics are doing so in the context of Responsible Conduct of Research (RCR) programs. In these cases, RCR content is adapted to the specificities of data science by including discipline-specific case studies. While inheriting the strengths of the RCR approach, these programs also inherit many flaws that have led to criticism. These include promoting a 'compliance-based' approach to good conduct and emphasizing ethics as a bureaucratic activity.

In this article, we propose an ethics program for data scientists framed in contrast to traditional RCR programs. We outline a new framework for a non-traditional data science RCR based on the language of virtue ethics and the notion of microethics. This new framework can support the aspirational nature that motivated the creation of RCR programs decades ago and foster contextual awareness and strategies of resilience amongst data scientists.

The structure of this article is as follows. Section 11.2 describes the nature of traditional RCR programs and their flaws (11.2.1). In addition, we show why data science results in new issues that require RCR training (11.2.2), and why we cannot use traditional RCR programs as blueprints (11.2.3). In Sect. 11.3, we describe a framework for a new RCR program tailored for data science. Finally, in Sects. 11.4 and 11.5, we describe how our RCR framework works by describing its application in five steps, exemplified by the Committee on Data Research and Research Data Alliance (CODATA-RDA) Schools for Research Data Science.

11.2 RCR Training and Data Science

11.2.1 RCR Programs and Their Flaws

RCR is a widespread institutional training that aims to familiarize scientists with ethical, legal, and societal issues in scientific research. In the United States context, RCR training was mandated by the NIH in 1989 and by the NSF in 1997 (Chen 2021). RCR training utilizes a framework that outlines areas in which researchers have responsibility. These include mentoring, authorship, conflicts of interests, protection of human subjects, use of animals in research, as well as "FFP" (fabrication, falsification, and plagiarism) misconduct (Shamoo and Resnik 2015). While the RCR discourse has been spelled out in various ways, especially in the research context of the United States, principlism has been seen as a means of structuring ethical discussion (Shamoo and Resnik 2015). Despite the varieties of ways in which these programs are designed, the philosophy and the roll-out of RCR have been associated with several tensions. Chen (2021) elaborates on these issues, which we summarize as follows.

The first issue is that mandated RCR training has not been accompanied by mandated funding allocations, meaning that institutions dedicate varying amounts of resources to the development and roll-out of training. This has led to considerable variation in the way the content is delivered. In most cases, well-established RCR programs reflect institutional champions and supportive leadership rather than the size of institutions and institutional budgets.

The variations in RCR roll-out can be seen as a side-effect of a lack of clarity on what RCR should accomplish. As Chen (2021) shows, an assessment of current RCR training shows high variability in expected outcomes – from rule compliance to character change. Similarly, methods for evaluating RCR training outcomes are highly variable, making cross-institutional comparisons challenging. Moreover, responsibility for RCR training, the position of RCR training in institutional budgets, and accountability for long-term assessment differ across institutions.

The roll-out of RCR training, challenging as it is, masks several tensions and paradoxes that persist within RCR as a subject. Many scholars recognize the mandatory nature of RCR training as a primary source of problems. On the one hand, RCR programs were motivated originally by appealing to an aspirational component, "seeking to promote research integrity through changing behavior, character, and judgment" (Chen 2021, p. 232). This is consistent with an 'authenticity paradigm of professional ethics' (Kelly 2018): the efforts to initiate and develop RCR training exemplifies ongoing 'existential' discourses of scientific communities, aiming at self-understanding, most notably in the form of delineating their own 'ethos', goals, and relations with the public. On the other hand, the mandatory nature of RCR programs has promoted a 'compliance paradigm' (Kelly 2018) that conflates with a 'rule-following' dimension. This can lead researchers to interpret RCR as "meaningless boxes to be checked, (...) byzantine and vaguely threatening bureaucracies" (Chen 2021, p. 231). This bureaucratic interpretation of RCR training misrepresents ethics as external to scientific practice and not as something that can be learned in an analogous way in which scientific skills are learned. Such training represents an ethical researcher following well-defined rules, thus conflating unquestioning compliance and ethics. This approach contrasts with the original aspiration of promoting "deep and thoughtful exploration into the various contexts and factors at play in ethical issues" (Chen 2021, p. 231).

The challenges embedded within current RCR framings and roll-out has lessened the potential training to function as a transformative tool to build character. This lost opportunity, while often remarked on, is difficult to address within the current structures of RCR training.

11.2.2 Why We Need RCR Training for Data Science

Data science has recently emerged as an interdisciplinary endeavor. This rapidly evolving discipline integrates practitioners from different scientific and engineering traditions (e.g., statistics, computer science, mathematics) to generate, process, and visualize research data. As a result, researchers occupying data science roles are heterogeneous regarding disciplinary expertise and computing training. Because of this, it is difficult to understand the boundaries between data science and the other fields that are 'assembled'. Who, exactly, is a data scientist remains an open question.

Despite its heterogeneity in application, data science coalesces around many common practices. Using computer programming to conduct data analyses means that – regardless of the application – data scientists often use a common set of computer programs and programming languages. These common practices also give rise to a common 'way of doing', in that daily data science activities comprise many repetitive tasks such as cleaning, annotating, and visualizing data and writing code to assist these activities. Data science, it can be said, is both discrete and iterative.

There has been plenty of discussions about data science not just for its interdisciplinary and 'vague' (in the technical philosophical sense) nature. It has been overwhelmingly documented that data science raises many moral and political issues relating to data creation, analysis, and reuse. We understand 'moral' as how the design and use of data science tools influence data subjects' lives and their autonomous choices (Ratti and Graves 2021), while 'political' refers to how the design and use of data science tools end up amplifying those 'moral issues' at a societal/structural level. For instance, it has been shown that large language models (LLM) – used in heterogeneous contexts from designing chatbots to the early phases of drug design – can incorporate intrinsic biases in the forms of misrepresentation, underrepresentation, and overrepresentation (Bommasani et al. 2021). These intrinsic biases can potentially lead to extrinsic harms, which can be at the individual level and especially at the collective level, when, for instance, they result in representational harms affecting both performance disparities among groups and a technical standardization of biases (Bommasani et al. 2021). These problems have emphasized underappreciated, highly technical, and repetitive practices such as data curation and documentation or model tuning (Bender et al. 2021). These issues are particularly pressing, and they also apply to scientific research, particularly in fields such as climate science or biomedicine, where data science tools are spreading rapidly. The spreading of data science can be direct or sometimes indirect: using some of its tools or simply taking advantage of the infrastructures (e.g., databases) this new field has made possible.

Two fundamental issues are key to understanding data science tools' ethical, political, and social implications. The first is the realization that all algorithmic systems, as products of human action, are *value-laden*. This links to the problem of algorithmic bias (Fazelpour and Danks 2021). 'Bias' here "simply refers to deviation from a standard (…) [where] there are many types of bias depending on the type of standard being used" (Danks and London 2017, p. 4692). Accordingly, an algorithm can be morally, statistically, or socially biased. Algorithms are optimized "for performance relative to a standard" (Fazelpour and Danks 2021, p. 3), where such standards are built upon preferences which, at their core, are values. To illustrate this point with a straightforward example, consider this. When data scientists are cleaning data sets used to train an algorithm, they often deal with OCR data, which are very time-consuming to prepare because they are full of gaps. For this reason, data scientists may prefer data gathered digitally. Behind this choice is a preference for speeding up routine operations like data cleaning and preparation motivated by efficiency and a preference for high-quality data, which can be interpreted as a 'desideratum' or a value for data. However, systematically using only high-quality data in the healthcare context may result in using data only from wealthy areas, where healthcare infrastructures can gather data in a more cutting-edge way. This will result in neglecting data from underserved areas, where data gathering is done by using more traditional means because of the lack of resources (Ratti and Graves 2021). This, long term may result in representational harms, especially in segregated countries like the United States. Recognizing that these biases are not necessarily the byproduct of malicious behavior is important. It is difficult to point to cases where the biases are intentional. They can happen independently of practitioners' intentions when, for instance, a particular framing systematically biases the entire system towards a certain value-laden direction (Selbst et al. 2019). In other words, algorithms are biased because they are value-laden, and as such, they can be socially and morally controversial. In addition to the example just mentioned, one can also consider other famous examples, such as the algorithmic tools to calculate the probability of recidivism (Angwin et al. 2016), which seem to incorporate specific notions of what is the goal of justice in the justice system (Pruss 2021). While algorithmic biases have been widely discussed in relation to society-focused computing, such as facial recognition software, there are additional potential biases within academic research. As a result, each data scientist can potentially integrate algorithmic biases into analysis and visualization software.

The second way data science tools can be morally or socially problematic is that they generate ethical 'effects' when deployed. For instance, when algorithmic systems are used and scaled up, they constrain decisions and capabilities (Ratti and Graves 2021). This can lead to unexpected downstream ethical consequences that may not match practitioners' intentions. In the cleaning data mentioned above, focusing only on high-quality data may promote the overrepresentation of certain social groups while promoting the underrepresentation of others. It is unlikely that a data scientist who is more comfortable dealing with high-quality data will promote such dynamics of under- and overrepresentation on purpose. In other words, ethical effects can be complicated to anticipate and connect to routine and technical practices involving data. We call these 'unanticipated effects'. Therefore, data scientists have been wondering how exactly to scale such tools to minimize such ethical effects (Bommasani et al. 2021).

The presence of values in algorithms and unanticipated effects pervade the daily activities of data scientists. It has been convincingly documented how micro- and incremental technical choices are inevitably intertwined with values and can have ripple effects, such as how health risks are managed at the population level (Obermeyer et al. 2019). This can happen in virtually any phase of the data science process, including the definition of a problem to be solved by the algorithm (Fazelpour and Danks 2021; Selbst and Barocas 2018), data acquisition and preparation (Ratti and Graves 2021), data analysis and modeling (Graves and Ratti 2021), and deployment (Wachter et al. 2021).

What is needed, then, is a means of teaching ethical awareness and sensitivity to data scientists that foregrounds the cumulative nature of data science. This means that data scientists need to recognize that ethically sound digital infrastructures and data analyses are influenced by their micro-technical decisions. There is a need to foreground the importance of scrutinizing these decisions and not simply consider the downstream consequences of data/digital development.

11.2.3 Finding the Right Blueprint or Rethinking RCR?

Given the potentially problematic status of data science practice, an appeal to create *ad hoc* training in data ethics is unsurprising. However, in many cases, this training was envisioned as an additional element within existing RCR programs. This positioning is logical in light of shared content and budgets. However, it is also potentially problematic in terms of content delivery. Current RCR programs may not be more reliable blueprints for data science, at least for two reasons.

First, the paradoxes and tensions noted above may lead RCR programs for data scientists to present ethics as a sterile and bureaucratic ritual. Using current RCR programs as models for data science ethics training may mean wasting a unique opportunity to design a different approach to RCR training. Data science's value-ladenness and unanticipated effects provide fertile ground for foregrounding the aspirational attitude that motivated the creation of RCR programs decades ago. Because the daily activities of data scientists are to use Vallor's expression (2016), 'techno-moral', there is a critical need for RCR programs to empower data scientists to realize the aspirational expectations that motivate their research. Moreover, given the highly discretized and iterative nature of data science, linking ethical decisions to highly repeated tasks affords the opportunity to design training that supports the internalization of norms by means of habituation. This would address the often-cited (but questionably realized) aim of RCR to foster "research integrity through changing behavior, character, and judgment" (Chen 2021, p. 232). Ethical choices, whether practitioners are aware or not, ethical choices are made at every step of the research lifecycle. However, few ethics training curricula explicitly link content to these repeated daily actions and thus fail to provide the cognitive links between individual practice and the broader ethical and societal relevance of technical decisions.

In addition to the mismatch between aspirational motivation and compliance-based implementation, linking data ethics content to current RCR programs may not be ideal for data scientists for *structural reasons*. Recent iterations of RCR training use certain aspects of the framework to demonstrate ethical behavior in daily research practices. Utilizing these "daily research examples" is critical to fostering self-reflection amongst the learners. However, tend to focus on more traditional research ethics case studies such as working with human or animal subjects, or FFP misconduct. Many data scientists are removed from the collection of primary data, and as a result examples about human/animal research will have little resonance. Furthermore, as data science practitioners are highly heterogeneous in their background and roles (e.g., researchers, technical support, data curators), many will not be focused on producing academic writings as their primary outputs. This decreases the impact of FFP and responsible authorship training.

While this section notes considerable challenges for integrating data ethics into RCR training, it also draws attention to the opportunities inherent in this field of practice. Most importantly, it highlights that data scientists, regardless of their role within the research process, will be engaged in highly repetitive tasks while engaged in critical strategic decisions regarding the selection of data, the design of algorithms, and the development of computational models. In the next sections, we describe how these characteristics can inform innovative RCR training for data science practitioners.

11.3 What Should a RCR Program for Data Scientists Look Like?

The previous sections outline two critical requirements for an effective RCR program for data scientists.

First, the programs must orient around the aspirational motivations of RCR. This means that the opportunity of addressing ethical and societal issues raised by data science tools should not be noticed by superimposing old and ineffective RCR programs. What is needed is for RCR programs to be *transformational* in what students will learn. As a technical course in one of the aspects of data science might be transformative by teaching a skill and providing the bases to cultivate that skill further, so an RCR module can be transformative by providing the basis for cultivating skills or traits to identify the moral and political dimensions of micro-technical choices. The 'transformative' in RCR modules must relate to character development,[1] equipping data scientists with the instruments for effective decision-making.

The second important consideration is that any RCR program for data scientists should not create the impression that ethics is external to the practice of data science. While some of the ethical and societal challenges described in Sect. 11.2.2 are clearly internal to the practice of data science, the mode of their presentation can nonetheless fuel 'compliance' rather than critical self-reflection and awareness. Ideally, a successful RCR program teaching ethics as internal to data science practice will provide the basis for a 'data science practical wisdom' or 'data science

[1] We use the term "character development" from a virtue ethics perspective to denote the ability of an individual to cultivate traits that will enable them to identify and perform the right action within a specific context. Character development also includes developing a consistency in action, so that the individual will be able to apply the same reasoning (and enact the same virtuous behaviour) regardless of the context.

phronesis'[2], which augments the perception of the ethical imports of data science procedures well beyond the artificial settings provided by RCR classrooms where usually only cases that blatantly undermine research integrity are discussed. It is important to stress that we aim to provide just the 'basis' of practical wisdom. One may argue (and with good reason) that the limited duration of an RCR program will not allow the cultivation of something like practical wisdom, and this is because of the nature of virtues like phronesis. Notoriously, what counts as a virtue is a matter of much controversy in virtue ethics (Vallor 2016). Sometimes scholars refer to excellences, other times to character traits, and sometimes to reliable dispositions (Annas 2011). But all these accounts share the idea that a virtue is something stable that is cultivated through practice and by following exemplars and entrenched in the character of individuals. This means that to cultivate virtues, one needs time; hence, it is unlikely that one will cultivate a full-blown virtue in the space of one module. This is certainly true, but it is not a problem at all. The same challenge applies to skills acquired in the same way virtues are (Russell 2015). However, this is not seen as an argument against attending technical courses, which provide some technical background and lots of practical activities. While these courses are not meant to have the students fully cultivate technical skills, they are seen as a way to familiarize them with certain practices and to develop certain proto-virtues that we call 'preliminary abilities', which can become virtues with more practice. We think about RCR modules in the same way, namely as a strategy to familiarize students with the moral and political dimensions of seemingly neutral technical activities. This is why we have said 'the basis for practical wisdom', rather than full-blown practical wisdom.

These requirements delineate an ideal of RCR programs as 'transformational' and 'internal to the practice'. We propose an RCR program with some notable characteristics to make this ideal more concrete.

First, the program must be embedded (Grosz et al. 2019; Bezuidenhout and Ratti 2020; McLennan et al. 2022) in technical curricula. There is a well-documented and positive trend of developing such 'embedded' programs, with notable examples including Harvard University,[3] MIT,[4] Technion, University of Toronto,[5] and Stanford University.[6] Embedding ethics in technical curricula provides the opportunity to teach ethics exactly where the 'action' is taking place. By showing a more direct connection between technical choices and ethical relevance, ethics is internalized as part of the daily practice of data science. Embedding an RCR program into technical curricula requires that RCR modules be specific to the tasks that characterize the

[2] The concept of "phronesis" relates to the consistency in behaviour mentioned in footnote 3 above. Phronesis, or practical wisdom, is the trait cultivated by mature virtuous individuals and refers to their ability to identify the right action within specific context. Phronesis involves not only that the individual possesses the virtues necessary for the action, but that they have a critical understanding of the context in which they are so as to identify what is the "right" or "appropriate" action in that specific setting.

[3] https://embeddedethics.seas.harvard.edu/

[4] https://computing.mit.edu/cross-cutting/social-and-ethical-responsibilities-of-computing/

[5] https://www.cs.toronto.edu/embedded-ethics/

[6] https://ethicsinsociety.stanford.edu/tech-ethics/tech-ethics-center-initiatives

particular courses they are embedded in. One way to do this is by reflecting the highly discrete and repetitive nature of the data science pipeline, such that the modules make clear how responsible data science is implicated in operations such as formulating a data science problem, selecting data sets, preparing data sets, etc. More precisely, one can use the idealized schema formulated in (Ratti and Graves 2021) and think about the data science process as an iterative process consisting of seven stages grouped into three main phases. The first phase consists of stages involving problem definition and data acquisition. The second phase includes data understanding and preparation, data analysis and modeling, and validation and interpretation of the model itself. Finally, the last phase includes the deployment of the model and the evaluation of feedback. Ideally, one must be able to devise an RCR module for each of these stages.

Next, the embedded curricula should be *transformational* because they must be scaled up to all technical courses data scientists must take during an undergraduate or graduate curriculum.

Furthermore, RCR programs for data science must use ethics exercises that habituate and familiarize students with ethical reflections on responsible conduct within specific tasks. This habituation would involve developing the practice of considering actions from an RCR perspective until this practice becomes a habit and an unconscious part of planning research activities.

Fourth, the curriculum should equip students to deal with conducting data science research/practice in a variety of different contexts. Instead of exposing students to rules established within a specific institution, the curriculum should help them identify the moral and political relevance of various data science contexts. Students should be able to understand, and cope with critically, the context's influence on their ability to enact ethical decision-making and practice.

The connections between ethics, habituation and daily tasks lead to our proposal that RCR programs for data science will be optimally transformational and internal if oriented around the concept of *microethics* (Komesaroff 1995). This concept emphasizes the importance of incremental and micro decisions and their ethical relevance. By fostering self-reflexiveness on the ethical implications of each micro-task – introduced as a component of each course of a technical curriculum – microethics makes ethical reasoning a routine activity. This approach offers "a way of developing a multifaceted awareness of the contexts" (Bezuidenhout and Ratti 2020, p. 8) and cultivates a much-needed moral sensibility.

11.4 Case Study: CODATA-RDA Schools for Research Data Science

To demonstrate how our proposed curriculum of data science-focused RCR training curriculum can be operationalized, we include the case study of the CODATA-RDA Schools for Research Data Science (SRDS)[7] where both authors teach RCR. The

[7] https://www.datascienceschools.org/ (accessed 02/02/2023).

SRDS network was founded in 2016 to provide basic data science training to early career researchers (ECRs), many of whom are based in low/middle-income countries. Sixteen schools have educated over 500 alums through two-week residential schools held worldwide. The curriculum covers software/data carpentry (R, Git, data visualization), data security, neural networks, and computational infrastructures. It also includes research data management, Open Science, responsible authorship, and data science ethics (Bezuidenhout et al. 2020; Bezuidenhout et al. 2021).

The data science ethics training (RCR and Open Science) receives five hours of formal instruction during the curriculum. This means that, like many other RCR courses, this means there is little time for long-term engagement and in-depth ethics instruction. The challenge for the instructors is thus to introduce students to key ethical concepts and current ethical debates in data science to ensure they understand their ethical responsibilities and how they are enacted in daily research practice.

Unlike many other short ethics courses, we opted not to utilize high-level case studies as a primary means of instruction. While engaging for the students, we felt that these case studies often do not provide the cognitive pathway through which students can locate their actions within the evolving ethical landscape. Presenting students with the 'big picture', or asking them to occupy hypothetical roles can provide the misimpression that ethical dilemmas only operate on societal levels or at higher levels of the career trajectory.

The SRDS curriculum utilizes the formal teaching time to introduce ethics, RCR and Open Science to the students and to highlight the key responsibilities and characteristics. A later session also introduces topical ethical issues within data science. The different themes of the lectures are then practically introduced through a series of ethics exercises linked to the technical module content. These ethics exercises are run throughout the school after each module and not only foreground individual responsibility but also iteratively repeat the concepts from the formal lectures. Together, we call this curriculum "open and responsible (data science) research" (ORR).[8]

Our idea of an embedded, transformational, incremental, and microethical RCR training is operationalized in the context of CODATA-RDA through five stages of cognitive linkage. First, instructors give students a general understanding of what ethics and moral abilities can possibly be. This general understanding should not be too technical regarding meta-ethics or moral philosophy, as the main goal is to communicate how moral and political aspects play leading roles even in technical decisions. For this reason, the next step is understanding how technical decisions and ethics are connected in the data science environment. This is done by problematizing the operations taught in the technical modules into microtasks, and understanding their connections to the proto-virtues like moral abilities we refer to above, and ethics: what is the relation between microtasks and virtuous behavior? The third

[8] An example of the 2022 programme for the SRDS school run at the ICTP in Trieste can be found here https://indico.ictp.it/event/9806/other-view?view=ictptimetable

step is to connect the microethics of technical tasks to the working environment. In this way, virtuous and vicious behavior is not only analyzed in terms of isolated microtasks; rather, the environment's role in shaping individuals' behavior is accounted for. In this context, vicious behaviour is understood as undermining the researcher's or the research's integrity. After taking stock of the complex relationship between microtasks, the environment, and virtuous/vicious behavior, the next step is to discuss how microtasks and the environment can be modified - in other words, the fourth step is to develop resilience strategies. Finally, the fifth step is to discuss how to identify mentors and exemplars in one's environment and how to develop strategies of resilience to cope with (un)expected environmental challenges to open and responsible data science practice. In the next sections, we describe these five steps in detail.

11.4.1 First Step: A General Understanding of Ethical Issues in Data Science and Open Science

The first step in ORR training is to introduce students to ethical concepts and issues surrounding data science, RCR, and the culture of open science. The topics covered in these formal lectures are illustrated in Fig. 11.1. The first two topics provide an introduction to RCR and Open Science. These concepts are linked by illustrating how open research practices address vital activities identified in RCR, which is an essential means of ensuring that students do not compartmentalize the different movements, meaning that they should not think of them as unconnected or independent of each other. Both the RCR and Open Science topics highlight not only the responsibilities of the individual researcher but also the roles that institutions and national and international infrastructures play in safeguarding research integrity.

The third topic brings the responsibilities identified in RCR and Open Science under the concept of 'open and responsible science citizenship'. The concept of science citizenship has proven a valuable means of ensuring that students can personally contextualize the earlier topics. Most of the students taught on the SRDS have no prior ethics training and little knowledge of Open Science. Because of this, these topics and the extensive lists of expectations outlined can seem overwhelming. Bringing the discussion on responsibilities back to citizenship is helpful, as all learners have some understanding of the concept. We use this to discuss community membership, community rules, and the rights, freedoms, and duties associated with citizenship. This is a means of discussing various duties, such as community rules (i.e., crediting and citation), duties (i.e., peer review and contributing to community activities), and community maintenance (safeguarding open resources and whistleblowing).

The last section taught in the formal lectures applies the concept of science citizenship more broadly to look at science in society. This section covers topics such as bias and value-ladenness in algorithms, data visualization, and the problems of

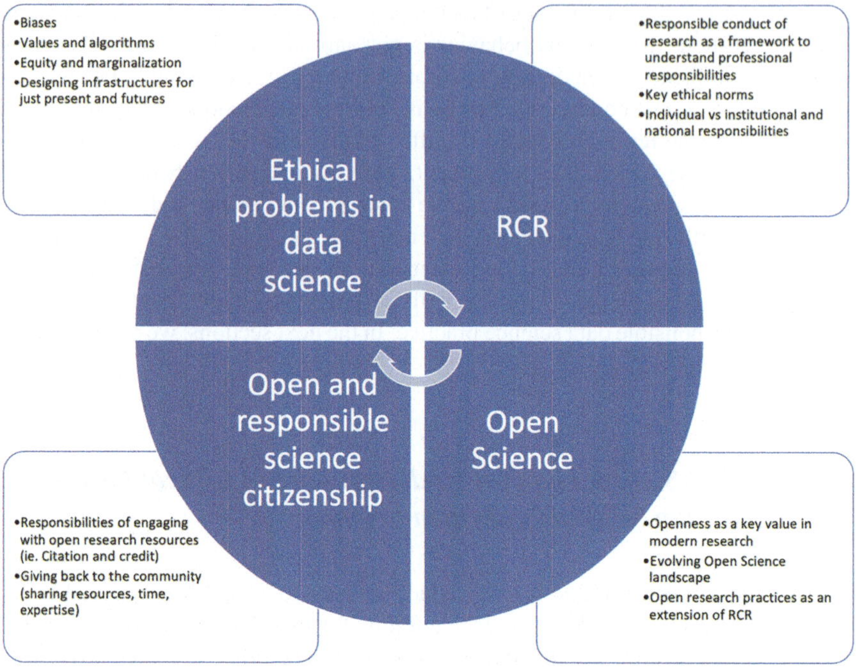

Fig. 11.1 Overview of topics taught in formal ethics lectures at SRDS

marginalization and misrepresentation through data re-use, data infrastructures, and information and communication technologies. This section serves to orient the students toward the societal impact of data science and highlight their responsibility for engaging (as data experts) in these discussions.

11.4.2 Second Step: Linking Ethical Content to Data Science Tools

The second step is understanding the connections between ethics and technical decisions. In the SRDS, we run a series of ethics exercises through which students can connect daily research decisions and the technical tools used in these everyday practices with RCR ethics and the "big picture" ethics discussions. This is an essential means to iteratively revisit ethics throughout the summer school, but also to ensure that students do not compartmentalize ethics as a "stand-alone" topic unrelated to the technical subjects being taught at the school.

The ethics exercises used to understand the connections between ethics and technical decisions consist of problematizing the practice of data science into discrete microtasks that use data science tools. The ORR microethics exercises are designed to accompany each technical module of the SRDS curriculum and foster reflection on using the data science tool and the broader implications of misapplication. A complete list of ethics exercises utilized at SRDS is published online,[9] with an example given below.

GitHub is a platform for (amongst other things) sharing software and managing version control. Data science students will interact with GitHub regularly as they share code, reuse code and engage with scholarship in their field. Nonetheless, the use of GitHub is reasonable. As demonstrated in Fig. 11.2, GitHub is blocked in countries against which the US holds sanctions. GitHub thus provides a useful way of linking issues of inclusion and exclusion to daily research practice, and the future use of the platform provides a means through which students can regularly question whether they are fulfilling their duties towards the inclusivity and universality of research. When students use the platform, they are encouraged to think about whether there are alternative actions that they can undertake to minimize the harm caused by the geoblocking of GitHub.

Students undertaking the GitHub ethics exercise illustrated in Fig. 11.2 are asked to vote, choosing between the following options:

- no, it's not my problem
- no, the blocking is probably justified
- yes, but only if I am collaborating with researchers in those countries
- yes, it is unjust to shut researchers out of Open Science platforms

The results of the voting provide a basis on which to start a discussion about critical issues, such as limits of openness, equity, and marginalization, political influence on research and non-maleficence in research. This provides the students with a

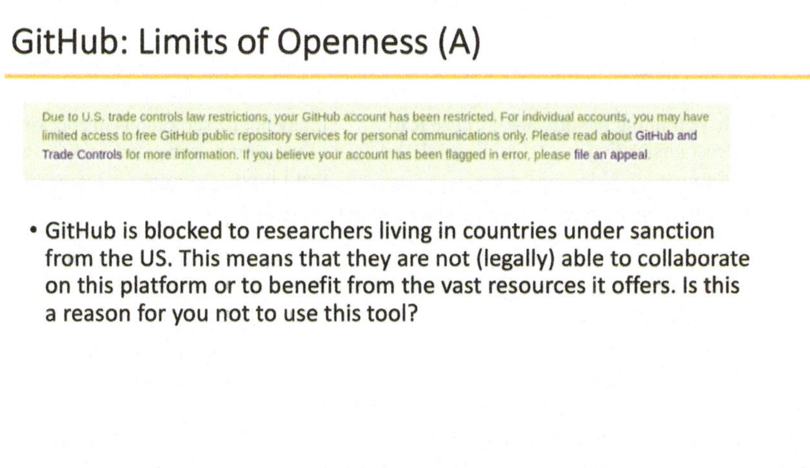

Fig. 11.2 Example microethics exercise from GitHub module

[9] https://doi.org/10.5281/zenodo.6874958.

cognitive link between the ethics content from the lectures and its applicability to the data science tools used (Table 11.1).

In addition to providing the cognitive links between daily research practice and broader ethical discussions, the microethics exercises offer an important means of distinguishing between good and poor conduct. In the discussions, it is also possible to critically unpack what 'misbehavior' would look like within a specific microtask. The identification of this poor conduct enables the students to identify what the right action would be. Table 11.2 illustrates how poor conduct can be identified from the specific microtask under discussion within the ethics exercise.

As will be apparent to many, these discussions are underpinned by the virtue ethics traditions of virtuous and vicious behavior. Misbehaviors lie at two extremes - omission and commission - Understanding what poor conduct would look like within a specific context. Consequently, 'good behavior' requires finding the midpoint of good behavior between these poles. Using these vices of omission and commission as prompts within discussion assists the lecturer in broadening the students' understanding of poor research conduct.

Highlighting the distinction between doing too little or doing too much (as demonstrated by Table 11.2) is an important contribution to the ethics exercise discussions. Indeed, it contrasts with the highly bureaucratic approach of other RCR training, which provides a list of rules to follow but does not encourage discussion around different ways of enacting poor conduct around a single action.

This strategy will be familiar to all scholars using virtue ethics or virtue language. Indeed, following the virtue ethics tradition, the discussion aims to provide

Table 11.1 Sample of instructor sheet with prompts for guiding behaviour outlining potential vicious behavior associated with a microtask

If I'm prevented from doing my task well, what happens?	Microtask	If I am under pressure to finish the task quickly, what happens?
If I don't have training or software, I may construct misleading visualizations	I use R to visualise data	I may not be scrupulous in checking my visualization or may not seek the advice that I need

By vicious behaviour we refer to poor conduct within the situation

Table 11.2 Extended mapping of virtue ethics guiding discussions on microethics

If I'm prevented from doing my task well, what happens?	Vices (omission)	Microtask	Virtues	If I am under pressure to finish the task quickly, what happens?	Vices (commission)
If I don't have training or software, I may construct misleading visualizations	Slothfulness, Apathy	I use R to visualize data	Temperance, wisdom	I may not be scrupulous in checking my visualization or may not seek the advice that I need	Recklessness, Lack of judgment

students with the tools to identify the *right action* to take in relation to the use of tools within their research context. In effect, for students to get used to identifying the "golden mean" of good behavior when they utilize data science tools or engage in other research practices.

Table 11.2 provides even more context underpinning these discussions, and is used as a train-the-trainer tool for teachers undertaking to teach Open and Responsible Research at the SRDS. As is evident, the virtues associated with the "right" execution of the microtask are counterbalanced by the vices of omission and commission linked to the misbehaviors identified in Table 11.1.

By following the structure of the ethics exercise discussions, the instructor is thus able to introduce the students not only to virtues but also to the process of identifying the "golden mean" between vices of omission and commission that is suitable for a specific context and action. It is important to mention that the formal language of virtue ethics is rarely used within the SRDS school. As most students come from natural and life science backgrounds, introducing a list of new terms can cause cognitive dissonance and detract from the learning process. Rather, virtuous behavior and associated vices are termed using the conventional "right behavior" and "misbehavior", more generally providing consistency between our teaching and RCR instruction.

It is also important to spend a few words on the process of identifying the 'golden mean'. This is connected to a central trait in our RCR training: the virtue of phronesis. Phronesis has many names, including prudence or practical wisdom. It is a virtue or habit not just limited to the Western tradition starting from Aristotle, but it is also recognized in the Confucian moral tradition as well as in Buddhism (for a detailed comparison of Aristotelian phronesis with these traditions see Vallor 2016). Phronesis is defined as the necessary virtue to identify the right action within a specific context, in the sense that 'controls' the "enactment of a person's individual moral virtues, adjusting their habitual expression to the unique moral demands of each situation" (Vallor 2016, p. 19). Phronesis is deliberative in the sense that it is operationalized via prudential judgments. But despite the importance of a skill such as 'prudential judgment', phronesis is not only the ability to "judge and choose a course of action" (Vallor 2016, p. 106); it is much more, as it also presupposes other moral virtues, but at the same time the other virtues cannot be properly coordinated and modulated without phronesis. When describing 'technomoral' wisdom, Vallor notices that (practical) wisdom is not a specific excellence like other virtues but a general condition of possibility for the other virtues to be exercised – it is unifying other virtues. The goal of the ethical exercises is to put students on the right path of moral self-cultivation that can lead to practical wisdom; this 'wisdom' will allow them to aim at the correct goods (i.e., the ones which lead to human flourishing within a particular context) and for the right motives. While we are certainly aware that a school cannot result in cultivating such an important and controversial virtue, we do believe we should start from somewhere, and SRDS can be a much needed starting point.

11.4.3 Third Step: Recognizing Context

In order to assist the students in understanding the role of the research context on their conduct, the ORR curriculum includes a session dedicated to discussing the challenges to responsible research conduct that students identify in their home environments. The object of this session is twofold. First, to demonstrate to the students that they are not alone in having problems and concerns about implementing open and responsible research practices. Second, many problems can be addressed using national and international infrastructures, tools, and communities.

In structuring these discussions, we draw on the Institutional Analysis and Development Framework developed by Hess and Ostrom (2007). This framework analyzes institutional structures based on biophysical conditions, community attributes, and rules-in-use. This enables institutional outcomes to be understood on a multidimensional level and draws attention to how these different elements shape action situations and outcomes.

This exercise is introduced by the instructor, who explains the different aspects of the institution that can present challenges to open and accountable research practices (using prompts such as those outlined in Table 11.3). The students are then invited to anonymously contribute to a shared document in which they list their concerns according to three categories:

- institutional/cultural issues
- infrastructural issues
- personal concerns

Discussion of these concerns forms the basis of a broader discussion about challenges and existing opportunities for their amelioration. As mentioned above, a key objective of the discussion is to empower the students to realize that they are not alone in their concerns and that existing structures can offer support. To reiterate that all researchers have concerns, students are also introduced to the findings of many international surveys, such as the State of Open Data run annually by FigShare.

Table 11.3 Sample of instructor sheet with prompts to introduce the exercise on identifying contextual challenges to open and responsible research practice

Context category	Sub-category	Description	Characteristics of own institution
Biophysical		Access to resources	
	Common resources	What resources do you have access to in order to do your work? Is anything missing?	
	Public resources	How is the provision of the internet, power, working environment in your institution?	
	Personal goods	Do you have access to a mentor, support system at work?	

Table 11.4 Sample of concerns (column 2) raised by students in 2022 SRDS schools and group discussion on potential tools to ameliorate concerns (column 3)

	Concerns	Possible Resources to Use
Institutional/ cultural concerns	Wide mentoring gap between professors and young researchers	Look for mentors outside of institutions. This could include independent networks such as AuthorAid, OpenLifeScience, disciplinary communities, or international organizations such as the Research Data Alliance
Infrastructural concerns	Lack of infrastructures and institutional support on Data Archiving	Understanding international repository landscape can assist (i.e., re3data), as well as understanding what constitutes a *trusted digital repository* and open licensing for research resources Many institutions are developing institutional repositories. Enquire from your library services about current developments
Personal concerns	Limited knowledge of open science infrastructures	Considerable Open Science online training such as Open Science training handbook, Open Science MOOC, and FOSTER. Other useful courses through data.europa academy

Examples of the content discussed in these exercises are available online,[10] while a sample of the concerns and possible solutions are presented in Table 11.4.

As will be evident from Table 11.4 and the extended document, we strongly encourage our students to seek advice, develop support strategies, and identify mentors. Identifying networks of support and trusted advisors is key to developing the ability to identify the correct action within a specific context. Crucially, the exercise highlights the importance of looking for support beyond their researchers' institutions. A range of local/national/international communities provide these levels of interpersonal contact and having the students understand these opportunities is vital for their ability to withstand the environmental challenges to responsible and open research practices that will inevitably occur during their careers.

11.5 Strategies of Resilience

The different steps of the ORR curriculum are outlined in Fig. 11.2. In contrast to other RCR curricula, our training focuses on linking ethical principles and current broader ethical concerns in data science to the daily research practices of the individual researcher. This empowers students to realize that their actions cumulatively play a role in safeguarding beneficial research and just digital futures. The curriculum also focuses on problematizing the research contexts in which the students are embedded, focusing on helping them to find the *right action within the context.* This marks a change from RCR education that utilizes rules or codes of conduct to outline good conduct. The focus

[10] https://zenodo.org/record/6875154#.Yt-qOC98rBI

on right action in context reflects our awareness that RCR readiness differs significantly between institutions and that there cannot be a "one size fits all" approach to outlining the best course of action within a research context. Instead, the ORR curriculum focuses on getting students to be mindful of their responsibilities and critically find tools to help them implement them within their research context while offsetting negative environmental influences' impact (Fig. 11.3).

The discussions about research contexts and the discussions associated with the microethics exercises also serve another vital purpose. By encouraging students to discuss the challenges they face in their research environments, we create a space in which students recognize that their concerns are in no way unique. Students worldwide need help with similar issues relating to implementing responsible and open research practices. In this way, we draw the students' attention to the importance of integrating into local/national/international research communities and other support organizations. This is reflected in the sample answer outlined in Table 11.4.

Taken together, the ORR curriculum offers students a pathway towards developing open and responsible research practices and strategies of resilience in the face of (un)expected challenges. This pathway is outlined in Fig. 11.4. Providing students with these strategies of resilience is an essential contribution towards long-term ethical behavior, and ensures that students do not become discouraged as they

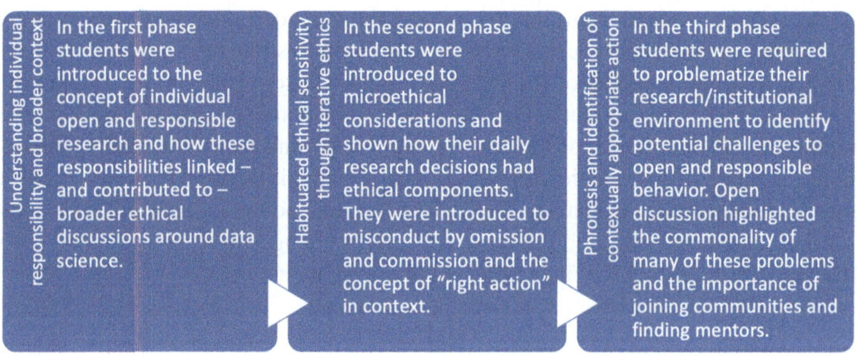

Fig. 11.3 Building resilience through ethics training

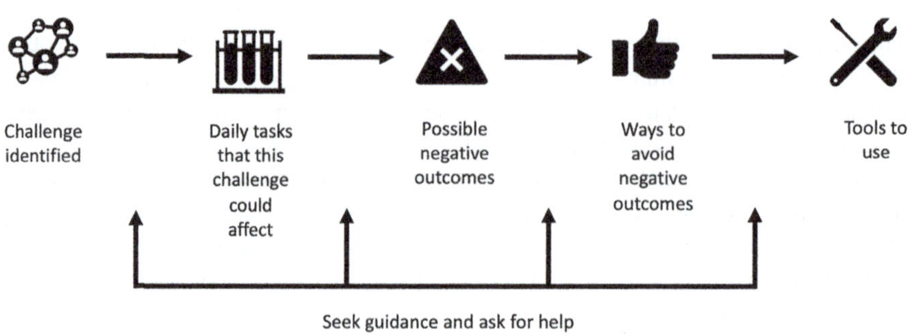

Fig. 11.4 Resilience strategy logic

go forward as early career researchers in environments that are perhaps not optimally equipped to support an "ideal" of RCR practice.

Developing the ability to identify "right action" and developing strategies of resilience to deal with (un)expected challenges are an important counter to the rule-focused approach of many RCR training. If students are simply taught to follow rules – and that this exemplifies good conduct – they are ill-equipped to deal with situations in which these rules cannot be followed in how they were taught. As few students will remain in their original institution for the duration of their careers and often move between countries or sectors, this should be reason for concern.

11.6 Concluding Comments: Leveraging Data Science to Foster Ethically Robust Digital Systems

The expansion of ethics training, such as ORR, that focuses on contextual awareness and strategies of resilience has the potential to impact data systems and digital infrastructures positively. Fostering data scientists habituated to scrutinizing the ethical impact of the smaller units of digital infrastructures and the data use practices will ensure that the developments within the emerging digital/data landscape are continually monitored.

We outline this aspiration in Fig. 11.4. We suggest that training data scientists in ethical awareness and resilience will support them in becoming virtuous and able to identify right action in context. This will lead to the development of ethically robust data systems that, in turn, influence users (Fig. 11.5).

Together, we may globally be able to ensure that our digital futures reflect the ethical principles that underpin research.

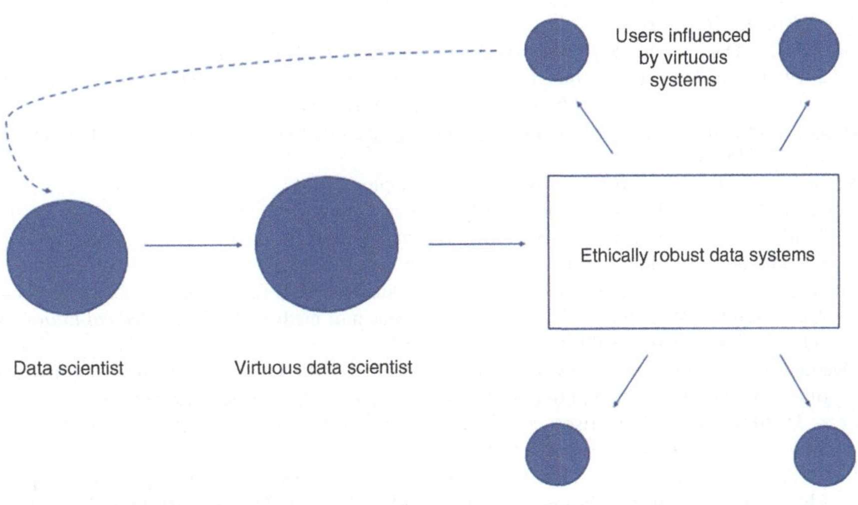

Fig. 11.5 Developing ethically robust data systems

References

Angwin, J., J. Larson, S. Mattu, and L. Kirchner. 2016. Machine bias: There's software used across the country to predict future criminals. And it's biased against blacks. *ProPublica*. https://www.propublica.org/article/machine-bias-risk-assessments-in-criminal-sentencing.

Annas, J. 2011. *Intelligent virtue*. Oxford University Press.

Bender, E.M., T. Gebru, A. McMillan-Major, and S. Shmitchell 2021. On the dangers of stochastic parrots: Can language models be too big? *FAccT 2021 - Proceedings of the 2021 ACM Conference on Fairness, Accountability, and Transparency,* 610–623. https://doi.org/10.1145/3442188.3445922

Bezuidenhout, L., and E. Ratti. 2020. What Does It Mean to Embed Ethics in Data Science? An Integrative Approach Based on Microethics and Virtues. *AI and Society* 0123456789. https://doi.org/10.1007/s00146-020-01112-w.

Bezuidenhout, L., R. Quick, and H. Shanahan. 2020. "Ethics When You Least Expect It": A Modular Approach to Short Course Data Ethics Instruction. *Science and Engineering Ethics* 26: 2189–2213. https://doi.org/10.1007/s11948-020-00197-2.

Bezuidenhout, L., S. Drummond-Curtis, B. Walker, H. Shanahan, and M. Alfaro-Córdoba. 2021. A School and a Network: CODATA-RDA Data Science Summer Schools Alumni Survey. *Data Science Journal* 20 (1): 10. https://doi.org/10.5334/dsj-2021-010.

Bommasani, R., D.A. Hudson, E. Adeli, R. Altman, S. Arora, S. von Arx, M.S. Bernstein, J. Bohg, A. Bosselut, E. Brunskill, E. Brynjolfsson, S. Buch, D. Card, R. Castellon, N. Chatterji, A. Chen, K. Creel, J.Q. Davis, D. Demszky, and P. Liang. 2021. *On the opportunities and risks of foundation models*. http://arxiv.org/abs/2108.07258

Chen, J.-Y. 2021. Integrating Virtue Ethics into Responsible Conduct of Research Programs: Challenges and Opportunities. In *Science, Technology, and the Good Life: Perspectives on Virtues in Modern Science and Technology*, ed. E. Ratti, and T. Stapleford

Danks, D., and A.J. London. 2017. Algorithmic Bias in Autonomous Systems. In *IJCAI International Joint Conference on Artificial Intelligence, IJCAI*, 4691–4697. https://doi.org/10.24963/ijcai.2017/654

Fazelpour, S., and D. Danks. 2021. Algorithmic Bias: Senses, Sources, Solutions. *Philosophy Compass 16* (8). https://doi.org/10.1111/phc3.12760.

Graves, M., and E. Ratti. 2021. Microethics for healthcare data science: Attention to capabilities in sociotechnical systems. *The Future of Science and Ethics* 6: 64–73. https://doi.org/10.53267/20210106.

Grosz, B.J., D.G. Grant, K. Vredenburgh, J. Behrends, L. Hu, A. Simmons, and J. Waldo. 2019. Embedded ethiCS: Integrating ethics broadly across computer science education. *Communications of the ACM* 62(8): 54–61. http://arxiv.org/abs/1808.05686

Hess, C, and Ostrom, E. 2007.Understanding knowledge as a commons: From theory to practice. The MIT Press.

Kelly, T. 2018. *Professional Ethics*. New York: Lexington Books.

Komesaroff, P. 1995. From Bioethics to Microethics: Ethical Debate and Clinical Medicine. In *Troubled Bodies – Critical Perspectives on Postmodernism, Medical Ethics, and the Body*, ed. P. Komesaroff, 62–86. Durham: Duke University Press.

McLennan, S., A. Fiske, D. Tigard, R. Müller, S. Haddadin, and A. Buyx. 2022. Embedded ethics: A proposal for integrating ethics into the development of medical AI. *BMC Medical Ethics 23* (1): 10.1186/s12910-022-00746-3.

Obermeyer, Z., B. Powers, C. Vogeli, and S. Mullainathan. 2019. Dissecting racial bias in an algorithm used to manage the health of populations. *Science*. https://www.science.org

Pruss, D. 2021. Mechanical jurisprudence and domain distortion: How predictive algorithms warp the law. *Philosophy of Science* 88 (5):1101–1112.

Ratti, E., and M. Graves. 2021. Cultivating moral attention: A virtue-oriented approach to responsible data science in healthcare. *Philosophy and Technology 34* (4): 1819–1846. https://doi.org/10.1007/s13347-021-00490-3.

Russell, D. 2015. Aristotle on cultivating virtue. In *Cultivating Virtue – Perspective from Philosophy, Theology, and Psychology,* pp. 17–48. Oxford University Press.

Selbst, A.D., and S. Barocas. 2018. The intuitive appeal of explainable machines. *Fordham Law Review* 87 (3): 1085–1139. https://doi.org/10.2139/ssrn.3126971.

Selbst, A.D., D. Boyd, S.A. Friedler, S. Venkatasubramanian, and J. Vertesi. 2019. Fairness and abstraction in sociotechnical systems. In *FAT* 2019 – Proceedings of the 2019 Conference on Fairness, Accountability, and Transparency*, 59–68. https://doi.org/10.1145/3287560.3287598

Shamoo, A., and D. Resnik. 2015. *Responsible Conduct of Research.* Oxford: Oxford University Press.

Vallor, S. 2016. *Technology and the Virtues – A Philosophical Guide to a Future Worth Wanting.* Oxford: Oxford University Press.

Wachter, S., B. Mittelstadt, and C. Russell. 2021. Bias preservation in machine learning: The legality of fairness metrics under EU non-discrimination law. *West Virginia Law Review* 123 (3): 735–790.

Chapter 12
Encouraging Transparency in Lab Safety via Teachable Moments and Positive Feedback

Melinda Box and Maria Gallardo Williams

Abstract Transparency is an essential part of developing and maintaining an ethical culture and a safety culture. We found two practices that significantly contributed to both - capitalizing on teachable moments that stemmed from safety incidents and the active provision of detailed positive feedback on laboratory inspections. When we applied these concepts to instances of evaluating performance, we found that they facilitated communication and supported the inquiry necessary to develop and sustain openness. While enforcement, an alternative approach, may evoke rapid compliance, it may also tend to reduce discussion. By contrast, we found that considering the vulnerability of those in non-compliance tended to ease the exchange and support the robust relations needed to create and maintain safe conditions. Appreciation for the state of intense interest and hunger for information characteristic of these teachable moments led to providing specific guidance that recipients were ready to use immediately. In addition, reinforcement of existing desirable practices via specific positive comments gave recipients the guidance necessary to build on their existing foundation of skills, knowledge, and strengths. This chapter combines research in behavioral psychology, principles of management and evaluation, and our own observations to illustrate how performance evaluations and incident reports can be leveraged to achieve an ethically responsive culture.

Keywords Positive feedback · Inspections · Compliance improvement · Safety · Culture · Teachable moment

---

M. Box (✉)
Department of Chemistry, Elon University, Elon, NC, USA
e-mail: mbox@elon.edu

M. Gallardo Williams
Office for Faculty Excellence, North Carolina State University, Raleigh, NC, USA
e-mail: mtgallar@ncsu.edu

E. Hildt et al. (eds.), *Building Inclusive Ethical Cultures in STEM*, The International Library of Ethics, Law and Technology 42,
https://doi.org/10.1007/978-3-031-51560-6_12

12.1 Introduction

Providing working conditions that are safe is generally accepted as an essential ethical priority. However, the means of achieving this goal may not be so obvious or widely understood. One example is academic chemistry research laboratory settings. Although generally considered lower-risk environments relative to their industrial counterparts (Álvarez-Chávez et al. 2019; Taylor and Snyder 2017), harmful events in recent decades have made it clear that this is not necessarily the case and that a culture of safety must also be prioritized in these environments to maintain continuous operations (Hill Jr 2019). Smaller scale work and the absence of profit pressures have not necessarily reduced the risk of adverse events and so alternate strategies must be employed.

Conditions that have made achieving a culture of safety in academic labs difficult include a decentralized organizational structure, a wide scope of methods and materials, high personnel turnover, and variable levels of abilities and backgrounds of those working in these research settings (Fivizzani 2016; Reniers et al. 2014). Within all this variability, student researchers are working on acquiring mastery of the essential safety skills of recognizing and managing the risks associated with their lab work; these skills cannot be taken for granted, but instead, the institution itself must be structured to provide a framework for making well-reasoned, ethical, safety-related decisions and support for carrying them out (Hill Jr 2019).

Creating that structure through an enforcement-type approach can risk evoking an "eyes glazed over" response from information overload involving communicating such things as requirements, regulations, rules, and restrictions (Hover and Schneider 2019). Instead, providing information to recipients when they are prepared for it can result in more meaningful comprehension and bring about more long-term, fundamental improvement. In fact, providing timely information is an essential part of safety engineering. This approach involves designing a feedback loop to provide cues to operators about where the boundaries of safe behavior are by allowing them to occasionally and manageably cross those boundaries. Analogously, a newly licensed vehicle driver might find the safest braking operation by initially pressing too hard or too lightly on the brake pedal. The repercussions are unpleasant but, most often, are not irreparable. Through a similar arrangement of mild experimentation and consequence, systems operators develop their mental models of how their system works and maintain their skill to run it safely (Leveson 2011).

In the same vein, when accidents and mistakes happen in labs, they can be utilized to help guide future decision-making. Acknowledging and treating these undesirable events as learning opportunities can facilitate appreciable changes in culture. The accompanying feelings of vulnerability, helplessness, and apprehension can lead to intense interest in finding solutions (Garrison 1997; McBride et al. 2003). Because these destabilizing conditions lower people's defenses against uncomfortable insights, they increase willingness to consider the need for change, and as a result, have been referred to as "Teachable Moments" (Hansen 1998). Responding to incidents as teachable moments systematically can contribute significantly to developing a safety culture because doing so helps participants personalize their understanding of how to achieve improved working conditions (Reason 1998).

In the Department of Chemistry at North Carolina State University, we leveraged teachable moments to create occasions of learning through the use of Lessons Learned memos. Summaries of incidents in story format with useful takeaways connected with readers' vicarious sense of vulnerability. This, in conjunction with a second initiative that involved the provision of positive feedback in lab safety inspections, extended our safety efforts beyond legal requirements into the realm of worker interests and concerns. In the process, we discovered that by addressing workers' needs, we had encouraged a culture of transparency, which is both an essential component of a culture of ethics and of safety.

This transparency in a lab setting involved workers revealing uncertainties, shortcomings, and workplace issues. Since all of these can be viewed as deficiencies, they can be difficult for workers to disclose. Nonetheless, these conditions are at the heart of workers' needs so encouraging openness about them is essential. To do that, an atmosphere of security must be established, and positive feedback, such as was provided in safety inspections, helped. It inspired the self-confidence necessary to admit to mistakes, reach out for help, and express concerns. This, in conjunction with the supportive sharing of lessons learned, reduced the isolation that might otherwise accompany the experience of making mistakes. This sense of connectedness stimulated sharing in confidence and thus encouraged users to communicate their experiences.

Our initiatives encouraged transparency by unifying two ways of approaching compliance. The two ways are at the ends of a spectrum from oversight to self-evaluation. At one extreme, knowledgeable safety professionals monitor practitioners and pass on the benefits of their expertise. They provide feedback that can be either corrective or affirmative. At the other end, practitioners monitor themselves by applying their own safety knowledge. We encouraged the growth of that knowledge when we shared information about incidents via the Lessons Learned memos. These passed on what went right, what went wrong, and what could be done differently to prevent a similar incident in the future.

As part of our professional practice in recent years, we have observed significant changes in engagement and safety compliance following these two active efforts to turn teachable moments into occasions of learning and to include positive feedback in environmental health and safety inspections (Box et al. 2020, 2021). Inspired by our own experiences, we have assembled in this chapter descriptions of our initiatives along with explanations of our observations based on principles of organizational management, evaluation tool design, educational practice, and behavioral psychology.

12.2 Teachable Moments

Since a teachable moment involves intense interest caused by need, doubt, and/or uncertainty, it can be both vulnerable and valuable (Garrison 1997; McBride et al. 2003). These moments are pivotal because they can generate both protectiveness and a strong desire to learn. Therefore, if inclinations to hide, blame, or defend are met with demonstrations of social safety and suitable information, recipients hungry for solutions can promptly incorporate those into their understanding with an amplified appreciation for the context in which they apply (Hansen 1998).

In general, these moments are circumstances of deficiency. One example might be getting caught in non-compliance, such as improper waste management. While that is generally not considered desirable, it can be an opportunity on the part of the one who discovers the infraction to convey social safety. Letting others know they have help and support in making needed changes can lead recipients to learn why those changes are required and how to make them effectively.

Confusion is also a state of deficiency, so expressing it might be avoided, particularly in the lab safety setting, where it may imply pre-existing conditions of non-compliance or, even worse, skeptical resistance. To illustrate, a lab worker may need clarification about chemical or waste labeling requirements. If they ask an inspector about this, it may prompt a closer look and discovery of deficiencies. If they ask their principal investigator about it, they may be viewed as someone seeking to do as little as possible. However, if this confusion is, instead, met with listening and emotional support, then recipients may feel freer to ask more questions and thereby leverage their confusion to get the input they most need when they are most interested in getting it.

And lastly, expressing concern can convey recognition of another's deficiency in some sense. Expressing uncertainty about work conditions can risk evoking defensiveness from representatives of the institution responsible for addressing or modifying those conditions. For example, if, in a lab setting, health symptoms are experienced, they could potentially be due to ventilation infrastructure issues. Investigation of this can be daunting, and the remedy can be extremely disruptive and expensive. As a result, weighing the balance between concern and remedy may result in management responding with skepticism, dismissiveness, or negation. However, if a concern of this type is, instead, met with openness and curiosity, apprehension can be leveraged for mutual gains in the broader understanding of systems, practices, and priorities. In addition, meeting these concern-based needs can lead to recipients reciprocating with an amplified commitment to transparency and compliance (Reason 1998).

From another angle, one might also recognize a teachable moment by staying alert to temptations related to finding these deficiencies, such as feeling victorious in finding errors, feeling superior in demonstrating personal knowledge, or feeling determined to quash defiance. Pursuing those, unfortunately, evokes compliance by fear rather than by understanding and so misses out on the benefits of a teachable moment.

Turning a teachable moment into an occasion of learning involves first engaging in ethical listening. This means taking in a speaker's message with an open mind, i.e., being able to hear the good, the bad, and the ugly without pre-judgment or mental counterargument. As a result, optimizing these opportunities begins not with the instruction of the evaluee (or of the incident victim) but with that of the evaluator (or incident investigator). To be certain that knowledge supplied by the evaluator or investigator is meaningful, one must first obtain a complete picture of the context. To obtain that, the evaluator or inspector must openly show deficiencies in knowledge by demonstrating curiosity (Wagner and Ash 1998). The inquirer must also demonstrate openness to input from multiple sources by pursuing a conversation with them and must demonstrate exposure to a new perspective by working to imagine what it would be like to do the same task, with the same resources, in the same space (Reason 1998).

Another key to turning a teachable moment into an occasion of learning is making sure that the information provided is applicable to the user's needs, useful to them in the context of their work, and available in manageable amounts. Framing related communications as stories distill all those necessities into one format. Responding to incidents by compiling details into a story connects with the user by using a timeless structure that evokes empathy and amplifies meaning.

Rather than a list of facts, story for a lab incident must include all of the standard components of storytelling. For example, establishment of setting might be by reference to where and when work was being done or an event occurred. Relevant locations might include at a fume hood, at a lab bench, or in proximity to a particular piece of equipment. Setting might also include conditions that contributed and time of day of the events that transpired. Characters might be those who experienced the incident firsthand, those who were contacted for assistance, and others who were involved or affected. Character types could be indicated by job role. Details of action could be those that will later be identified as correct or incorrect or as targets for change in the interpretation that follows the narrative. Action could unfold chronologically or in a circular fashion to explain a major event (Yellowlees 2022).

The familiarity of having an introduction, some action, a significant event, and a resolution creates anticipation of direction and a sense of purpose that focuses attention and prepares recipients to learn. Aspects like a protagonist who experienced a challenge in a particular setting seamlessly fall into resonant familiarity with plot and story. Therefore, the impact of the information is strengthened because using this structure taps into shared understanding and capitalizes on universal expectations.

Regardless of whether the story is conveyed informally or formally, verbally or in writing, the wider the distribution, the greater the range of comprehension, imaginative appreciation, and the capacity to learn for both the teller and the listener (Woodhouse 2011). In addition, storytelling facilitates transformation from the bottom-up by opening spaces for the expression of a wide range of social groups, including those that might be marginalized or perceived to be of low standing, such as gender and racial minorities (Woodhouse 2011).

However, realizing this potential requires a mindful approach to a teachable moment. Consideration of the role of emotions and relationships is essential in achieving the transformational learning needed (Baumgartner 2012; Taylor and Cranton 2012). Taking a punitive tactic can result in recipients feeling isolated and ashamed, leading to withdrawal and hiding of perceived failings or limitations. Examples of this approach might be reprimanding, controlling, quashing, advising, and lecturing. Alternatively, if a supportive approach is used, recipients may feel secure and understood and more inclined to be open about their shortcomings, which are essential ingredients in facilitating this transformative process (Vella 2000). Examples of positive tactics could be curiosity, empathy, and connecting with physical conditions (Johnston 1995).

12.3 Generating Lessons Learned Memos

In the North Carolina State University Department of Chemistry, we began an initiative in 2017 of documenting incidents in a "Lessons Learned" format. In it, we used story structure to provide anonymized accounts followed by identification of what went right, what went wrong, and what preventative measures could be taken. These formal accounts were generated through a consistent process of investigation that first began with requesting to learn more from those involved, followed by site visits to better understand the arrangements involved in the incident. Next, a draft summary was generated and sent to the individuals involved for their input regarding accuracy and completeness. Finally, a revised draft was provided to the Chemistry Department Safety Committee for their input, and from this, the formal account in the form of Lessons Learned memo was made and distributed via email to all members of the department (Box et al. 2020). If helpful, illustrative photos and/or additional resources were also provided in the memo and the accompanying email.

The following is a sample of one of those communiques:

What happened?

An undergraduate researcher set up a diethyl ether distillation. During the process, he saw the level of diethyl ether in the source flask run low. In response, he lowered that flask from the condenser to refill it. In this effort, some ether vapor or liquid got either onto or into the hotplate, where it caught fire, which then caught a solvent squirt bottle on fire. The student pushed this bottle away from the hotplate and off to the side. There it caught a lab coat on fire that was hanging on the external hood fixtures. Another lab member quickly responded with a fire extinguisher, but in attempting to pull the pin, broke it in the handle and so could not use it. A second fire extinguisher was found and used successfully. However, the sprinkler system activated during the time it took to get that one, producing an accompanying flood.

What went right?

- *The student was wearing safety glasses, gloves, long pants, and closed-toed shoes.*
- *The student had completed related university-mandated safety training.*
- *The student had also completed fire extinguisher training.*
- *Others present in the lab responded quickly and appropriately to put out the fire.*

What went wrong?

- *The student set up and performed the procedure without the knowledge or oversight of senior group members.*
- *Volatile solvent transfer on or near a hotplate risked combustion of vapor or liquid, since hotplates have thermostats in them that produce open sparks.*
- *A heated process was done with combustible solvent sitting nearby and with that solvent in a squirt bottle.*
- *A lab coat was hung next to a setup that involved heat and flammable, volatile liquids.*
- *The fire extinguisher handle was squeezed while also trying to pull the pin.*

What could be done differently?

- *Seek a review of a senior group member who has appropriate experience for any new set-up.*
- *With flammable liquids, use a non-sparking heat source, such as a heating mantle, instead of a hotplate.*
- *Store squirt bottles of flammable liquids away from sources of heat and ignition.*
- *Store lab coats away from potential sources of ignition.*
- *Hold the top or bottom half of a fire extinguisher handle when pulling the pin.*

Our memo writing efforts have likely been a greater investment of time and energy than a traditional enforcement approach might have been. However, we found that it successfully shifted previous behavioral norms and expectations toward more openness and voluntary sharing. In that sense, we saw that our commitment to leverage teachable moments through the use of story structure, including positive feedback in the interpretative portion, supported cultural change toward a more mindful and open value system. From there, we branched out to also make positive feedback an important part of our laboratory inspection process.

12.4 Positive Feedback in Safety Inspections

Checklist-guided inspections often achieve an evaluation of safety efforts. This checklist format is in widespread use because it makes content readily accessible, organizes knowledge in a way that encourages systematic evaluation, and reduces the frequency of omission errors (Scriven 2005). In addition, it standardizes the evaluee's expectations, thereby facilitating proactive preparation and compliance for regular inspections. However, checklists do have their limitations. The unavoidability of variation in results caused by such things as the inspector's time restrictions, limitations in training, and individual interpretation can lead to certainty in only the deficiencies found (Drury 2001). In other words, if something is recorded as satisfactory, the question will remain whether the check was thorough enough or done at all. As a result, feedback in the form of a checklist can, unfortunately, tend to be experienced as an itemized record of shortcomings.

Recognition of these limitations for us necessitated reconsideration of the checklist-only approach. As part of that reevaluation, we wanted to begin with a good foundation by establishing the goals of our inspection process. Our intention was to ensure that the evaluation tool would be designed so that the purpose of the inspection would be widely understood and accepted (Davidson 2005). Some of our goals were to improve working conditions, to increase compliance with institutional requirements, to strengthen preparedness for external inspections, and to increase workers' awareness and understanding of hazards and regulations. In all of these, the aim of the feedback would be to either prompt change or strengthen existing favorable conditions. We recognized that while checklists provided some guidance for change, they did not bring about that change and were very limited in conveying what favorable conditions to strengthen (Buckingham and Goodall 2019; Folkman and Zenger 2019; Porath 2016). As a result, we turned to including positive feedback via descriptive comments to help us fulfill these particular goals.

In our department, we began an initiative in 2016 to actively and regularly provide positive feedback with annual lab inspection reports. This was added to the standard inspection process, which involved the university Environmental Health and Safety (EH&S) representative recording observations based on a checklist. These were later sent to the safety plan holder, the group's safety officer, and the department's safety officer. Additional corrections and/or clarifying comments may have been included in the checklist reports.

The positive feedback that accompanied was compiled by the department safety officer, approved by the assigned inspector, and delivered by email to each research group's safety representative and Principal Investigator, as well as to the department administration. Acknowledging these successes to those in oversight positions conveyed the value and significance of the feedback, as did prompt delivery (Benson and Associates 2012). Descriptive feedback also included specific details, which amplified meaningfulness (Lippman 2015; Boyatzis 2011).

The following is a sample email sent by the department safety officer:

"Hi, Dr. XXX and XXX,

I am emailing your research group to document my observations of positive conditions found during EH&S inspections in an effort to begin a process of recording these successes. My goal is to have positive feedback become a part of EH&S's routine recordkeeping.

These observations have been reviewed and approved by XXX, head of our recent inspections:

- *consistent use of PPE when in lab spaces.*
- *chemical inventory in room temp storage well organized and maintained.*
- *good interest in using respiratory protection with silica gel transfer.*
- *good organization and signage in waste submission area.*
- *good interest in useful ventilation of rotovaps.*
- *good labeling and containment of squirt bottles.*
- *good use of recirculation pumps to cool condensers.Let me know if you have any questions about this feedback,"*

The results we saw over 2 years were that over two-thirds of groups in the first year either improved or remained the same, with equal or fewer issues than the year before. This impact was particularly noted by the long-time inspector of the lab spaces for our department. What he observed was that, in general, the number of violations had declined, and in fact, the average had decreased by over 15% per group in that first year. Then it declined by almost 10% more the next year. Thus, adding this descriptive positive feedback to a standard checklist was both a quantitative and a qualitative success. Not only did it yield improved responsiveness in the first year, but it also brought about extended responsiveness and compliance in the second year. In addition, we observed more initiative in requesting assistance and in addressing issues from the lab groups that received this active, positive feedback. Overall, the time and effort required to make and provide this record of desirable ongoing activities in the labs was a worthwhile investment relative to the benefits achieved (Box et al. 2021).

In implementing this initiative, we also made an additional serendipitous discovery. By including specific detail in the positive feedback, it became more evident to recipients that the record reflected one person's perspective. By association, it also acknowledged the same was true for the checklist portion of the record, and so improved the overall inspection recipient experience.

12.5 Why Positive Feedback Worked

The success we observed with positive feedback is counterintuitive since negative feedback can more readily achieve compliance in recipients. However, positive feedback increases openness to learning and consideration of a broader array of solutions (Fredrickson 2001). As a result, both are needed.

This is what Baumeister and Bratslavsk found in their 2001 review article. Titled "Bad is Stronger than Good," the authors found one exception to the generalization of their title and that was whether the *content* of negative feedback would be remembered. It turned out that the feedback recipients wouldn't remember as well was negative content delivered by itself. However, if it was accompanied by sufficient and meaningful positive feedback, recollection of the specifics of the negative feedback measurably improved.

In the case of the exceptional study utilizing positive feedback with negative feedback, a fixed ratio of three positives to every one negative was employed (Baumeister and Cairns 1992). To identify a more optimal ratio, a study of workplace team performance that found similar results might be useful. In that related study, it was observed that 5.6 positive comments were exchanged for every one negative in high-performing teams. Whereas in medium-performing teams, there were 1.9 positive comments for every negative, and the ratio flipped in low-performing teams, with more negative comments exchanged than positive ones (Losada 1999).

To explain why positive feedback is so important, it can be useful to turn to realms of study beyond safety and ethics, in particular evaluation tool design, organizational management principles, and neuroscience. From organizational management, one established principle is that positive feedback develops trust and rapport between the giver and the receiver (Boyatzis et al. 2012). In conjunction with that, evaluation tool design recognizes that this strengthening of trust facilitates revelation (Stufflebeam and Coryn 2014). To elaborate, if recipients' strengths have been acknowledged, they feel an added sense of security, and this security can contribute to revealing more about their work conditions and practices. Organizational management principles note that this added security also amplifies the evaluee's confidence to build on their existing strengths. In other words, it empowers them to perform at a higher level (Buckingham and Goodall 2019).

In business management, it is also recognized that positive feedback conveys to recipients' acknowledgment of how they are achieving excellence and, by association, what they are doing correctly and, therefore, what they need to keep on doing (Folkman and Zenger 2019; Porath 2016). In addition, it acknowledges what is often most valuable and meaningful to the recipient, i.e., whether or not the evaluee has succeeded at something they were striving for (Dasborough 2006).

To explain how all of these benefits of positive feedback lead to transparency, it can be helpful to first look at the effects of negative feedback for comparison. Psychological studies have shown that negative feedback encourages adherence to standards but also reduces the range of behavioral responses (Losada and Heaphy 2004). This reduction in the thought-action repertoire can be evolutionarily attributed to a response to a perceived threat. By contrast, as studies in behavioral psychology have also shown, positive feedback expands the spectrum of responses because the perceived risk of punishment is lower. As a result, people consider a more comprehensive array of solutions and try new things (Fredrickson 2001). They become more open to learning and more engaged with others and their work effort. Positive feedback adds to a recipient's security, amplifies their confidence to build on their existing strengths, and empowers them to perform at a higher level (Buckingham and Goodall 2019). As a result, they develop greater security to reveal ignorance and deficiency and so transition to a state of greater transparency.

The impact of positive feedback, then, is so significant that there may be better strategies than waiting until the spontaneous urge to record it develops. Instead, actively identifying the conditions and actions that are praiseworthy can ensure that the benefits of positive feedback are noticed. Making sure that feedback is timely, significant, specific, and sincere can also ensure that the benefits of positive feedback are not lost (Lippman 2015).

12.6 Best Practices

Although every situation is different, we have come to understand that a sustainable safety culture can be tailored to meet an organization's unique requirements while being flexible enough to adapt to evolving needs. Using teachable moments and positive feedback as a framework for developing such cultures nurtures the conditions for a bottom-up approach, where all stakeholders participate in the creative endeavor. In our experience, the following categories must be considered: communication, learning, mindset, and institutional structure.

12.6.1 Communication

The communication of safety practices and reporting incidents are central to establishing safety cultures (Reason 1998). This information must be conveyed in a timely fashion to ensure that users receive that information in connection with a salient concern and can take advantage of information that could lead to a teachable moment (Box et al. 2020). It must also be communicated using non-punitive language, contributing to the openness required for transformative learning (Vella 2000; Vesel 2020; Box et al. 2021). By using elements of storytelling to summarize incidents, their meaning can be amplified, and recipients might construct their own understanding. Communicating openly and using positive language can maximize community-wide awareness so that teachable moments can contribute to an informed culture (Box et al. 2020, 2021).

Specificity in the feedback improves the experience of inspection feedback for recipients for two reasons. One is that descriptive specificity could highlight things that could not be captured in a checklist. Examples include the use of clever and efficient strategies, noteworthy demonstrations of risk recognition, hygiene, and recordkeeping, and exemplary efforts to overcome obstacles in solving safety issues. Another reason is the implicit acknowledgment of the limitation of an inspection record to one person's perspective. Since the descriptive feedback reflects an individual's view, accompanying checklist feedback with it conveys that the checklist part of the feedback had been affected by that, as well.

12.6.2 Learning

Providing inspection-related positive feedback encourages dialog and relationship building, both of which pave the way for a greater understanding of follow-up priorities, such as how frequently, how thoroughly, and how soon to make changes. As an example, a checklist may document that chemical storage is excessive. Yet, inspection recipients might need to learn how their inventory crosses that line. Although specific descriptive comments may provide some clarification, in-person interaction bridges the gap to an even more concrete understanding of the changes needed. In this way, the relationships built by positive feedback can facilitate more in-depth learning and comprehension of findings from an inspection.

Leveraging teachable moments via Lessons Learned memos moderates the unpleasant feelings associated with incidents by turning them into community-based learning events (Merriam and Bierema 2014). Providing active learning experiences through discussion instead of lecturing or giving advice can help to create feelings of community, open-mindedness, and a healthy curiosity so that more learning can follow an incident (Vella 2000). In addition, institution-wide learning can develop from sharing a teachable moment widely, thereby building institutional memory and long-term changes in safety culture (McBride et al. 2003; Reason 1998).

12.6.3 Attitude/Mindset

Individual attitudes play an important role in responding to a teachable moment. Modeling a supportive approach to learning from our mistakes can help others to feel safe in such situations, to be open to suggestions, and to admit to confusion. These items are critical elements in transformative learning (Vella 2000). In addition, encouraging curiosity about the particulars of a safety-related event (including details that enable the listener to imagine doing the same task, with the same resources, in the same space) is essential in obtaining an accurate understanding of contributing factors, which is necessary for meaningful follow-up and the recognition of possibilities for change and remedy (Wagner and Ash 1998; Reason 1998). This growth mindset in responding to a teachable moment is essential to encouraging candor and willingness in others to learn (Wagner and Ash 1998).

12.6.4 Institutional Structure

To take full advantage of teachable moments, support structures must be in place to provide centralized oversight. A designated individual, most likely the unit or departmental safety officer, can be responsible for the investigation, follow-up, and communication on safety-related matters. With experience and training, this individual can be prepared to optimize teachable moments as they arise and can be ready to share this knowledge with all relevant parties (Haug 2014). This role requires institutional support and continuity to contribute to a culture of safety that benefits all (Zohar 1980).

12.7 Conclusions

Designating a champion, such as a departmental safety officer, to model and lead in efforts to capitalize on teachable moments and actively provide valuable and relevant positive feedback is essential in a decentralized organizational structure such as that of academic lab research. If we can turn teachable moments into occasions of learning, then we can help recipients make significant gains in the contextual knowledge they need for self-evaluation. This progression is essential for being able to recognize safety needs at the source and address them proactively. By the same token, if we actively note the successes of those being evaluated, we can also help them build the confidence they need for self-revelation. Allowing evaluees' strengths to shine gives them a firmer foundation from which to show the vulnerability of their safety issues.

Subsequently, with the active inclusion of positive feedback, external evaluation is more likely to be invited, and transparency is achieved. This transparency can increase continuity in knowledge transfer which is essential in a setting with high personnel turnover and variable levels of abilities and backgrounds. Together these initiatives can also spur a positive, open-minded, ethical culture that spans a broad scope of methods and material-handling needs, thereby addressing the other common safety culture challenge in academic lab research. As a result, the collective, therefore, will have the tools it needs to grow in exercising ethical consideration and in their culture of safety. These recommended best practices, then, tap into innovative communication channels by sourcing from users and moving up the chain of command, a learning process that allows for a holistic movement in safety culture.

References

Álvarez-Chávez, Luz S., Karla Perez-Gamez Marín, Mariona Portell, Luis Velazquez, and Francisca Munoz-Osuna. 2019. Assessing college students' risk perceptions of hazards in chemistry laboratories. *Journal of Chemical Education* 10: 2120–2131. https://doi.org/10.1021/acs.jchemed.8b00891.

Baumeister, Roy F., and Ellen Bratslavsky. 2001. Bad is stronger than good. *Review of General Psychology* 5: 323–370. https://doi.org/10.1037/1089-2680.5.4.323.

Baumeister, Roy F., and Kenneth J. Cairns. 1992. Repression and self-presentation: When audiences interfere with self-deceptive strategies. *Journal of Personality and Social Psychology* 62: 851–862. https://doi.org/10.1037//0022-3514.62.5.851.

Baumgartner, Lisa M. 2012. Mezirow's theory of transformative learning from 1975 to present. In *The handbook of transformative learning: Theory, research, and practice*, ed. Edward W. Taylor and Patricia Cranton, 99–115. San Francisco: Jossey-Bass.

Berson & Associates. 2012. *Berson & Associates Unlocks the secrets of effective employee recognition*. Cision PR Newswire. https://www.prnewswire.com/news-releases/bersin%2D%2Dassociates-unlocks-the-secrets-of-effective-employee-recognition-158548395.html. Accessed 25 May 2020.

Box, Melinda, Ciana Paye, and Maria T. Gallardo-Williams. 2020. Creating a positive, community-based learning environment in a Chemistry Department. *ACS Chemical Health & Safety* 27: 341–345. https://doi.org/10.1037/acs.chas.0c00033.

———. 2021. Positive feedback via descriptive comments for improved safety inspection responsiveness and compliance. *ACS Chemical Health & Safety* 28: 397–401. https://doi.org/10.1021/acs.chas.1c00009.

Boyatzis, Richard. 2011. *Neuroscience and leadership: The promise of insights*. https://iveybusinessjournal.com/publication/neuroscience-and-leadership-the-promise-of-insights/. Accessed 17 Dec 2020.

Boyatzis, Richard E., Melvin L. Smith, and Alim J. Beveridge. 2012. Coaching with compassion: Inspiring health, well-being, and development in organizations. *The Journal of Applied Behavioral Science* 49: 153–178. https://doi.org/10.1177/0021886312462236.

Buckingham, Marcus, and Ashley Goodall. 2019. The feedback fallacy. *Harvard Business Review*. https://hbr.org/2019/03/the-feedback-fallacy. Accessed 17 Dec 2020.

Dasborough, Marie T. 2006. Cognitive asymmetry in employee emotional reactions to leadership behaviors. *The Leadership Quarterly* 17: 163–178. https://doi.org/10.1016/j.leaqua.2005.12.004.

Davidson, E. Jane. 2005. *Evaluation methodology basics: The nuts and bolts of sound evaluation*. Thousand Oaks: Sage.

Drury, Colin G. 2001. Human factors and automation in test and inspection. In *Handbook of industrial engineering*, ed. Gavriel Salvendy. New York: Wiley.

Fivizzani, Kenneth P. 2016. Where are we with lab safety education: Who, what, when, where, and how? *Journal of Chemical Health and Safety* 23: 18–20.

Folkman, Joseph, and Jack Zenger. 2019. Resolving the feedback quandary. *Chief Learning Officer* 18: 44–53.

Fredrickson, Barbara L. 2001. The role of positive emotions in positive psychology: The broaden-and-build theory of positive emotions. *American Psychologist* 56: 218–226. https://doi.org/10.1037/0003-066X.56.3.218.

Garrison, Jim A. 1997. *Dewey and Eros: Wisdom and desire in the art of teaching*. New York: Teachers College Press.

Hansen, Edmund J. 1998. Creating teachable moments...And making them last. *Innovative Higher Education* 23: 7–26. https://doi.org/10.1023/A:1022916412432.

Haug, Berit S. 2014. Inquiry-based science: Turning teachable moments into learnable moments. *Journal of Science Teacher Education* 25: 79–96. https://doi.org/10.1007/S10972-013-9375-7.

Hill, Robert H., Jr. 2019. Recognizing and understanding hazards – The key first step to safety. *Journal of Chemical Health and Safety* 26: 5–10. https://doi.org/10.1021/acs.chas.8b26306.

Hover, Kenneth C., and Michael J. Schneider. 2019. Safety in educational and research laboratories: Seizing the opportunity. *Concrete International* 41: 34–40.

Johnston, Neil. 1995. Do blame and punishment have a role in organizational risk management? *British Flight Deck* 15: 33–36.

Leveson, Nancy G. 2011. *Engineering a safer world*. Cambridge: MIT Press.

Lippman, Victor. 2015. *Management feedback is vital – Why is it so often given badly?* Forbes. https://www.forbes.com/sites/victorlipman/2015/09/02/management-feedback-is-vital-why-is-it-so-often-given-badly/?sh=4ab811d2695a. Accessed 17 Dec 2020.

Losada, Marcial. 1999. The complex dynamics of high performance teams. *Mathematical and Computer Modelling* 30: 179–192. https://doi.org/10.1016/S0895-7177(90)00189-2.

Losada, Marcial, and Emily Heaphy. 2004. The role of positivity and connectivity in the performance of business teams. *American Behavioral Scientist* 47: 740–765. https://doi.org/10.1177/0002764203260208.

McBride, Colleen M., Karen M. Emmons, and Isaac M. Lipkus. 2003. Understanding the potential of teachable moments: The case of smoking cessation. *Health Education Research* 18: 156–170. https://doi.org/10.1093/her/18.2.156.

Merriam, Sharan B., and Laura L. Bierema. 2014. *Adult learning: Linking theory and practice*. San Francisco: Jossey-Bass.

Porath, Christine. 2016. Give your team more-effective positive feedback. *Harvard Business Review*. https://hbr.org/2016/10/give-your-team-more-effective-positive-feedback. Accessed 17 Dec 2020.

Reason, James. 1998. Achieving a safe culture: Theory and practice. *Innovative Higher Education* 23: 7–26. https://doi.org/10.1080/02678379808256868.

Reniers, Genserik L.L., Koen Ponnet, and An Kempeneers. 2014. Higher education chemical lab safety interventions: Efficacious or ineffective? *Journal of Chemical Health and Safety* 21: 4–8.

Scriven, Michael. 2005. Checklists. In *Encyclopedia of evaluation*, ed. Sandra Mathison, 53–59. Thousand Oaks: Sage.

Stufflebeam, Daniel, and Chris L.S. Coryn. 2014. *Evaluation theory, models, and applications*. San Francisco: Jossey-Bass & Pfeiffer Imprints.

Taylor, Edward, and Patricia Cranton, eds. 2012. *The handbook of transformative learning*. San Francisco: Jossey-Bass.

Taylor, William D., and Lori A. Snyder. 2017. The influence of risk perception on safety: A laboratory study. *Safety Science* 95: 116–124. https://doi.org/10.1016/j.ssci.2017.02.011.

Vella, Jane. 2000. *A spirited epistemology: Honoring the adult learner as subject*. San Francisco: Jossey-Bass.

Vesel, Crista. 2020. Agentive language in accident investigation: Why language matters in learning from events. *ACS Chemical Health & Safety* 27: 34–39. https://doi.org/10.1021/acs.chas.0c00002.

Wagner, P. Susan, and Katherine L. Ash. 1998. Creating the teachable moment. *Journal of Nursing Education* 37: 278–280.

Woodhouse, Howard. 2011. Storytelling in university education: Emotion, teachable moments, and the value of life. *The Journal of Educational Thought* 45: 211–238.

Yellowlees, Duncan. 2022. *3 story frameworks you can use to make your research presentations*. https://medium.com/@info_98779/3-story-frameworks-you-can-use-to-make-your-research-presentation-49567d7c53d6. Accessed 13 Dec 2022.

Zohar, Dov. 1980. Safety climate in industrial organizations: Theoretical and applied implications. *Journal of Applied Psychology* 65: 96–102. https://doi.org/10.1037/0021-9010.65.1.96.

Chapter 13
In Situ Ethics Education Within Research Laboratories: Insights into the Ethical Issues Important to Research Groups and Educational Approaches

Kelly Laas, Christine Z. Miller, Eric M. Brey, and Elisabeth Hildt

Abstract This chapter describes the development of a workshop series focused on helping students develop research lab ethics guidelines. The workshop was developed through a National Science Foundation-funded project that situates ethics education within the research environment. Students in four departments at a private research university were recruited to join a Student Ethics Committee that collaboratively developed context-specific codes-of-ethics-based guidelines for their departments. These bottom-up developed guidelines were revised in an iterative process, including feedback from faculty, other graduate students, and the original Student Ethics Committee. The goal of the workshops was to promote the cultivation of an ethical culture in experimental laboratories by integrating research stakeholders in a bottom-up approach to developing context-specific guidelines and to identify factors that students consider relevant to the ethical conduct of research. After describing the workshop series, this chapter presents a qualitative content analysis of the guidelines. The guidelines developed by the Student Ethics Committees reflect the current culture of experimental labs in these departments

K. Laas (✉)
Centre for the Study of Ethics in the Professions, Illinois Institute of Technology, Chicago, IL, USA
e-mail: laas@iit.edu

C. Z. Miller
Savannah College of Art and Design, Savannah, GA, USA

E. M. Brey
Department of Biomedical Engineering, AET, The University of Texas at San Antonio, San Antonio, TX, USA
e-mail: eric.brey@utsa.edu

E. Hildt
Center for the Study of Ethics in the Professions, Illinois Institute of Technology, Chicago, IL, USA
e-mail: ehildt@iit.edu

E. Hildt et al. (eds.), *Building Inclusive Ethical Cultures in STEM*, The International Library of Ethics, Law and Technology 42,
https://doi.org/10.1007/978-3-031-51560-6_13

and the ethical issues that students think are important. The content of these guidelines and the ethical issues the students focused on contrast in some key ways with what ethical issues faculty who head research groups saw as paramount. The chapter describes the first two iterations of the ethics education workshop module, a qualitative analysis of the first draft of these guidelines, and finally reflects on future work in this area of in situ ethics education.

Keywords Ethics guidelines · Responsible conduct of research · STEM ethics

13.1 Introduction

This chapter aims to describe a model teaching method that involves students in active learning by asking them to analyze existing codes of ethics and ethical guidelines relevant to their research environment and then use their new knowledge of research ethics to craft draft guidelines relevant to their personal research situation. The chapter first describes our initial six-week workshop series with four groups of graduate students who formed four Student Ethics Committees and developed a set of guidelines specific to their discipline. The chapter then provides an analysis of the contents of these codes. We end with reflections on topics and teaching strategies stressed in the final workshop module developed and outline possible iterations of the guidelines workshop module.

Integrity in the scientific research process is integral to the education of graduate students in the STEM fields. Since the U.S. National Institutes of Health began requiring Responsible Conduct of Research (RCR) education in 1989 (NIH 1989), debates have continued about what should be included in the research ethics curriculum. As Michael Kalichman notes in his history of RCR education, a major question dominates the field. "*Which* of the vast number of potential ethical challenges faced by scientists should be addressed and *how* [these issues should be addressed] are not made clear by the general injunction to teach RCR." (Kalichman 2013).

The goals of research ethics education do seem to be slightly more clear, though this, too, is a moving target. At its most basic level, one goal is to help researchers avoid research misconduct. However, leaders in this field have been working for years to slowly move the focus of RCR from a focus on compliance to actively embracing practices that build a culture of ethical practice and inclusiveness. Michael Kalichman, in his 2013 article, outlines these goals as" Instead, the focus is something much more aspirational: is to increase student's awareness of the ethical challenges inherent in research, prepare them to meet these challenges, and also persuade them "that surmounting these challenges is worth the trouble." (Kalichman 2013: 381). Another eminent scholar in the field, Allison L. Antes, in her overview of RCR education, adds that another goal is to foster a community of social responsibility (Antes et al. 2009: 398).

In 2014, the National Science Foundation (NSF) changed the focus of its Ethics Education in Science and Engineering Program (EESE) that supported the development of ethics courses, case studies, and teaching modules. This change was in response to many studies finding that commonly used instructional practices had limited effectiveness in meeting the main goals of ethics education (Waples et al. 2009; Cech 2014). This new program funds research projects that identify (1) factors that are effective in the education of ethical STEM researchers and (2) approaches to developing those factors in all the fields of science and engineering that NSF supports. The named Ethics and Responsible Research (ER2) program[1] solicits research proposals exploring the following questions: 'What constitutes responsible conduct for research (RCR), which cultural and institutional contexts promote ethical STEM research and practice, and why?'

Against the background of an approach that considers the broader cultural and institutional contexts that promote ethical STEM research and practice, this project described in this chapter seeks to answer the "which" and the "how" that Michael Kalichman raises in his 2013 article – *which* of the huge number of potential ethical issues should be addressed, and *how*.

One of the key components of this project is to expand the conversation around research integrity and the responsible conduct of research in a way that engages students and other stakeholders in bottom-up approaches to building ethical cultures. As those applying for research funding, supervising students, and producing quality research in a "publish or perish" environment, faculty have certain ethical issues and areas in research that they consider extremely important. As individuals new to the research environment, working toward their degree, and juggling academics, family, teaching assistantships, and in the case of some international students, visa issues, students have other ethical topics that they feel should receive greater consideration. Though there is much overlap, the differences between these two groups are telling and worthy of further exploration. Approaches to ethics education and building cultures that foster scientific integrity must address the concerns of both sets of stakeholders. To understand the "which" question around what ethical issues should be addressed in STEM ethics education, the project team surveyed and observed graduate students and STEM faculty at one private technical university to help identify factors relevant to ethical STEM.

In answering Michael Kalichman's "how" question, how should ethics be taught, we sought to situate ethics education within university research groups rather than the classroom. Reasoning that graduate students who had previously been engaged in research would have the requisite experiences to draw on, we recruited students to take part in a six-week workshop during which they would form ethics committees divided by discipline. Over two semesters, two student ethics committees worked in parallel with each other – the biology and biomedical engineering ethics committees in semester one and the physics and chemical and biological engineering committees during semester two – to engage in active discussions about ethical issues they had encountered in the lab and to develop guidelines that could serve as a guide for addressing or mitigating these issues in the future. As Dena Plemmons and Michael Kalichman state:

[1] https://www.nsf.gov/pubs/2019/nsf19609/nsf19609.htm

> ...it makes sense to consider the research environment itself—where learning one's science happens where one also engages in social interaction around that science—as offering a specific, key, and underutilized arena in which to actively think about ethical behavior as it relates to scientific practice. (Plemmons and Kalichman 2018, 209).

Situating ethics education in the research environment and tailoring it to address the ethical issues that are important for the discipline and the social interactions that happen around the research process helps make ethics education both pertinent and timely. Our goal was to develop a flexible teaching approach that could engage students in active learning about research ethics and help empower participants by helping them become aware of the – sometimes unacknowledged – rules and norms that exist in research groups and giving students a chance to be a part of discussions and negotiations around formalizing these rules and norms. Recognizing that most research groups do not often have written documents outlining best practices for research and the responsibilities of student lab members and faculty lab leaders, we decided this would likely be a fruitful approach.

Our approach of asking students to develop ethical guidelines for their lab groups allows them to ask questions, challenge assumptions, and engage in discussions on a slightly more equal footing. The bottom-up approach of this project is also integral in shifting the power dynamic that often exists in ethics education and academic research. In many approaches, students are informed of what the regulations and policies are and may engage in some discussion about fictional or historical case studies. Afterward, orientations for student researchers may touch upon issues such as lab safety, authorship, and managing data but may have limited interactive discussions about science's norms.

The approach presented here helps to slowly push the discussion about the responsible conduct of research from compliance to building ethical cultures by engaging students in these discussions, fostering decision-making skills, and developing the research culture in a way that helps cultivate open discussion around scientific integrity. The guidelines discussed below both help identify which ethical issues the Student Ethics Committees saw as important in their research context and represent one approach that answers how we can help researchers develop practices that reinforce integrity in science and foster discussions that ultimately should help improve the environments in which research is done. In our view, this educational approach is a beneficial approach to complement the more traditional ethics education students engage in throughout their academic careers.

We expect this kind of education to be most effective when it includes a discussion of ethical issues in situ in the research group (rather than in the classroom) and includes all stakeholders, including students' peers, mentors, and supervisors (Watts et al. 2017). Finally, ethics education was observed to be most impactful when it occurs within the respective *institutional culture,* which comprises both the *organizational context* and the *peer environment* (Holsapple et al. 2012).

13.2 Project Background

Our project sought to situate ethics education within research laboratories to help students address ethical challenges in a more informed way that improves the ethical culture of their research environments. Funded by a three-year National Science Foundation grant, this project sought to identify what university students and faculty saw as important ethical issues in STEM through a series of surveys and observations during the first year.[2] During the second year, the project team recruited four groups of graduate students, biology and biomedical engineering during the first semester, and physics and chemical and biological engineering during the second semester, to join a student ethics committee and take part in a multi-week workshop. The workshop consisted of six sessions. It actively engaged graduate students in reflecting on and articulating what they believe are the critical ethical issues they encounter in their research environments. After discussing and reflecting on existing ethics codes, participating students were tasked to develop guidelines that would be useful in addressing ethical issues in their respective research environments.

Before running the workshops, the project team hypothesized:

- Student-developed guidelines will differ in ethical topics covered based on discipline and student experiences in the lab.
- Student-developed guidelines will focus on issues that impact them directly.
- Student-developed guidelines will reflect the current culture of research groups and call attention to areas students see as areas of concern or needing clarification.

These guidelines draw on two major areas of information. First, they draw on and complement existing professional codes and organizational guidelines/regulations. Secondly, they seek to address ethical issues that the authors think need more guidance and reflection. We assume that direct involvement in developing ethical guidelines may positively influence researchers' understanding of ethical research and practice issues, their handling of these issues, and promoting an ethical culture in the respective lab. These guidelines also add the benefit of shedding light on ethical areas that the authors believe are important to clarify and discuss.

The workshop series (described in the next section) was developed as the basis for a flexible teaching strategy that could be used in a classroom setting, research group, or a two or three-day workshop series. The project team tested different variations of this module in the settings described below.

The final version of this workshop series (currently available online, including a slide deck and teaching notes: see https://ethics.iit.edu/cultivatingcultures) reflects the findings of a background study we performed at our home institution, the Illinois Institute of Technology, consisting of qualitative and quantitative analysis (Hildt et al. 2022, Laas et al. 2022.) Our research suggests that research ethics extends

[2] For more information see Hildt et al. 2019, 2022 and Laas et al. 2022.

beyond the commonly covered responsible conduct of research (RCR) topics: plagiarism, falsification of data, and treatment of human and animal subjects to issues that involve social interaction, power relations, gender, and other forms of difference. Peer-to-peer, supervisor-to-lab member, and PI-to-student relationships are significant and provide the foundation for ethical research practice. We have found that creating opportunities for students to interact as peers allows for frank and open discussion about issues they encounter that might otherwise not be put into words or raised. The research also supports the need for bottom-up approaches to ethics education that actively engage students in learning about best practices in research and in discussions around setting expectations, standards, and guidelines for the ethical practice of research in their labs.

13.3 Overview of Initial Workshop Series

During the first half of the project, the team worked on developing a six-session interactive workshop that was co-led by project team members and graduate student leaders who were recruited and trained by the project team before running the workshops. The goal was to develop a module that could be adapted for use in different settings that seeks to introduce students to the intricacies of research ethics and may also be useful as part of professional development training for ethics instructors or future mentors.

During the Fall 2017 and Spring 2018 semesters, we recruited graduate students from four departments at the university in four Student Ethics Committees: Biology (BIO), Biomedical Engineering (BME), Chemical and Biological Engineering (ChBE), and Physics (Phys). Students were nominated to participate by their faculty advisor or responded to an email sent out to all research-active graduate students by their department coordinators. Students who responded to the invitation volunteered to participate in six 90-min workshop sessions. The module sessions actively engaged graduate students in reflecting on and articulating what they believe are the critical ethical issues that they encounter in their research environments. After discussing and analyzing existing professional ethics codes and university-specific guidelines (such as those around authorship, departmental guides for graduate students, etc.), participating students were tasked to develop guidelines that would be useful in addressing ethical issues in their respective research environments. Students engaged in these workshops had at least 6 months of experience working in research laboratories. Research groups at the university tend to be relatively small, usually composed of a faculty PI, 3–8 graduate students, and in some cases, undergraduate students (Table 13.1).

The sessions were as follows:

- Introduction to ethics, ethics codes, and guidelines
- Discussion of issues encountered in research labs for participating students, analysis of discipline-specific ethics codes

Table 13.1 Number of students from each department in student ethics committees

Departments	Fall 2017	Spring 2018
Biology	8	
Biomedical Engineering	7	
Chemical and Biological Engineering		7
Physics		9

- Extended discussion of students' laboratory situation and ethical issues encountered
- Begin to draft guidelines
- Discussion of guidelines
- Refinement of 1st draft of guidelines

In designing these sessions, we worked with two student facilitators per workshop series – one from each department – who stood out as a leader in their community and had connections with many other graduate students in their department. The goal was to bring in the student experience before designing the workshop series. Engaging these students turned out to be particularly important. They brought in a fresh perspective and helped highlight often-overlooked issues in research ethics, such as special issues that some minority and international students face.

Student facilitators led discussions among members of the Student Ethics Committees. The project team's role was limited to presenting educational materials, such as information on what ethics is, an introduction to RCR, and the role codes of ethics play in shaping professional practice.

One of the key goals of the first session was to begin building a community where members felt safe and supported and potentially willing to share some of the ethical issues they have encountered in research labs. Building this community of trust hinged on having strong student leaders who help plan and run the workshop sessions. A project team member co-led these sessions with the two student leaders and shared information on the role codes of ethics and guidelines can play in building ethical research environments and the main goals of the workshop series.

In the second and third sessions, student participants were asked to talk about ethical issues faced in their own lab experience that they were willing to share. These discussions happened without faculty present to help make sure students' comments were kept confidential outside of the student ethics committee. Following this, the students analyzed existing codes of ethics and guidelines from their discipline to see if these codes potentially addressed some of these issues or provided helpful guidance. In sessions three and four, students chose the most important ethical issues they had discussed and jointly wrote the first draft of a set of guidelines that tried to address these issues in a way that would be helpful to their research groups. Team leaders provided examples of other codes to look at and helped students locate existing guidelines and resources provided by the university that could be referred to in the student guidelines. During the fifth session, students put their guidelines on a screen and jointly edited and discussed them as a group. During the sixth session, each departmental group shared its guidelines with all workshop participants. Students found this particularly helpful in looking at new ways to frame these ethical issues.

At the end of the sixth session, students provided the project team with the first draft of their ethical guidelines that reflected their group understanding of ethical issues important in their research environment, as informed by discussions and readings completed during the workshop. These guidelines were edited for grammar and punctuation but otherwise represent a direct view of the students' current research environment and what students considered important ethical issues and areas of concern. Copies of these guidelines can be found as part of the workshop module teaching guide (http://ethics.iit.edu/cultivatingcultures).

13.4 Analyzing the Ethics Guidelines

To complete a qualitative content analysis of these four sets of guidelines, the project team developed a taxonomy for coding the guidelines that are based on an expanded understanding of the ethical topics important to the research environment (i.e., beyond just RCR topics).[3] To begin with, we drew from the nine areas of research ethics (United States, NIH, 2009) and added codes from our experience coding 30 different graduate student interviews about their experiences in research labs. Ultimately, our taxonomy included 61 codes (See Appendix). The three research team members coded each set of guidelines using the software MaxQDA, and then differences were reconciled through discussion (Table 13.2).

Table 13.2 Summary table of taxonomy themes

Main theme	Example codes within theme
Interpersonal Relationships	Equitable treatment Working in diverse groups, Workplace norms, Respect for personal property, Relationships among lab members
Management of Guidelines	Approval of guidelines, Ethics education, Living document/ reviewed routinely
Responsible conduct of research topics/Good research practices	Animals in research, Collaborative science Conflict of interest, Data management, Human subjects, Peer review, Publication practices, Research misconduct
Role of PI	Clearly defined expectations, Fairness, Meetings with PI, Recommendation letters
Work Environment	Acceptable workplace norms/behaviors, Competition within lab, Funding, Lab meetings, Work hours/vacation policies

[3] For more information on discussions of the importance of covering issues beyond traditional RCR topics, see Hildt et al. 2022, and Hildt et al. 2019.

13.5 Overview of the Guidelines

The first draft of the guidelines developed by the Student Ethics Committees draws on two major sources of information. First, they draw on and complement existing professional codes and organizational guidelines/regulations (avoiding reinventing the wheel). Secondly, they seek to address ethical issues where the authors thought more guidance and reflection were needed. We assume that the topics that ultimately were included in these guidelines represent the issues that the Student Ethics Committee members saw as the most important in their department/research group and that having these topics highlights areas of concern or areas where more clarity is needed in each respective discipline. Regarding learning goals, we assume that direct involvement in developing ethical guidelines may positively influence researchers' understanding of ethical research and practice issues, their handling of these issues, and the promotion of an ethical culture in the respective lab.

Each of the four guidelines developed by the Biomedical Engineering (BME), Biology (BIO), Chemical and Biological Engineering (ChBE), and Physics (Phys) Student Ethics Committees were different. However, one could see a cross-pollination of ideas happening between groups. The Biomedical Engineering guidelines focused on the responsibilities and students in research groups and were extremely transactional in their focus, and students should do x, faculty supervisors should do y. It sought to lay out general responsibilities but did not provide any other details about handling ethically questionable situations or building stronger research communities.

The Biology group, in contrast, was described by some readers as a kind of "Ten Commandments" and laid out precisely what PIs and students should and should not do. These guidelines focused on issues of responsibilities, relationships, and conduct. Sections were labeled "Laboratory Conduct," "Student-Student Relationships," "Mentor-Mentee Relationships," and "Papers and Authorship" and were preceded by proscriptions of what individuals should and should not do.

In the second iteration of the workshop series during the second semester, based on lessons learned during the first semester, the physics and ChBE student leaders and their colleagues turned more toward issues specific groups of students faced. The ChBE guidelines focused on the culture of research groups, outlining the role free discussion and questions have in good scientific practice, the need for respectful dialogue, confidentiality, and the need for clear policies to ensure the fair treatment of students, regardless of background. Some sections of the guidelines seek to address issues faced by international students. The guidelines also included specific sections on authorship, data and resource management, and patents.

The guidelines developed by the Physics Student Ethics Committee were the most detailed of any of the guidelines developed. In response to a discussion about changes to the American Physical Society's Guidelines for Professional Conduct that was being discussed in 2017 around adding a section that specifically addressed issues of explicit, systemic, and implicit bias,[4] the students chose to focus attention

[4] See APS Guidelines on Ethics, section III. https://www.aps.org/policy/statements/guidlines-ethics.cfm

on issues of harassment in the workplace and diversity in the discipline. While the first sections of the guideline concentrate on issues of authorship, transparency, and reproducibility of experiments (both subjects that were discussed in detail during the workshop sessions), the second two-thirds of the document are titled "social" and provide detailed guidance, educational resources, and examples of unwanted behavior including sexual harassment and bullying. Some sections talk about issues faced by minorities and women, individuals with mental and physical disabilities, family status and its implications for students in research, religion, and international students and collaborators.

13.6 Detailed Results

13.6.1 Themes Across All Four Guidelines

There were some clear themes across the four sets of guidelines, regardless of discipline.

All four guidelines spoke about the role of the principal investigator in the research environment (14), with the Biology group mentioning it seven times. The Biology Student Ethics Committee outlined specific responsibilities, for example, "PIs shall set aside a period of at least 3 hours once every two weeks to meet with students," and required the PI to abide by common workplace norms. In comparison, Physics only mentioned this topic once, outlining the role of the PI to either adopt or expand a version of these guidelines into a formal, written policy for the research group (Table 13.3).

The second most prevalent code was the need for clearly defined expectations and responsibilities (10). This call for clarity applied to everything from vacation policies to outlining the specific responsibilities of the PI and graduate student. Three of the four guidelines (BIO, ChBE, and Physics) speak of the need for written contracts that clearly outline what is expected of graduate research students. The ChBE guidelines require the following information to be included: "...working hours, compensation, duration, task description, and qualitative measures to determine completion of the job."

Table 13.3 Themes across four guidelines

Main themes across four guidelines	Bio	BME	ChBE	Physics	Total
Role of principal investigator	7	2	4	1	14
Clearly defined expectations and responsibilities	1	1	3	5	10
Work hours/vacation policies	2	4	2	1	9
Good research practices	3	1	1	3	8
Publication and authorship	1	1	1	2	5

All four guidelines also mentioned the need for work hours/vacation policies (9) to be included. In some cases, these policies were justified as a way to "avert the occurrence of inconsistent behavior between specific groups. For example, international and domestic students should be treated equally in respect to U.S. national holidays." (ChBE). In other instances, students tried to balance the time requirements of laboratory research with what students saw as an acceptable work schedule. The BIO guidelines gave a justification for this limit. "Although certain laboratory work requires time exceeding that of a normal work-week (more than 40 hrs/week or working weekends), principal investigators shall not force students to work more hours than what is considered a typical full-time work schedule (40hrs/week)." (Emphasis included in the original.)

The 4th and 5th most prevalent codes that showed up across all four guidelines had to do with good research practices (8 total) and publication and authorship practices (5 total). While both of these subjects were mentioned in all of the codes, the length of the mentions varied substantially. "Good research practice" referred to sections of the guidelines where the authors detailed what they saw as the norms of science, and many of these sections coded under "Good research practice" also refer to traditional RCR themes like data management (not excluding data that does not fit a hypothesis) or avoiding misconduct. All four guidelines dedicate a relatively substantial section of their guidelines outlining practices they see as adding to the integrity and transparency of science in their own discipline. For instance, the Physics guidelines include almost a page of guidance about the transparency and reproducibility of experiments and the need to add comments to computer code written by students, using open-source software and version control to maintain this code.

Most discussions about publication practices dealt with questions of authorship, with some specifically claiming the power for an equal say, "If there is a question regarding whether or not an individual should be considered an author, all other authors shall take a vote to decide. This decision should not be left solely to the principal investigator." (BIO). Other guidelines closely follow the recommendations of the International Committee of Medical Journal Editors, which are widely used by many disciplines and journals (ChBE). In the two Student Ethics Committees held in the Spring 2018 semester, both of these guidelines also provide links to the institutional Authorship Guidelines that the university's Faculty Council drafted.[5] Interestingly, none of the students in the Student Ethics Committees knew about these authorship guidelines before a faculty member brought them to the project team's attention. When made aware of this, the students added links to this and other institutional policies to their guideline drafts.

---

[5] https://web.iit.edu/sites/web/files/departments/general-counsel/faculty_handbook/Appendix-S.pdf

13.6.2 Themes Across Three Guidelines

Some themes were not present across all four guidelines but only across three guidelines. This is an indication of differences between departments and also may be due to the various individual ethics committee members influencing the content of the guidelines (see Discussion part).

Fairness and discrimination appeared 14 times overall in three guidelines: BIO, ChBE, and Physics, and were discussed in various ways. BIO included six sections that were coded as "Fairness" by the project team, ChBE included three sections that specifically addressed issues of equity and discrimination, and the Physics guidelines mentioned this issue five times and at far greater detail in the "Social" section of their guidelines. Other issues that came up across three of the four guidelines include expectations ("...students should be able to express their scientific opinions freely," ChBE) and responsibility, where the guidelines specifically detail the responsibilities of individual lab members ("Graduate Students [will] be primarily responsible for the successful completion of their own research projects. / Research advisors [will] provide supervision for the research and a safe, productive research environment." BME). The remaining four topics mentioned in three of the four guidelines set up expectations for safety training and ensuring the lab's safety, access to shared resources such as lab equipment, expectations for meetings between the student/PI, and issues of accountability for lab members (Table 13.4).

13.6.3 Traditional RCR Topics in the Guidelines

After surveying the most-mentioned topics in the guidelines, the project team decided to see if some codes could be grouped under traditional responsible conduct of research (RCR) topics, such as animal and human subjects research, authorship, data management, etc. Codes such as publication ethics and animal subjects were relatively straightforward, but in other instances, it became necessary to group some codes together. For example, when looking at the topic "mentor/mentee relationships" we grouped the following codes under this category

Table 13.4 Major themes across three guidelines

Main themes across three guidelines	Bio	BME	ChBE	Physics	Total
Fairness/discrimination	6	0	3	5	14
Expectations	4	0	2	1	7
Responsibility	0	4	1	1`	6
Shared resources (lab equipment)	2	2	2	0	6
Safety issues/safety training	1	1	0	2	5
Meetings with PI	1	3	1	0	5
Accountability	1	2	0	1	4

- Role of PI
- Expectations (relating to both roles)
- Clear Expectations and Responsibilities (when relating to PI/Student relationship directly)
- Responsibility (relating to both roles)
- Meetings with PI

Example of quotes falling under this category include this one from the Physics Student Ethics Committee,: "When a new member joins the research group, it is the responsibility of the PI to review the group's ethics policy with the new member, to ensure that all parties have a clear understanding and to create an initial set of responsibilities and expectations". Similarly, we grouped the codes data management, data sharing, and data collection under data management.

These groupings help provide an overview of the RCR themes that were covered in the four sets of guidelines. Overall, you can see that the frequency with which issues of mentor/mentee relationships were mentioned was extremely high in all four guidelines. Publication practices and data management were discussed but to a smaller extent, and the other topics received much fewer mentions, with conflict of interest and research misconduct each mentioned in only one set of guidelines (Fig. 13.1).

The guidelines focused on many of the details of the mentor role, including the need to set aside time to regularly meet with students (three guidelines), provide documentation (such as work contracts, four guidelines), and provide opportunities to discuss expectations and responsibilities for students in the lab (three guidelines),

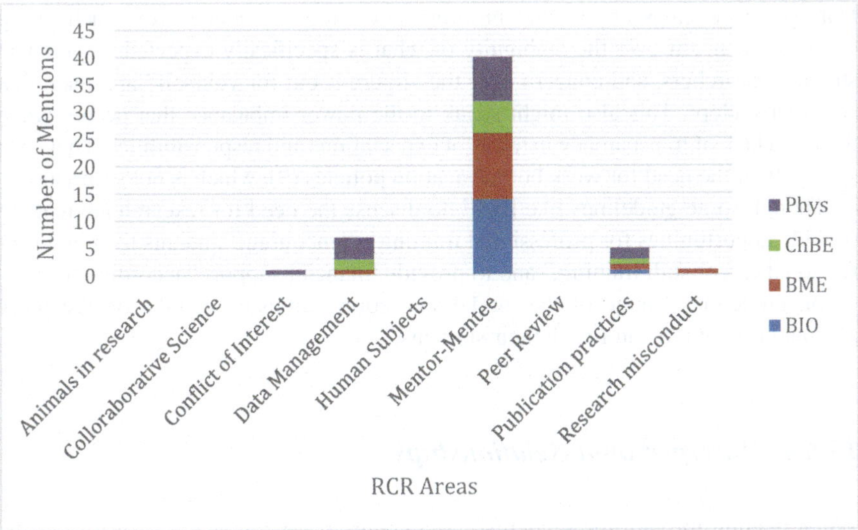

Fig. 13.1 RCR themes across disciplines

Beyond these similarities, the guidelines varied widely in how they addressed the role of the PI.

In the mentoring relationship and the student/supervisor relationship, the issue of power imbalance plays a large role. This relationship can be abusive and supportive, and members of the Student Ethics Committees were eminently aware of this issue before the workshop began. The BIO guidelines specifically mention that principal investigators should not assign the same work to two different students to force unnecessary competition between lab members. Alternatively, in the ChBE guidelines, the PI is called upon to exercise their power, both in acknowledging and making final decisions on authorship issues and in deciding if the outcome of research being done in the lab can be shared with third parties. BME, which chose to organize its code by listing the responsibilities of PIs and students respectively, seems to be the most adamant, meticulously detailing what it means to be committed to trainee graduate students' training and scientific education. These guidelines detail how the PI should help the students navigate institutional and departmental requirements, provide financial assistance for tuition and travel expenses for conferences, and urge the student to attend professional skills training. Overall, three of the four guidelines (excluding physics) clearly sought to delineate some sort of boundaries around the role and power PIs have in the research group, or at least lay some ground rules for how these relationships play out.

There were some clear topical themes across the four guidelines, regardless of discipline. Echoing findings from 30 graduate student interviews analyzed by the project team (Hildt et al. 2022), all four guidelines spoke at length about the role of the principal investigator (14). The second most prevalent code was the need for clearly defined expectations and responsibilities (10). This related to everything from vacation policies to outlining specific responsibilities of the PI and graduate student. During the meetings of the student ethics committees, students from each group expressed concern over the ambiguity of what is specifically expected of graduate student researchers, and concern over the "figure it out for yourself" approach that some labs adopt. This also might speak to the power imbalance that occurs when there is a lack of transparency in terms of expectations and responsibilities. They also spoke about the need for work hours/vacation policies (9), which is not surprising.

Though some guidelines like BME do discuss the need for research advisors to provide opportunities for professional training – "encourage students to attend professional career skill training," and to provide financial support "according to institution guidelines," none of the guidelines require advisors to take on the more personal role of a mentor with their students.

13.6.4 Interpersonal Relationships

When reading through the guidelines, the coders were somewhat surprised at the widespread attention paid in the guidelines to interpersonal relationships beyond mentions of relationships between supervisors and students. The group of codes that emerged, which we classified as "interpersonal relationships", focused on relationships among lab members, how to be a good lab citizen, and discussed relationships outside of the lab that could impact a member's position or work. In session two of the workshop, where students were asked to share stories from their own experiences, students spent a good amount of time discussing issues of equitable treatment, cases they had either experienced or witnessed and the need for more discussion around these issues. Codes that fell under the interpersonal relationship category (excluding the relationship specifically between PI/Student) included:

- Good lab conduct/how to be a good lab citizen
- Working in diverse groups
- Dealing with issues of sexual harassment and discrimination
- Equitable treatment (international students, minority groups)
- Workplace norms
- Respect for personal property
- Relationships among lab members

Some examples of specific topics that fell into this category include outlining details about professional dress (Bio), avoiding disrespectful or disparaging language (ChBE), respecting the privacy of fellow lab members unless disclosing that information addresses a legitimate safety question (ChBE), and providing individuals with different religious beliefs the time and space appropriate for daily prayers (Phys).

This focus on interpersonal relationships rather than traditional RCR topics is somewhat telling. Students seem to see the research environment as a socially complex enterprise, and the guidelines try to set some boundaries on how research group members interact with one another. Three of the four guidelines discuss how lab members should share resources such as equipment and supplies, reflecting some departments' limited budget and space. Two codes specifically discuss the need to keep personal information about fellow lab members confidential. Other examples, such as a quote from the Biology guidelines, speak to the power differential mentioned above and issues of respect for personal property/factors outside the lab: "Students shall not be requested to use personal money for laboratory-related expenses."

Similarly, ChBE includes a discussion of how junior and senior research members should freely discuss their scientific opinions:

> This requires an atmosphere where junior members of the group are able to question authority with "fact-based arguments" and/or "demonstration of judgement-based arguments". In other words, the methods, approaches, and techniques used in research should not be limited to a specific set, preferred by the senior researchers. On the other hand, the junior members of the group should welcome sincere critics, have enthusiasm to learn best practices, and value collaboration with different components of the team.

This somewhat differs from the perspective of faculty who, having a greater level of autonomy and power in these environments, focus on ethical issues that directly impact the quality of research.

The focus of the student-written guidelines on issues of discrimination, diversity, and inclusion were in part shaped by guidelines read by the students in preparation

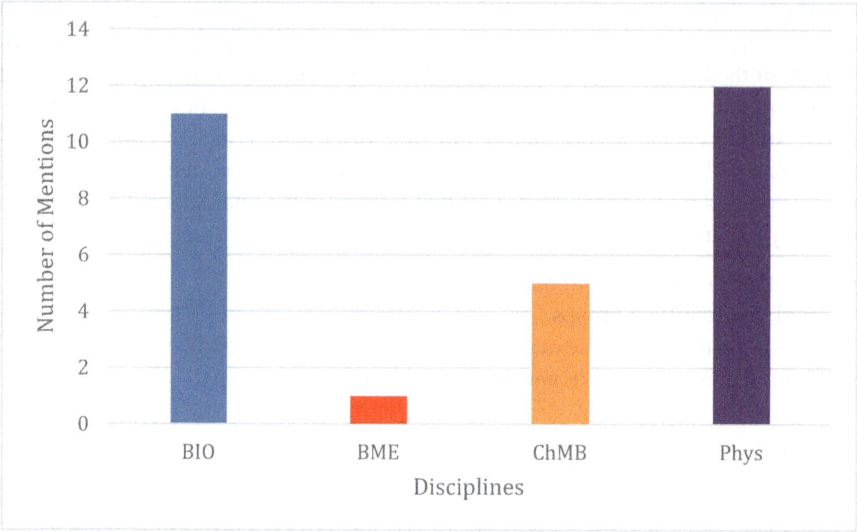

Fig. 13.2 Codes discussing interpersonal relationships by discipline

for drafting their guidelines, as well as recent events in the news. Most codes of ethics reviewed by members of the Student Ethics Committees included sections discussing sexual harassment and discrimination issues. News coverage of the #Me Too movement in 2017 that focused attention on issues of sexual abuse and harassment, as well as increased coverage of issues faced by women and minorities in the research environment also likely contributed to the students' focus on these issues (Fig. 13.2).

13.6.5 Relevant Topics in the Guidelines

13.6.5.1 Social Interactions, Equity, and Equality

In February 2022, the National Institutes of Health published "FY 2022 Updated Guidance: Requirements for Instruction in the Responsible Conduct of Research" (NIH 2022). In discussing the subject matter, the notice states, "Developments in the conduct of research and a growing understanding of the impact of the broader research environment have led to a recognition that additional topics merit inclusion in discussions of the responsible conduct of research." The notice goes on to highlight several new and expanded topics, including safe research environments (e.g., those that promote inclusion and are free of sexual, racial, ethnic, disability, and other forms of discriminatory harassment), collaborative research, conflicts of interest and conflicts of commitment, in allocating time, effort, or other research resources, and secure and ethical data use, to name a few.

Though this study predates the NIH updated guidance by several years, the Student Ethics Committee members we worked with recognized the importance of including the ethics of social interaction in responsible research guidelines. While some topics, such as equity and inclusion, were introduced to the students by the student discussion leaders and the project team, all four student ethics committees committed a good portion of the guidelines outlining positive social practices to help improve interactions in the lab.

13.6.5.2 Supervisor as a Role Model and Mentor

The supervisor as a role model is an issue that comes up in the literature about mentoring (Bird 2001; Weil 2009) and in the guidelines produced by the Student Ethics Committees. In discussions, students saw the PI as setting the overall tone of the research group and recommendations for how a PI should lead.

Supervisors also have a duty to not only model but also articulate the values and standards of the research community and help students understand discipline and lab-specific practices. (Bird 2001). The call for PIs to introduce these guidelines to research group members, review the guidelines periodically, and provide opportunities for discussion goes some way to requiring this (ChBE, Physics).

Reasons for the focus on mentor-mentee relationships are multiple. In this context, we are using the terminology "mentor/mentee relationships" in the way utilized by the National Institutes of Health to include ethical issues and responsibilities that exist between student researchers and their supervisors/advisors/PIs. Stephanie Bird, in her 2001 article on mentoring, very rightly points out that while the terms 'mentor' and 'research supervisor' are often used interchangeably, a true mentor is someone who has a personal interest in the student they are supervising and has an ongoing interest in the student's professional development and career advancement (Bird 2001). A Principal Investigator/advisor can have a mentoring relationship beyond the required academic, research, and administrative responsibilities, but this is not always the case. In large labs, it is likely impossible for a lab supervisor to effectively have a mentoring relationship with every student under their supervision.

In all four departments that participated in the study, all students came from relatively small research groups, usually composed of a faculty PI, 3–8 graduate students, and in some cases, undergraduate students. While the supervision responsibilities of the lab students usually fell upon the PI, at least one group of students talked during the workshop sessions about how senior graduate students sometimes took on supervision responsibilities for their junior colleagues or undergraduates in the lab.

13.6.5.3 Limited Focus on Good Research Practices

The 4th and 5th most relevant codes across all four guidelines had to do with good research practices (following the scientific method, 8) and publication practices (5 total). While both were prevalent across all codes, the students dedicated less space to discussing these issues.

Authorship's important role in the success of graduate students is well known, with graduation and earning one's first job contingent on getting a first-author paper (Borenstein and Shamoo 2015: 272). Students from all four departments included both discussion and mention of authorship norms in the workshops and the guidelines produced. Data management was also seen as paramount, with Physics including a detailed section on how to keep accurate and transparent records of research results.

13.6.6 Student Influence on the Guidelines Content

When viewed as a whole, there were some extremely interesting differences in the tone and approach of the four different sets of guidelines. Changes in the tone likely reflect both developments in how the workshops were led by student facilitators and new content added to the second iteration of the workshop based on experience. The first two sets of guidelines developed by the BIO and BME departments are relatively transactional in nature, with the BME guidelines outlining what a research advisor should do, followed by a list of what graduate students should do (follow good research practices, participate in mandatory safety training, etc.).

The BIO code is organized with a section outlining laboratory conduct and what each lab member shall do in the student-student relationship and the mentor/mentee relationship. It ends with a section on papers and authorship and a miscellaneous section about safety training, respect for colleagues, and other points the Student Ethics Committee was unsure where to include.

During the second iteration of the workshop, the project team received feedback from faculty about the draft guidelines developed by the BIO and BME committees and incorporated some notes as to the future and reception of the guidelines when they were shared with the wider department. In response to this, the guidelines produced by the Physics and the ChBE Student Ethics Committees both provide a preface outlining the students' vision of how the guidelines should be used in the future as well as recommendations to form some kind of ethics committee that includes student and faculty representatives that could help facilitate discussion of ethical issues and play some role in resolving issues or questions that might arise.

The differences between the guidelines were also illuminating, as codes ranged from being a kind of "ten commandments" to the expansive guidelines of the Physics Student Ethics Committee, which sought to educate lab members on building an equitable lab environment.

Overall, these guidelines help cast light on the issues that individual students see as important. While some faculty may see research ethics as focusing on more traditional RCR issues such as data management, publication ethics, and animal and human subjects research, these guidelines shed light on ethics topics that students see as socially complex, areas where more guidance is needed, or areas where no guidance is provided currently.

Discussion about issues of fairness and discrimination were undoubtedly linked to a change in focus between the fall 2017 and the spring 2018 Student Ethics Committees. Beyond the potential influences of news coverage and mention of these issues in the guidelines read, this was almost entirely student-driven and represented an interesting finding of this study. During the first two Student Ethics Committees that involved the BIO and BME students, some limited discussion occurred around discrimination and equity issues in research groups.

In the second semester, the student facilitators of the ChBE and Physics Student Ethics Committees raised the issue in a meeting of the project team that one of the major issues they saw as important, based on their own experience, were ethical issues around fairness and equity faced by female students, international students, and students of color. In response to this, the project team worked with the student facilitators to choose two short case studies to help facilitate discussion around these topics. Members of both committees shared many stories about the discrimination they have faced in their current or former lab groups. In response to these discussions, the ChBE guidelines include three sections that specifically address issues of equity and discrimination, and the Physics guidelines mention this issue five times within the "Social" section of their guidelines which runs to one and a half pages detailing information about the historical discrimination of women, minorities, and individuals with disabilities in science, recommendations for handling issues of harassment and bullying, and even links to educational material and definitions of terms like "tokenism."

13.7 Comparison with Other Study Results

The approach taken in the analysis of these student-generated guidelines reflects the findings of a background study we performed at our home institution that includes both qualitative and quantitative analysis (Hildt et al. 2022; Laas et al. 2022). Our research suggests that ethics topics important in the research environment extend beyond the topics commonly covered by Responsible Conduct of Research education, such as plagiarism, falsification of data, and treatment of human and animal subjects. The student-developed guidelines highlight the importance of including ethical issues involving social interactions, power relations, discrimination, and sexual harassment, as well as issues of fairness in how opportunities, resources, and research projects are allocated. Peer-to-peer, supervisor-to-lab member, and PI-to-student relationships play a significant role in the creation of ethical research environments and provide the foundation for ethical research practice.

In part, there is an overlap between what students interviewed in an interview study said and the topics discussed in the guidelines. Echoing findings from graduate student interviews (Hildt et al. 2022.), the guidelines focus on providing a structure where principal investigators and students can build relationships built on trust and mutual understanding, where expectations for both roles are clear, and where two-way communication is maintained. In situations such as crediting authors, providing travel and conference funds, or allocating other resources, the student-developed guidelines stressed transparency and fairness.

Focus on these topics and practices also mirrors surveys of exemplary lab managers who identified some best practices in running lab environments. Much like the guidelines, these studies focus on holding regular meetings (Antes et al. 2016, 415; Antes et al. 2019a; Antes et al. 2019b, 11; Resnik et al. 2021,12), providing sufficient oversight (Antes et al. 2019a, 11; McIntosh et al. 2020, 264) and having open-door policies that allow for two-way communication (Antes et al. 2016, 415; Resnik et al. 2021, 12), and the need for the PI to lead by example (Antes et al. 2019a, 13; Resnik et al. 2021, 12). Similarly, these studies focus on the need to provide adequate training for students (Antes et al. 2019b, 11; Resnik et al. 2021), and establishing and following clear operating standards (Antes et al. 2019b, 11; McIntosh et al. 2020, 264; Resnik et al. 2021, 12).

Based on findings from these studies that focus on both students and exemplar lab managers, there seems to be a set of shared topics, values, and practices that both parties feel are necessary for building a strong, supportive lab environment. Creating these guidelines provides opportunities for students to interact as peers, raise important questions and concerns, and allow for frank and open discussion about issues they encounter that might otherwise not be put into words or raised. They also provide a mechanism for sharing these concerns with faculty supervisors and other research group members on a more level playing field than in other educational contexts.

13.8 In Situ Ethics Education on Ethical Issues in STEM Labs

The guidelines developed through this workshop module series are potentially more important as a catalyst for starting ethical conversations than in their final form. Though useful as a starting point for the orientation of new lab members and as a view into the ethical issues and questions important to a specific lab environment, the real power of the guidelines lies in their ability to spark discussion between students and their peers in a safe environment, and then help carry that conversation to lab supervisors and department chairs. The workshop module series immerses students in a series of active-learning activities where students apply their own experiences in the lab to evaluate case studies, existing ethical guidelines, and ultimately the stories shared by fellow members of the Student Ethics Committee during the group meetings. Students can listen to the experiences of students with different backgrounds and hopefully become more aware of some of the specific challenges faced by international students, women, and minority students.

In developing the guidelines, students brainstorm potential solutions to the ethical questions and issues they have discussed. During the sessions completed by our project team, the student participants became aware of existing institutional policies and resources that they were unaware of and began building social networks and contacts with individuals they could contact for further information and assistance – such as faculty interested in ethics and the institution's Patent Office. Throughout the workshop module, students became more aware of the rules and sometimes unacknowledged norms in research groups.

Ultimately, one of the main goals of the workshop module is to empower students by giving them the confidence, tools, and skills needed to speak up on ethical issues that can profoundly impact their educational careers.

13.9 Developing a Flexible Workshop Module

One goal of this project is to develop a flexible workshop module that could be integrated into various contexts. Beyond the two six-week workshop series described in this chapter, we also taught two different versions of the module, the first with students from a single department in a similar, six-week workshop series and the second with students enrolled in an introductory-level graduate course during three class sessions.

In the summer of 2019, an adapted version of the module described above was run with 13 graduate students enrolled in the Biomedical Engineering program at the University of Texas, San Antonio. Similar to the initial workshop series run at Illinois Tech, this iteration was held over six sessions and featured similar content. When developing their guidelines, the students chose to focus on a wide range of topics, including issues that come up with undergraduate students working in labs along with graduate students, animal husbandry, pressures that minorities, international students, and women face in research, and overloaded schedules of teaching assistants.

In the fall of 2021, two project team members worked with a colleague from Illinois Tech's Biomedical Engineering department to co-teach a shortened workshop module to 18 students enrolled in an introductory graduate class. The students were all pursuing a master's degree in biomedical engineering but had a mixture of experiences working in research labs or industry, while a handful of students had no research experience. The workshop module was held over three 100-minute sessions, and students worked in teams, so each group had at least one student with experience in hands-on research. The students developed rough drafts of their chosen guidelines during the final two sessions of the workshop and provided relatively robust drafts of the guidelines despite the shortened length of the module. The students covered data management, fair and equitable use of research equipment, authorship, and developing shared expectations between students and advisors.

Overall, these examples show the flexibility of the general approach developed in the initial workshop series. The longer version of the workshop series (six sessions) is extremely useful in providing students with a more nuanced understanding of ethics in the research context and in developing stronger relationships between students through prolonged discussion and guideline development. The shorter version of the workshop series (2–3 sessions) introduces students to key ethical issues, resources for solving potential issues, and the importance of strong interpersonal relationships in research settings. A teaching guide, slide deck, and resources can be found at Illinois Tech's Center for the Study of Ethics in the Professions (https://ethics.iit.edu/cultivatingcultures).

13.10 Conclusion

We think that the guideline-developing approach is useful in providing in situ ethics education to graduate students. Compared to more traditional classroom-based courses, this workshop module asks students to reflect on their experiences in the lab, discuss potential solutions, and develop context-specific guidelines. These guidelines can be utilized to help orient new students to the lab environment and serve as a catalyst for ongoing ethical discussions among their colleagues and principal investigators. The workshop modules can also be used to complement existing teaching approaches, either by inserting the module in an existing introductory course or to serve as reinforcement after the student has gained a working knowledge of the main ethical issues in research ethics.

The guidelines can also serve as a tool to shed light on the situation in specific research labs and on ethical issues graduate students consider important in a research lab or have encountered. Department chairs, lab supervisors, or other administrators can use the topics covered in the guidelines to begin working on strategies to address systemic issues identified in the guidelines and engage students in the process by using their developed ideas as a starting point.

Finally, the guidelines developed through the workshop series discussed above reinforce the need for research ethics education to help students and faculty supervisors pay closer attention to mentor-mentee relationships and interpersonal relationships in the lab. We believe that expanding the umbrella of what is considered research ethics is necessary; not only that students understand and practice responsible research, but also that they comprehend the important role team dynamics play in building a strong, supportive research environment and put these standards into practice in their own work.

Acknowledgement This work was part of the research project "A Bottom-up Approach to Building a Culture of Responsible Research and Practice in STEM", funded by the National Science Foundation Award # 1635661.

Appendix

Taxonomy Utilized in Quantitative Analysis of Codes, grouped by Themes

Interpersonal Relationships
Acceptable workplace norms/behavior
Accountability
Collaboration
Common language
Conflict resolution
Cultural barriers/exchanges
Equitable treatment
Gender
Inclusion
Morale
Opportunities to create/build shared understanding
Personal property/respect for/theft
Relationships among lab members
Sexual harassment
Socializing outside of lab with members
Working in diverse groups
Management of Guidelines
Approval of guidelines
Ethics education
Ethics review/ethics committee
Living document/reviewed routinely
Responsible conduct of research/good research practices
Animals in research
Collaborative science
Conflict of interest
Data acquisition
Data management
Documentation
Good research/science practices
Human subjects
Intellectual property/patents
Mentor/trainee
Peer review
Publication practices
Reproducibility
Research misconduct
Whistleblowing

(continued)

Role of PI	
	Clearly defined responsibilities and expectations
	Discrimination and/or abusive behavior
	Expectations
	Fairness
	Meetings with PI
	Orientation for new lab members
	Power imbalance issues
	Professional development/skills
	Recommendation letters
	Safety training
Work environment	
	Acceptable workplace norms/behavior
	Competition within the lab
	Freedom to explore new ideas
	Funding
	Lab meetings
	Non-work activities in lab
	Opportunities to explore new ideas
	Shared resources
	Time pressure
	Transparency
	Work hours/vacation policies
	Work-life balance

References

Antes, A.L., S.T. Murphy, E.P. Waples, M.D. Mumford, R.P. Brown, S. Connelly, and L.D. Devenport. 2009. A meta-analysis of ethics instruction effectiveness in the sciences. *Ethics and Behavior* 19 (5): 379–402.

Antes, A.L., A. Mart, and J.M. DuBois. 2016. Are leadership and management essential for good research? An interview study of genetic researchers. *Journal of Empirical Research on Human Research Ethics* 11 (5): 408–423. https://doi.org/10.1177/1556264616668775.

Antes, A.L., A. Kuykendall, and J.M. DuBois. 2019a. Leading for research excellence and integrity: A qualitative investigation of the relationship-building practices of exemplary principal investigators. *Accountability in Research* 26 (3): 198–226. https://doi.org/10.1080/08989621.2019.1611429.

———. 2019b. The lab management practices of "research exemplars" that foster research rigor and regulatory compliance: A qualitative study of successful principal investigators. *PLoS One* 14 (4): e0214595. https://doi.org/10.1371/journal.pone.0214595.

Bird, S. 2001. Mentors, advisors and supervisors: Their role in teaching responsible research conduct. *Science and Engineering Ethics* 7: 455–468.

Borenstein, J., and A.E. Shamoo. 2015. Rethinking authorship in the era of collaborative research. *Accountability in Research: Policies & Quality Assurance* 22 (5): 267–283. https://doi.org/10.1080/08989621.2014.968277.

Cech, E.A. 2014. Culture of disengagement in engineering education? *Science, Technology, & Human Values* 39 (1): 42–72.
Hildt, E., K. Laas, C.Z. Miller, S. Taylor, and E.M. Brey. 2019. Empowering graduate students to address ethics in research environments. *Cambridge Quarterly of Healthcare Ethics* 28 (3): 542–550.
Hildt, E., K. Laas, C.Z. Miller, and S. Taylor. 2022. Student views on the culture of STEM research laboratories: Results from an interview study. *Accountability in Research.* https://doi.org/10.1080/08989621.2022.2109018.
Holsapple, M.A., D.D. Carpenter, J.A. Sutkus, C.J. Finelli, and T.S. Harding. 2012. Framing faculty and student discrepancies in engineering ethics education delivery. *Journal of Engineering Education* 101 (2): 169–186.
Kalichman, M. 2013. A brief history of RCR education. *Accountability in Research* 20: 380–394. https://doi.org/10.1080/08989621.2013.822260.
Laas, K., S. Taylor, C.Z. Miller, E.M. Brey, and E. Hildt. 2022. Views on ethical issues in research labs: A university-wide survey. *Accountability in Research* 29 (3): 178–201. https://doi.org/10.1080/08989621.2021.1910503.
McIntosh, T., C. Sanders, and A.L. Antes. 2020. Leading the people and leading the work: Practical considerations for ethical research. *Translational Issues in Psychological Science* 6 (3): 257–270. https://doi.org/10.1037/tps0000260.
National Institutes of Health. 2022. FY 2022 Updated Guidance: Requirement for Instruction in the Responsible Conduct of Research NOT-OD-22-055. https://grants.nih.gov/grants/guide/notice-files/NOT-OD-22-055.html. Accessed 30 Mar 2023.
National Institutes of Health. Alcohol, Drug, and Mental Health Administration. 1989. Requirement for programs on the responsible conduct of research in National Research Service Award Institutional Training Programs. *NIH Guide for Grants and Contracts* 18: 1.
Plemmons, D.K., and M.W. Kalichman. 2018. Mentoring for responsible research: The creation of a curriculum for faculty to teach RCR in the research environment. *Science and Engineering Ethics* 24: 207–226. https://doi.org/10.1007/S11948-017-9897-Z.
Resnik, D.B., E. Lee, B. Jirles, E. Smith, and K. Barker. 2021. For the "good of the lab": Insights from three focus groups concerning the ethics of managing a laboratory or research group. *Accountability in Research*: 1–20. https://doi.org/10.1080/08989621.2021.1983799.
United States, National Institutes of Health. 2009. NOT-OD-10-2019: Update on the requirement for instruction in the responsible conduct of research. https://grants.nih.gov/grants/guide/notice-files/NOT-OD-10-019.html. Accessed 28 Mar 2023.
Waples, E.P., A.L. Antes, S.T. Murphy, et al. 2009. A meta-analytic investigation of business ethics instruction. *Journal of Business Ethics* 87: 133–151. https://doi.org/10.1007/s10551-008-9875-0.
Watts, L.L.K.E., T.J. Medeiros, L.M. Mulhearn, S. Connelly Steele, and M.D. Mumford. 2017. Are ethics training programs improving? A meta-analytic review of past and present ethics instruction in the sciences. *Ethics & Behavior* 27 (5): 351–384.
Weil, V. 2009. Mentoring: Some ethical considerations. *Science and Engineering Ethics* 7: 471–482. https://doi.org/10.1007/s11948-001-0004-z.

Chapter 14
Engineering an Ethical Ethos: Reframing Ethics Education for Engineers and Researchers

Juhi Farooqui, Sarah Dawod, Erinn M. Grigsby, Josep-Maria Balaguer, and Devapratim Sarma

Abstract Engaging with ethics is a critical component of working in science and engineering, from institutional approvals to research training to ethics coursework. However, in order for ethics to be more than simply a requirement or constraint in the research environment, ethics education must be designed to foster deep engagement. In this work, we present our approach to developing an *ethical ethos*, which we define as a research culture in which ethics is an integral part of all stages of research. To cultivate this ethos, we develop a monthly semi-structured discussion

J. Farooqui (✉)
Rehab Neural Engineering Labs, University of Pittsburgh, Pittsburgh, PA, USA

Neuroscience Institute, Carnegie Mellon University, Pittsburgh, PA, USA

Center for the Neural Basis of Cognition, Pittsburgh, PA, USA
e-mail: jfarooqu@andrew.cmu.edu

S. Dawod
Center for Bioethics and Health Law, University of Pittsburgh, Pittsburgh, PA, USA

E. M. Grigsby
Rehab Neural Engineering Labs, University of Pittsburgh, Pittsburgh, PA, USA

Center for the Neural Basis of Cognition, Pittsburgh, PA, USA

Department of Physical Medicine and Rehabilitation, University of Pittsburgh, Pittsburgh, PA, USA

J.-M. Balaguer
Rehab Neural Engineering Labs, University of Pittsburgh, Pittsburgh, PA, USA

Center for the Neural Basis of Cognition, Pittsburgh, PA, USA

Department of Bioengineering, University of Pittsburgh, Pittsburgh, PA, USA

D. Sarma
Rehab Neural Engineering Labs, University of Pittsburgh, Pittsburgh, PA, USA

Center for the Neural Basis of Cognition, Pittsburgh, PA, USA

Department of Mechanical Engineering, Carnegie Mellon University, Pittsburgh, PA, USA

E. Hildt et al. (eds.), *Building Inclusive Ethical Cultures in STEM*, The International Library of Ethics, Law and Technology 42,
https://doi.org/10.1007/978-3-031-51560-6_14

seminar series designed to critically engage researchers and engineers with ethical principles by connecting with their research and their own deeply held values. Through this approach, we empower researchers to take ownership of ethical principles and ultimately apply these principles in the day-to-day work of research.

Keywords Ethics · Education · Research cultures · Engineering · Neuroscience · Neural engineering

Engaging with ethics is critical for scientists and engineers operating in a complex world. From Institutional Review Boards (IRBs) to Institutional Animal Care and Use Committees (IACUCs) to the professional ethics norms that undergird research, engagement with ethics is unavoidable in science. However, ethical engagement can go much further than the procedural mechanisms that researchers are required to interact with. Researchers who want their work to make a positive impact on individuals, communities, and society can derive greater fulfillment from their work when it commits to ethical principles that align with their values. Ethics education that prompts researchers to reflect on their own values and goals and connects ethical principles with these values can foster deep engagement and encourage ethically conscientious science and engineering. To facilitate such reflection, conscientious growth, and fulfillment, our group at the Rehab Neural Engineering Labs (RNEL) at the University of Pittsburgh set out to make ethics engagement an intrinsic part of the lab culture by creating what we call an *ethical ethos*. Over the last four years, students, post-docs, staff, and faculty in our lab have implemented this ethos through a series of interactive seminars designed to encourage personal reflection through small-group discussion.

In the following sections, we lay out the context of our development of the ethical ethos, our approach, and our observations regarding this process. We begin in Sect. 14.1 by describing the state of existing ethics education for members of RNEL prior to the implementation of our interactive seminars. Section 14.2 defines our understanding of the ethical ethos and the philosophical theory that drives it. In Sect. 14.3, we describe our approach to cultivating an ethical ethos at RNEL. Section 14.4 discusses the intersection of ethics with inclusion and equity work and our approach to making this intersection explicit. Section 14.5 presents feedback received on the approach. We conclude in Sect. 14.6 with a discussion of the practical and philosophical outcomes of applying an ethical ethos approach.

14.1 Existing Ethics Education at RNEL

RNEL is a consortium comprising 10 faculty members and their respective research groups, including graduate students, postdocs, and dedicated engineering, clinical, and administrative support staff, totaling nearly 80 members. Lab members come from multiple departments and institutions, often working closely across research groups. This creates a densely collaborative environment well-suited for joint critical engagement with complex topics. However, the diversity of departments and institutions represented also means that there is no cohesive, shared institutional ethics education standard for lab members. Instead, lab members receive ethics training and education from a variety of sources.

The most common sources of ethics training for lab members at all levels are online training modules that are required by the university and/or funding agencies in order to conduct research. These modules cover basic tenets of research ethics ranging from plagiarism to animal research to the Belmont Report addressing human subjects research. Most students at RNEL are also members of the Center for the Neural Basis of Cognition (CNBC), a center that provides educational support and training for faculty members and trainees engaged in neuroscience research at the University of Pittsburgh and Carnegie Mellon University. During their graduate career, CNBC student members must attend three "Ethics Roundtable" talks delivered by faculty members whose work brings them in contact with the topics of animal research ethics, human subjects research ethics, and professional ethics. While these talks provide a starting point for discussion of the three topics they cover, they are necessarily limited in scope, and students have reported in post-attendance feedback surveys that they do not feel grounded in a rigorous ethical education framework.

In addition to online training and ethics roundtable talks, most students at RNEL are required to receive ethics training as part of their graduate degree programs, usually delivered in the second or third year. The two most common sources of departmental ethics education are: (1) a one-semester ethics seminar course delivered by the University of Pittsburgh Department of Bioengineering and (2) a seminar series on the Responsible Conduct of Research (RCR). The bioengineering ethics course consists of 15 weekly three-hour discussion sections. The course primarily focuses on current-event case studies related to industry-level medical device research and commercialization. Each week, class time is divided into a brief introduction to an ethics principle, an overview of the case, a small group discussion, and finally a large group discussion. The RCR seminar series consists of 10–12 one-hour sessions taken over the course of one semester, covering high-level research ethics topics such as responsible authorship, data sharing and confidentiality, conflicts of interest, IRB review, and research misconduct. These educational efforts are well-intentioned and seek to engage students in discussions of ethical principles and questions. However, because they cater to a broad audience and are delivered during a constrained time period early in a student's graduate career, they are limited in their capacity to relate continuously with students' research. In departmental surveys within the Department of Bioengineering, students were more likely to report that they would not recommend the ethics seminar course than any other course, citing the course's lack of relevance to their research and prescriptive approach to ethics training.

Beyond academic requirements, researchers are subject to several ethics requirements during the process of research. For researchers at any stage in their career, addressing ethical considerations when seeking approvals from IRBs or IACUCs is unavoidable, as is keeping up-to-date with ethics certifications delivered online.

Research ethics driven solely by regulatory oversight and enforced compliance has two major consequences: first, researchers develop an adversarial relationship with moral oversight and identify the guidelines simply as constraints on their work (Melo-Martín et al. 2007). Second, because standardized regulation must account for a breadth of disciplines, the guidelines become high-level checklists, which trivializes ethical questions and enables researchers to disassociate from them (Antes et al. 2009; Johnsson et al. 2014). Based on RNEL members' experiences of ethics, as described above, both of these consequences are relevant concerns.

Taken together, the state of ethics engagement and education for RNEL members mirrors the broader scientific research landscape. Ethics engagement is often characterized by didactic and/or regulatory approaches, ranging from online ethics training to protocol approval by federal or institutional bodies. This can lead researchers to view ethics as a regulatory step, and, at worst, as an obstacle in the path to pursuing their research (Melo-Martín et al. 2007). Formal ethics education for researchers, which often focuses on analytical reasoning skills and professional codes of conduct, does not lend itself to deep engagement (Antes et al. 2009) or an effective integration of ethical principles into research.

Moreover, the culture of engineering, like other science, technology, engineering, and mathematics (STEM) fields, is often one that implicitly or explicitly deprioritizes ethics, treating it as irrelevant to core engineering work or relegating it to the domain of ethicists (Martin et al. 2021). Nevertheless, based on informal feedback, comments, and requests from local research community members, we perceived a definite hunger for ethics engagement among our faculty and student colleagues. For example, post-attendance survey responses from the ethics roundtable talks frequently included explicit requests for speakers to engage the audience and provide more time for discussion. This suggested to us that researchers *want* to personally engage with the ethical implications of their work, consider the role of their research in society, and understand how to conduct more ethically grounded research.

One approach to better ethics integration is ethics consultation (Cho et al. 2008; Fisher et al. 2015; Melo-Martín et al. 2007), wherein researchers work with ethicists to better inform the ethical dimensions of their specific projects during the course of the projects themselves. At the strong end of the ethics consultation model is the embedded ethicist model (Goering and Klein 2020), wherein an ethicist is part of the core research team on a project. At the other end, an ethicist may be available for specific, more limited consultation on issues, questions, or problems that arise. In any case, ethics consultation requires that the researchers themselves have sufficient awareness to notice the issue/question/problem and seek the consultation. Moreover, these approaches are still project-specific and rely on an external expert to provide ethical guidance. Our approach aims to go one step beyond this model, to deeply engage individual researchers *themselves* in the ethics of their research – not to replace the subject-matter expertise of a consulting or embedded ethicist, but rather to more fully integrate the researcher with the process of ethical consideration. To do this, we formulated the concept of ethical ethos central to our approach.

14.2 Ethical Ethos – Philosophical Underpinnings

In *Ethics in Conflict*, Dzeng and Wachter argue that "moral distress and professional ethical dissonance" are major sources of burnout among medical professionals (Dzeng and Wachter 2020). To resolve this, they propose an ethics-rooted approach that enables physicians to act in alignment with their values. At a lab like RNEL, which brings together clinicians and engineers to bridge these two disciplines, Dzeng and Wachter's framework resonates. Both disciplines are shaped by the distinct pressures and incentives created by academic norms and industry pressures, but are also driven by their responsibilities to patients, participants, and potential users (Harris Jr. et al. 1996). This leads us to believe that an ethics-rooted approach like that suggested by Dzeng and Wachter can also resonate with researchers, especially in labs subject to similarly conflicting interests.

A challenge emerges within engineering disciplines, where ethics frameworks are often regulatory and compliance-based, resulting in engineers feeling disassociated from or even adversarial toward ethics (Melo-Martín et al. 2007). Moreover, the culture of engineering often treats ethics as a secondary or low-priority concern, to be offloaded to others (Martin et al. 2021). Such challenges suggest the need for a fundamental shift in culture to view ethics with nuance instead of as a nuisance. Our solution is to develop an ethical ethos.

We define an *ethical ethos* as a research culture in which ethics is a deeply integrated part of the research environment. In such a culture, people are encouraged to consider ethics at every stage of the research process, from project planning to implementation to dissemination of the ultimate product (e.g., knowledge, technology). As ethics becomes integrated, it becomes natural or "second nature" for researchers to consider ethics at every stage; thus, an ethical *ethos* develops and becomes a part of the research culture.

The need for an ethical ethos grows from a natural consequence of action theory, also known as philosophy of action, wherein an action can be defined in terms of both the *agent* who produces the action and the *situation* in which the action is performed (Argandoña 2008). For our purposes, we can understand the *agent* as the individual researcher and the *situation* as the overarching institutional, societal, and historical context in which the research takes place. Korsgaard defines a genuine or authentic action as one that reflects the lived values and morals of the individual, whereas disingenuous actions are driven by external pressures (Korsgaard 2009). Perhaps for this reason, compliance-based ethics approaches alone are often less effective than values-based approaches (Weaver and Treviño 1999). Therefore, to enable researchers to act genuinely in their work, we must support them in illuminating their beliefs and values about a situation and understanding how their professional actions can reflect those beliefs. This can empower researchers to act authentically, which Korsgaard connects with agents' self-conception of their own identities. Inspired by Miller et al.'s approach, we posit that authenticity engendered by the integration of practical work and ethical reflection can enable researchers to develop an ethical identity by which they can expand their understanding and self-definition of their roles (Miller et al. 2022). An approach to ethics education that supports such authenticity and value-alignment can in turn foster a greater sense of commitment and integrity (Weaver and Treviño 1999).

In an action theory framework for ethics, a researcher (agent) must learn to examine their own moral code *and also* the context (situation) of the research and the ethical question (Argandoña 2008). Common approaches to ethics education in such a framework can be analyzed in terms of how they address each component of an action (agent or situation). For example, one common approach involves presenting abstract moral dilemmas, prompting the learner to consider how they might respond to a hypothetical scenario under imagined circumstances. This exercise can develop the learner's own moral code and reasoning about it (Argandoña 2008; Martin et al. 2021), and can be viewed as primarily developing the agent.

Educators can incorporate the situation by guiding evaluation of ethical questions through analysis of the situation. One such approach centers on nuanced case study repetition, prompting learners to synthesize ethical themes in presented case studies (Martin et al. 2021; Vertrees et al. 2013). Through this process, learners develop ethical sensitivity: an awareness of ethical implications within a given context. Additionally, cognitive decision-making, an approach to ethics that emphasizes logical reasoning based on a deep analysis of the situation, can enhance case study analysis by requiring learners to incorporate contextual elements such as relevant history, participant-researcher relationships and trust, and experimental design (Martin et al. 2021). Failure to account for any of these features can result in unethical research conduct, even in the absence of malicious intent or regulatory noncompliance (Johnsson et al. 2014; Kass et al. 1996; Tubig and McCusker 2021).

Our approach seeks to incorporate both components of the action theory framework. We present researchers with complex ethical questions grounded in their research area to build exposure to and familiarity with such questions, encouraging analysis of both their own moral judgments and their understanding of the context. Furthermore, researchers work together in small groups to logically analyze the ethical questions presented. This process is supported by the existing culture of the engineering lab, wherein logical problem-solving as a team is a critical skill. We intentionally separate participants into small groups for these discussions, creating space for participants to collaboratively break down and analyze the questions they are presented with. Importantly, in our neuroethics discussions, facilitators encourage participants to relate these analyses to their own deeply held values, deepening their understanding of themselves as an *agent*. To bolster this collaborative analysis, we explicitly present connections to the ethical *situation*, for example, by introducing participants to relevant historical or societal contexts.

The ethical situation includes relevant historical context, the power and relational dynamics of the research environment, and the specific research considerations at play, among other factors. Our approach explicitly emphasizes awareness of the ethical situation by integrating sessions specifically aimed at learning about historical contexts or ongoing inequities that can have a bearing on the research environment (more concretely discussed in Sect. 14.4), as well as by consistently encouraging researchers to reflect on their relationship with research participants and their responsibilities with respect to both internal and external contexts.

By connecting researchers with relevant context for the ethical considerations of their research and enabling them to reflect on the relationship between *their* values and the values grounding research ethics, we create a space for researchers to examine their own beliefs and genuinely act on them. These genuine actions sometimes include updating or correcting their beliefs. This approach shifts the focus of ethics education from mere logical consistency and rule-following toward developing the intrinsic motivation to intentionally integrate ethical principles into research. It equips researchers to act authentically within their work, informed by their critical reflection on ethical arguments and their own values. We believe that such an approach would be transformative: being free to act congruously with one's own beliefs can engender greater curiosity, learning (Edmondson 1999), engagement (Christian et al. 2011; Rich et al. 2010), and sense of integrity (Weaver and Treviño 1999), which can, in turn, lead people to work more effectively and consistently with their own ethically warranted beliefs. This can further enable them to derive greater fulfillment from their work. Therefore, we posit that creating an ethical ethos can open the workplace to clarity of ethical and scientific purpose and create a genuine, empowering environment.

14.3 Cultivating an Ethical Ethos

To cultivate the ethical ethos, we developed an approach designed to allow researchers to contextualize their own research and values within a framework of ethics. Our approach (first described in Farooqui et al. 2021) is informed by the idea that interactive, experience-based ethical education can be more effective and engaging than classic didactic forms (Antes et al. 2009; Vertrees et al. 2013). We construct the process of developing an ethical ethos as a cycle with four major parts (Fig. 14.1): Topic Identification, Dialogue/Synthesis, Ownership, and Application.

We implement this approach through a monthly, semi-structured discussion-based seminar series, referred to here as the Neuroethics and Discussion Seminars. These seminars (described in more detail in the subsection titled "Dialogue/Synthesis") are led by the authors and other volunteers within the lab who are interested in advancing ethics education at RNEL. The seminars are optional, but faculty members actively encourage attendance. All lab members are invited to participate, including students, postdocs, faculty, and staff. Seminars were initially held onsite at the RNEL offices, but with the onset of the COVID-19 pandemic, seminars were offered in a virtual format. Even as our work has returned to the lab, we have persisted with a hybrid format to keep the seminars accessible to all attendees. This has also enabled us to make these seminars open to other members of the neuroscience and neural engineering community in Pittsburgh.

In the following subsections, we will lay out each stage of the cycle described above, including its role in the Neuroethics and Discussion Seminars as well as how we envision it manifesting in the context of a research culture with a persisting ethical ethos.

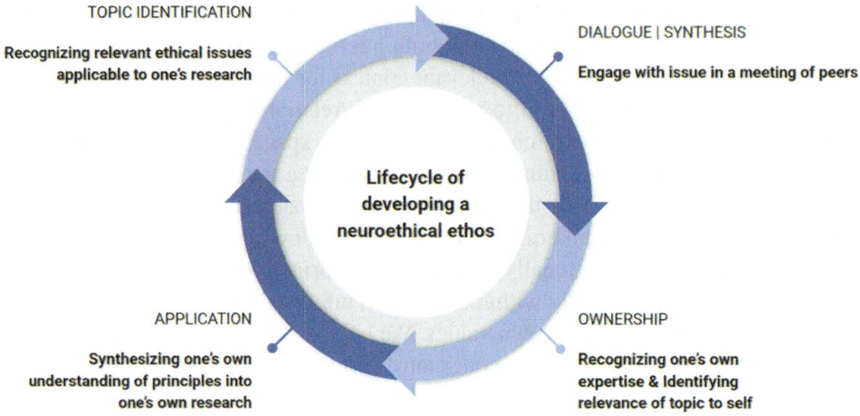

Fig. 14.1 Our approach to developing the ethical ethos can be conceptualized as a four-part cycle that is designed to self-perpetuate within the culture of a lab. The four major stages of the cycle are: Topic Identification, Dialogue/Synthesis, Ownership, and Application

14.3.1 Topic Identification

In the topic identification phase, lab members recognize and identify relevant topics, issues, and concepts in neuroethics that are applicable to their own research. In developing the Neuroethics and Discussion Seminars, the authors and a group of additional seminar volunteer organizers identify and research topics of interest to the lab in order to plan the seminars. To drive topic identification for the seminars, we focused around three questions:

- What are the questions the field is asking?
- What are the questions researchers are asking (formally or informally)?
- What are the issues that are relevant right now?

This led us to identify an initial set of topics in four major clusters: Rights and Experience of Participants, Equity and Structural Barriers, Translation and Dissemination, and Future Concerns of the Field. Each cluster consists of three to four sub-topics that are each explored in depth during a series of discussion seminars. The selected topic areas reflect concerns, issues, and questions that are often front-of-mind for researchers either in their professional or personal lives in ways that they might not even have recognized as being related to ethics. We formulated the topics to allow researchers to take a deep dive into their own perspectives, beliefs, and biases regarding research, technology, research participants, and equity.

However, this stage of the ethical ethos goes beyond seminar planning. Although this process is initially offloaded to the seminar organizers, we also encourage all lab members to engage in the process of topic identification during their day-to-day work. As an ethical ethos develops, all lab members can continue to engage in this process by identifying questions to address in future seminars and in their own research. For example, a lab member identifying an ethical principle or consideration that they need to account for while planning an experiment is an act of topic identification.

14.3.2 Dialogue/Synthesis

Inspired in part by the Scientific Perspectives and Ethics Commitments Survey (SPECS) project developed at the Center for Neurotechnology at the University of Washington (Tubig and McCusker 2021), our approach has dialogue at its heart. The core activities are monthly semi-structured discussions (the Neuroethics and Discussion Seminars). Lab members are invited to join a completely voluntary, open discussion of one of the identified topics. Each session includes a few key components which facilitate free and informed dialogue:

- Topic expert
  - an invited guest with relevant academic expertise to help frame the philosophical underpinnings of the issue and provide context (e.g., philosophical foundations of consent)
- Site expert
  - a lab member who interacts with the practical application of the concept to provide on-the-ground context (e.g., consent procedures at RNEL)
- Background material
  - relevant material for discussion preparation or further interest (e.g., academic article or explanatory video)
- Guiding Questions
  - a set of guiding discussion questions that provides structure and prompts participants to consider how the topic intersects with their research

The discussion session itself is designed as an open dialogue. Invited experts begin the session with a brief presentation including key background on the topic, and the bulk of the session is devoted to small-group discussions. All participants are treated as peers who can learn from one another on a level playing field. Lived expertise is given equal weight to academic expertise. The discussion carries no penalties or expectations, which allows researchers to participate fully and without fear of consequences. Critically, the discussion is conducted in groups of 4–8, enabling all participants to fully engage and have time to consider, formulate, and articulate their perspectives. Together with their small groups, participants have the space to think through ethical principles and problems deeply and freely, as well as a platform for constructive disagreement and debate.

We conceptualize these semi-structured small-group discussion sessions as critical to the development of an ethical ethos within the lab. However, in the broader context of an ethical ethos, this stage of the cycle can be understood as the organic, unstructured process of engaging in dialogue with colleagues and subject area experts.

14.3.3 Ownership

Our framework empowers researchers to reason about ethics on their own terms, challenging the perception of ethics as regulatory or didactic by nature. By supporting lab members in *articulating and exploring* the relationship between ethical principles and their own goals and values, this approach promotes the organic emergence of ethical principles from the beliefs and deeply held values of the individual. Initially, we introduce ethical principles (e.g., participant rights or the responsibilities of scientists in society) as part of the Neuroethics and Discussion Seminars, encouraging lab members to reflect deeply on them through the lens of the topics under discussion. Incorporating their own values and bringing their own professional experiences to bear on the discussion allows participants to take ownership of the ethical principles being discussed in light of their relationship with their own values and experiences. Moreover, the communal nature of these discussions leads to joint emergent insights, which can lead to shared principles and ethical beliefs and commitments. As participants move beyond the seminars, ownership can be seen in the way that individuals discuss and prioritize ethical principles in their work and interactions with other researchers. For instance, we have observed this in the form of lab members asserting the personal value of ethical principles during research discussions. We have also seen lab members take initiative to formalize some of these principles as core values in lab documentation.

14.3.4 Application

By allowing researchers to take ownership of their own beliefs, ideas, and convictions regarding ethical principles, we aim to create an environment in which thinking and talking about ethics in direct relation to research is normal and natural – i.e., an ethical ethos. This ethos, in turn, can empower researchers to apply their understanding of these topics in all stages of their research. When lab members internalize ethical principles as an element of their own value system, applying these principles becomes a genuine action rather than an act of rote compliance. Therefore, they organically begin to consider these principles in the day-to-day work of research, from developing research questions to designing and implementing experiments to analyzing and communicating results. This can also empower researchers to develop their own ethical questions and identify new topics to explore, thus completing our cycle.

In practice, outside of the constraints of our monthly seminar series framework, we can imagine the persistence of an ethical ethos to look something like this: in the act of conceptualizing or planning a new research project, researchers identify some potential ethical considerations (*topic identification*). They seek out expert opinions and discuss these considerations with their colleagues and area experts (*dialogue/synthesis*). Through this discussion, they strengthen the relationship between the consideration they identified and their own value system (*ownership*), leading them to structure their approach to the research project to be mindful of the consideration (*application*).

14.4 Justice, Inclusion, and the Ethical Ethos

In 2021, the International Neuroethics Society (INS) announced the theme of that year's Annual Meeting as "Social Justice at the Center: Shaping the Future of Neuroethics." Social justice was described in the INS Annual Meeting theme description as "the project of promoting equity across geographic, economic, social, and cultural boundaries" (International Neuroethics Society 2021). This theme choice reflected a growing recognition of the critical importance of the ideas of social justice and equity across disciplines, particularly within the discipline of neuroethics.

If an ethical ethos strives to integrate the moral values and lived expertise of researchers with their day-to-day research, then the conception and practice of ethics must reflect researchers' commonly held values. In the summer of 2020, amidst rising global demands for racial justice spurred by the murder of George Floyd by Minneapolis police, a group of students and postdoctoral fellows at RNEL mobilized. Led by members of the RNEL Neuroethics team in collaboration with other engaged students, this group was organized to demand greater action and accountability from RNEL to improve inclusion and equity efforts within the lab. The result was the establishment of a wide-ranging diversity, equity, and inclusion (DEI) effort at RNEL, including (1) a multi-pronged outreach arm aimed at delivering STEM education to underserved students and local communities, (2) several internal efforts to foster a more welcoming and inclusive work environment, and (3) a series of didactic and discussion sessions to provide education to lab members about history and concepts under the social justice umbrella. These persistent programs reflect a deep commitment and enthusiasm on the part of many lab members for creating a more equitable future in STEM.

As DEI efforts began to take hold at RNEL over the course of summer 2020, it became increasingly apparent that diversity, equity, and inclusion were deeply held values and commitments for many members of the lab. This presented an important opportunity for the lab's ethical ethos to develop toward authentically aligning with lab members' genuine beliefs. We initially approached this with a block of explicitly DEI-oriented ethics discussion sessions, with the goal of increasing awareness of particular topics and concepts and illuminating the link with ethics. This block started with a session defining and describing disparity. It was followed by sessions focused on implicit bias and perceived discrimination in healthcare, trust and risk as a function of race in medicine, and assumptions regarding accessibility and technology for disabled populations. These topics were chosen to track closely with issues of direct relevance for RNEL researchers, who work on clinical research and medical technology, primarily intended to benefit disabled populations.

This initial block of DEI-oriented ethics discussions was eagerly received in a work environment where people were actively seeking fulfillment of their own values of equity and justice in their workplace. Lab members were both challenged and edified by discussions that prompted a close examination of the motivations and intentions of neural engineering research as they relate to disabled people's lived experiences. These discussions also encouraged careful consideration of the power and trust dynamics in the relationship between research participants and researchers. During the discussions and in feedback provided after the session, attendees identified important considerations regarding the diversity of research participant pools and their own motivations for research. For instance, a discussion on trust and race in medicine and research prompted a seminar attendee to reflect on the racial disparities between the demographics of the city and the demographics of the research participant population and to suggest that the lab make a conscious effort to connect with underrepresented communities. Another discussion regarding disability and access prompted researchers to reflect on the assumptions underlying their interest in neurotechnology research and to recognize the need for engaging with disabled communities in identifying research directions.

The early success of this block of sessions inspired a new approach to ethics programming at RNEL. At the beginning of 2022, we merged the ethics working group with the DEI training and discussion subcommittee established in 2020, with the goal of converging on an equity-centered approach to neuroethics discussions. This equity-centered approach shows up in each of the four main stages of our ethical ethos cycle.

14.4.1 Topic Identification

Most of the topics we select to drive future discussions are chosen through a combination of attendee feedback, current events, and alignment with a justice and equity framework. By tying the topics to justice and equity in light of current events, we drive engagement through familiarity and timeliness. Even when developing discussions around fairly standard/classical ethics topics (e.g., consent or researcher responsibilities), it has become natural to identify the ways in which these topics align with considerations of justice and equity.

14.4.2 Dialogue/Synthesis

Prior to each session, guiding questions are formulated to direct the discussion toward the intersection of justice and equity with other ethical principles, issues, and concerns. In this way, even topics that are not explicitly or obviously oriented toward equity and justice can be connected with the social justice ideals with which lab members have already demonstrated familiarity and interest. This encourages a constant process of formulating connections between existing values and the ethical principles under consideration.

14.4.3 Ownership

By making explicit the relationship between lab members' expressed values and beliefs and the ethical questions and topics we are discussing, we strengthen the sense of ownership that lab members develop toward ethical principles. In effect, we empower lab members to internalize ethical principles as part of their individual value systems by highlighting the connections with principles of justice and equity that they already hold dear.

14.4.4 Application

Lab members who have internalized ownership of ethical principles manifest this in multiple ways. Already many lab members have applied their values by participating in outreach and other DEI-related efforts. As the connections between these values and ethical considerations in research become strengthened, we see lab members integrating these ideas toward initiatives like soliciting community feedback regarding research directions or working to better communicate the fruits of their research to members of underserved local communities.

14.5 Responses and Outcomes

In cultivating an ethical ethos, the first goal is to drive engagement with ethics. Deep ethical engagement leads to the internalization and application of ethical principles. Therefore, we have consistently collected feedback from discussion attendees to assess their engagement, familiarity with the topic, and the relevance of the topic to their research in a total of 15 sessions (five held in person prior to the pandemic, five held virtually due to the pandemic, and five hybrid sessions since the lab returned to mostly in-person work).

Our survey questions are designed to prompt discussion participants to reflect on the topic of the discussion, with a particular emphasis on how the discussion has impacted their perspective on the topic and the application of the topic to their research. We ask participants to reflect on their familiarity with the topic before and after the discussion, key insights from the discussion, how their values regarding the topic impact their research, and how they or the lab could better align with their values regarding the topic.

This form of assessment suffers from a common shortfall of post-event surveys: low response rate. As expected, only a small fraction of attendees completed the feedback form after any session. However, the responses serve as a sample of the kind of feedback that has been shared with us both in feedback surveys and in informal comments made by attendees.

Survey data suggest that the implementation of a neuroethical ethos impacts the ethical education of the participants and their research culture (Fig. 14.2). Most respondents reported that these discussions increased their familiarity with ethical considerations in their research. Interestingly, this increase in awareness was independent of previous experience related to the topic, suggesting that revising ethical topics according to our approach can reinforce and enhance existing familiarity. However, a few attendees reported a decrease in familiarity. Feedback from one such attendee indicates that they believed themselves to be very familiar with the topic prior to the session but that the discussion revealed to them that their prior understanding was flawed. This suggests that the discussion prompted critical reflection, leading them to acknowledge their flawed understanding. Although this resulted in the attendee feeling less familiar with the topic after the session, the discussion appears to have prompted them to reflect more deeply on the topic than they had prior to the session.

When asked to propose actions that they or the lab can take to better reflect their values, most respondents proposed enhanced ways to engage with research participants and to design clinical studies to reflect justice and inclusion values that emerged from the discussions. After a seminar on trust and race, one respondent emphasized that "having positive relationships [with marginalized communities] could improve overall trust towards the scientific research community." An attendee of a session on disability and access cautioned against presenting research as "some kind of 'gift to the disabled community'". Another session prompted an attendee to

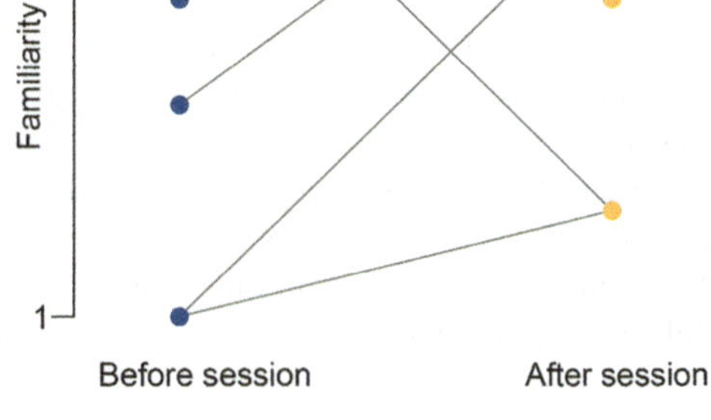

Fig. 14.2 Most participants reported an increase in familiarity after attending the session (yellow) compared with before the session (blue)

reflect that "we need a more comprehensive assessment of the moral implications of each technology [we develop]."

Most respondents highlighted that the structure and implementation of the discussion sessions were effective for them. Many commented on the value they gained from the opportunity to directly engage in conversation with an invited topic area expert and the effectiveness of breakout rooms (in a virtual environment) or simply small group discussions (when in person) for fostering insight and engagement. These elements contributed to a sense that these discussions felt like natural conversations with their research colleagues.

We did note one major area for improvement: session participants whose research involves human subjects reported benefiting more from the discussions, while others would sometimes comment that they didn't see the relationship to their work. As we continue to reflect on and refine our approach, it is critical that we more closely engage researchers who do not work with human subjects to recognize the ethical and societal implications of their work.

Overall, the implementation of an ethical ethos seemed not only to expand ethical awareness but also to strengthen the work environment. Discussion participants' feedback indicated that they feel better knowing that their place of work discusses ethics in research. This feeling encapsulates one of the most fundamental goals of the ethical ethos: for researchers to be embedded in a culture that values ethics and to feel that their workplace reflects their own values.

In effect, creating an ethical ethos contributes to a workplace culture in which researchers are free to act genuinely in accordance with their deeply held values. When people feel supported in their whole selves in the workplace, this can transform their work experience into one that feels more fulfilling. Moreover, in an academic landscape where researchers are often asked to situate their research in terms of its impact (e.g., through grant proposals and other funding mechanisms), an ethical ethos can support researchers' sincere aspirations toward socially impactful work and support the integration of the principles they espouse when they *communicate* their work with the day-to-day *reality* of the work.

14.6 Discussion and Takeaways

Our initial challenge was to overcome the constraints of regulatory/didactic ethics to make ethics a natural part of the research process. The resolution of this challenge lies in the development of what we have termed an ethical ethos.

In the last several years, various federal and institutional forces in the sciences and (of particular interest to our lab) in neuroscience and engineering have begun to prioritize ethics publicly. For example, the National Science Foundation explicitly requires impact statements (National Science Foundation n.d.) in funding applications, the Institute of Electrical and Electronics Engineers (IEEE) Brain Initiative has developed ethical guidelines for engineers, researchers, and neurotechnology companies (IEEE Brain n.d.), and the National Institutes of Health (NIH) Brain

Research Through Advancing Innovative Neurotechnologies (BRAIN) Initiative has funded neuroethics research projects and managed a working group of neuroethics and neuroscience experts to provide input on the direction of the program (Brain Initiative n.d.). All of these examples demonstrate that ethics and social impact are of critical importance to the landscape of science.

Critically, prior work (Johnsson et al. 2014; Tubig and McCusker 2021) has argued that deep ethical engagement by researchers can help to mitigate some of the valid mistrust that research participants or members of the public have toward scientists and researchers (Boulware et al. 2003; Crawley 2001; Sullivan 2020; Yeager et al. 2017). Trust in research, particularly medical research, is a complex phenomenon. On the one hand, there is a great deal of public and individual mistrust toward research and researchers. On the other hand, research participants and medical patients necessarily place a great deal of trust in researchers and providers when they engage with research or medical systems (Kass et al. 1996). It is, therefore, critical to ensure that trust in researchers is not merely bolstered but also warranted. Researchers and institutions must engage in work that *makes* them genuinely trustworthy. An ethical ethos is a step toward that ideal.

Conscious design is increasingly popular in the technology and research worlds, with concepts like value-sensitive design (Friedman and Hendry 2019) and equity design thinking prompting technologists and medical researchers to engage deeply and at an early stage with the implications of their work (The Equity Design Thinking Educational Series: Pitt Equity Design Thinking n.d.). The ethical ethos approach aligns with this movement, promoting the creation of cultures that make ethical and equity-centered design not only desirable but also natural.

Our approach to implementing an ethical ethos represents a grassroots, researcher-driven framework for ethical engagement. It centers the voices, lived experiences, and professional and personal expertise of the community of researchers in a way that empowers them to integrate ethical principles into their work. We provide a toolset for individual researchers to think proactively about ethical considerations and to design their research, teaching, and technology with ethics in mind.

Our experience over the last four years has led to some key insights into the implementation of programs like this. Firstly, these sessions are most successful and engaging when we intentionally focus on engaging participants as their whole selves. In particular, encouraging participants to actively reflect on the intersection of scientific ethics with their own lived values and experiences leads to greater engagement and interest, and can promote the integration of ethical principles into their own research. Secondly, it is critical to be thoughtful about how participants are likely to engage, and to provide frameworks to help shape that engagement. Finally, flexibility and adaptability are key. As the needs and interests of a group shift, ethics education must adapt with them.

This dialogue-centered approach has successfully engaged lab members and cultivated an ethical ethos at RNEL over the last several years. In fact, the success of this approach has generated interest even beyond the lab, drawing attendees from different departments and universities. These efforts continue to expand in the neuroscience and engineering community in Pittsburgh. Through normalizing the implementation of ethics in everyday research, this approach can support more conscientious research and technology development in Pittsburgh and beyond. The expansion of an ethical ethos across labs, companies, and institutions can foster a more hospitable ecosystem for thoughtful, ethically engaged work with positive resonances throughout scientific fields.

Acknowledgments This work was made possible by the generous financial support of the University of Pittsburgh Research, Ethics, and Society Initiative. The authors would also like to acknowledge the support they have received from the Rehab Neural Engineering Labs (RNEL) and the Center for Bioethics and Health Law at the University of Pittsburgh, as well as the University of Pittsburgh and Carnegie Mellon University's Center for the Neural Basis of Cognition (CNBC).

Special thanks to Dr. Lisa Parker (Center for Bioethics and Health Law, University of Pittsburgh), Dr. Lee Fisher (Department of Physical Medicine and Rehabilitation, University of Pittsburgh), and Dr. Douglas Weber (Department of Mechanical Engineering, Carnegie Mellon University).

References

Antes, A.L., S.T. Murphy, E.P. Waples, M.D. Mumford, R.P. Brown, S. Connelly, and L.D. Devenport. 2009. A meta-analysis of ethics instruction effectiveness in the sciences. *Ethics & Behavior* 19 (5): 379–402. https://doi.org/10.1080/10508420903035380.

Argandoña, A. 2008. Integrating ethics into action theory and organizational theory. *Journal of Business Ethics* 78 (3): 435–446. https://doi.org/10.1007/s10551-006-9340-x.

Boulware, L.E., L.A. Cooper, L.E. Ratner, T.A. LaVeist, and N.R. Powe. 2003. Race and trust in the health care system. *Public Health Reports* 118 (4): 358–365. https://doi.org/10.1016/S0033-3549(04)50262-5

Brain Initiative. n.d.. *Neuroethics | Brain Initiative*. https://braininitiative.nih.gov/research/neuroethics

Cho, M.K., S.L. Tobin, H.T. Greely, J. McCormick, A. Boyce, and D. Magnus. 2008. Strangers at the Benchside: Research ethics consultation. *The American Journal of Bioethics: AJOB* 8 (3): 4–13. https://doi.org/10.1080/15265160802109322.

Christian, M.S., A.S. Garza, and J.E. Slaughter. 2011. Work engagement: A quantitative review and test of its relations with task and contextual performance. *Personnel Psychology* 64 (1): 89–136. https://doi.org/10.1111/j.1744-6570.2010.01203.x.

Crawley, L.M. 2001. African-American participation in clinical trials: Situating trust and trustworthiness. *Journal of the National Medical Association* 93 (12 Suppl): 14S–17S. PMID: 11798059, PMCID: PMC2719992.

Dzeng, E., and R.M. Wachter. 2020. Ethics in conflict: Moral distress as a root cause of burnout. *Journal of General Internal Medicine* 35 (2): 409–411. https://doi.org/10.1007/s11606-019-05505-6.

Edmondson, A. 1999. Psychological safety and learning behavior in work teams. *Administrative Science Quarterly* 44 (2): 350–383. https://doi.org/10.2307/2666999.

Farooqui, J., S. Dawod, and D. Sarma. 2021. Towards a neuroethical ethos: A case study in reframing neuroethics education for engineers and researchers. The Neuroethics Blog, Emory University Center for Ethics. https://www.theneuroethicsblog.com/2021/02/towards-neuroethical-ethos-case-study

Fisher, E., M. O'Rourke, R. Evans, E.B. Kennedy, M.E. Gorman, and T.P. Seager. 2015. Mapping the integrative field: Taking stock of socio-technical collaborations. *Journal of Responsible Innovation* 2 (1): 39–61. https://doi.org/10.1080/23299460.2014.1001671.

Friedman, B., and D.G. Hendry. 2019. *Value sensitive design: Shaping technology with moral imagination*. MIT Press.

Goering, S., and E. Klein. 2020. Embedding ethics in neural engineering: An integrated transdisciplinary collaboration. In *A guide to field philosophy*. Routledge.

Harris, C.E., Jr., M. Davis, M.S. Pritchard, and M.J. Rabins. 1996. Engineering ethics: What? why? how? and when? *Journal of Engineering Education* 85 (2): 93–96. https://doi.org/10.1002/j.2168-9830.1996.tb00216.x.

IEEE Brain. n.d.. *IEEE Neuroethics Framework | IEEE Brain*. https://brain.ieee.org/publications/neuroethics-framework/addressing-the-ethical-legal-social-cultural-implications-of-neurotechnology/

International Neuroethics Society. 2021. *2021 annual meeting theme*. https://www.neuroethicssociety.org/2021-annual-meeting

Johnsson, L., S. Eriksson, G. Helgesson, and M.G. Hansson. 2014. Making researchers moral: Why trustworthiness requires more than ethics guidelines and review. *Research Ethics* 10 (1): 29–46. https://doi.org/10.1177/1747016113504778.

Kass, N.E., J. Sugarman, and J., Faden, and M. Schoch-Spana. 1996. Trust the fragile Foundation of Contemporary Biomedical Research. *Hastings Center Report* 26 (5): 25–29. https://doi.org/10.2307/3528467.

Korsgaard, C.M. 2009. Agency and identity. In *Self-constitution*. Oxford University Press. https://doi.org/10.1093/acprof:oso/9780199552795.003.0001.

Martin, D.A., E. Conlon, and B. Bowe. 2021. A multi-level review of engineering ethics education: Towards a socio-technical orientation of engineering education for ethics. *Science and Engineering Ethics* 27 (5): 60. https://doi.org/10.1007/s11948-021-00333-6.

Melo-Martín, I.D., L.I. Palmer, and J.J. Fins. 2007. Viewpoint: Developing a research ethics consultation service to foster responsive and responsible clinical research. *Academic Medicine* 82 (9): 900–904. https://doi.org/10.1097/ACM.0b013e318132f0ee.

Miller, M. H., E. Englehardt, M. Pritchard, S. LeFrancois G. Centeno, K. Reeves, and I. Roman, 2022, August 23. *Cultivating the ethical identities of STEM students through enhanced internships*. 2022 ASEE Annual Conference & Exposition. https://peer.asee.org/cultivating-the-ethical-identities-of-stem-students-through-enhanced-internships

National Science Foundation. n.d. *Broader Impacts | NSF National Science Foundation*. https://www.nsf.gov/od/oia/special/broaderimpacts/

Rich, B.L., J.A. Lepine, and E.R. Crawford. 2010. Job engagement: Antecedents and effects on job performance. *Academy of Management Journal* 53 (3): 617–635. https://doi.org/10.5465/amj.2010.51468988.

Sullivan, L.S. 2020. Trust, risk, and race in American medicine. *Hastings Center Report* 50 (1): 18–26. https://doi.org/10.1002/hast.1080.

The Equity Design Thinking Educational Series: Pitt equity design thinking. n.d.. https://www.pittequitydesignthinking.org/

Tubig, P., and D. McCusker. 2021. Fostering the trustworthiness of researchers: SPECS and the role of ethical reflexivity in novel neurotechnology research. *Research Ethics* 17 (2): 143–161. https://doi.org/10.1177/1747016120952500.

Vertrees, S.M., A.G. Shuman, and J.J. Fins. 2013. Learning by doing: Effectively incorporating ethics education into residency training. *Journal of General Internal Medicine* 28 (4): 578–582. https://doi.org/10.1007/s11606-012-2277-0.

Weaver, G.R., and L.K. Treviño. 1999. Compliance and values oriented ethics programs: Influences on employees' attitudes and behavior. *Business Ethics Quarterly* 9 (2): 315–335. https://doi.org/10.2307/3857477.

Yeager, D.S., V. Purdie-Vaughns, S.Y. Hooper, and G.L. Cohen. 2017. Loss of institutional trust among racial and ethnic minority adolescents: A consequence of procedural injustice and a cause of life-span outcomes. *Child Development* 88 (2): 658–676. https://doi.org/10.1111/cdev.12697.

Part IV
Introduction: New Approaches in Framing Ethical Issues

One extremely promising advancement in ethics education is helping students explore ethical questions from a new perspective. Educational strategies that break students out of their normal routines put them in the place of a designer or stakeholder or call upon skills that STEM students might often utilize in class can be helpfully disruptive.

Efforts to integrate the humanities into STEM ethics education have proven fruitful. The use of science fiction in teaching technology ethics has been a long one (Burton et al. 2018, Berne et al. 2005). There have been other excellent approaches that ask students to write personal ethics statements or engage in keeping a reflective journal in the form of an ethics autobiography wherein the students are asked to think about the ethical implications of their daily activities both in and out of the classroom (Zhu and Woodson 2020; Snieder and Zhu 2020).

Imagination and creativity can too often be an under-recognized tool for STEM students. Creative approaches to problem-solving are increasingly being seen as a key strength in the workforce, and it is clear that these skills are critical in finding solutions in designing for a sustainable future. Some educators argue that STEM education too often narrows students' focus to trying to find technological solutions to global problems (Smith and Watson 2020). Instead, there is a need to find educational approaches that open up students' curiosity, imagination, and creativity.

Another way of promoting changes in students' perspectives is by immersing them in different environments. Community outreach approaches move STEM ethics education outside the university and reinforce the main goals of STEM ethics education while also assisting students in engaging with the communities and environments in which they will work (Bielefeldt et al. 2016). Students learn how to interact with diverse stakeholders, balance the wants and needs of their community clients with the limitations of the project, and improve their communication skills when working with non-engineers.

This section of the volume looks at new approaches to framing ethics issues. The chapter describes using different lenses to help STEM students and practitioners look at their work in a different way. The approaches described build on what we know from neuroscience and psychology, use the power of narrative to help students explore their moral imagination and empathy, and integrate the direct experience of tinkering with technology in ethics education.

In the chapter by Mike T. Stuart and Hannah Sargeant, "Inclusivity in the Education of Scientific Imagination," the authors present the results of a 6-year ethnographic study exploring how scientists learn to use their imagination. After discussing the importance of imagination in scientific research and the implicit ways in which imagination is explicitly taught, the authors explore how individuals from traditionally marginalized groups often feel that imagination is less critical in science and have less faith in the strength of their own imagination compared to non-marginalized groups at the same career stage. The authors explore systemic reasons behind this. Dr. Sargeant provides a personal reflection on her journey of realizing that imagination is a critical scientific skill and her evolving relationship with imagination and curiosity. The authors conclude with a critical reflection on how educators might better support the development of scientific imagination for all students, including setting students tasks that require imagination, pointing to good role models, and supporting virtues that enable imaginativeness.

Storytelling is another beneficial approach. The chapter, "At the verge of 'is' and 'could be': storytelling as medium to develop critical ethical skills" by Marietjie Botes and Arianna Rossi, explores how storytelling can be effectively used to embed ethics into professional courses through its power to appeal to students' hearts and minds. The author examines how stories help to convey knowledge and develop students' empathy and understanding of different stakeholders' viewpoints. She then discusses the power of fictional narratives to act as a kind of sandbox where students can test open-ended scenarios and develop critical thinking skills. Dr. Botes and Dr. Rossi explore the benefits of using science fiction, digital storytelling, virtual reality, and video games in ethics education and reflect on best practices for embedding storytelling in STEM ethics training.

In the chapter, "Tinkering with Technology: How Experimental Engineering Ethics Pedagogy Can Accommodate Neurodivergent Students and Expose Ableist Assumptions," the authors Janna van Grunsven, Trijsje Franssen, Andrea Gammon, and Lavinia Marinfrom Delft University of Technology explore how educators can promote inclusivity at the content and form level. They describe an exercise where students were asked to critically engage with inclusivity-undermining ableist assumptions in technology development by participating in a hands-on "material-tinkering" workshop. In this workshop, student teams were asked to hack artifacts used in contexts of disability and healthcare with the values of inclusivity and accessibility at the forefront of their minds. The authors provide a detailed explanation and analysis of their tinkering exercise, photographs of the final alterations to the artifacts made by the students, and discuss the results of the triangulated research method used to access the pedagogical value of the exercise that included observations from the teachers, students, and observers of the exercise. Finally, the authors discuss the successes of this approach and suggest some areas for improving future tinkering exercises of this kind.

In "Philosophy in the Rainforest: Reflections on Integrating Philosophy and Fieldwork," author Clair Morrissey gives an overview of a long-term collaboration between Occidental College's Biology and Philosophy departments to develop a model for undergraduate ecological field research ethics education. After discussing how ecological field research often falls into an ethics "gap" regarding ethics oversight and normative guidance, the author describes how interdisciplinary collaborations between scientists and field philosophers can work, and how this model helps young ecologists understand ethics as intrinsic to their scientific practice. The chapter describes the details of Occidental College's field ecology-philosophy collaboration at the La Selva Biological Research Station in Puerto Viejo de Sarapiquí, Costa Rica, and at Occidental College's main campus. Morrissey reflects on the interdisciplinary work completed by the teams of philosophy and ecology faculty and students and the philosophy discussions and assignments completed by the student participants. She provides examples from reflection journals completed by the students, and discusses how this project helped students develop a habit of ethical reflection in their individual and collaborative work.

In the concluding chapter for this section, "Building Inclusive Cultures through Community Research," Jennifer F. Nyland, Timothy Stock, Michéle Schlehofer introduce the "Re-envisioning Ethics Access and Community Humanities (REACH) initiative at Salisbury University. Recognizing the need to reorient ethics teaching to include a more community-oriented approach, the authors implemented a community-source ethics case, "When to vaccinate, when to educate?" that focuses on the role of non-profit health workers and public health authorities as they tried to mitigate vaccine hesitancy during the COVID-19 pandemic. The project integrates community-based ethics research into the science classroom. It seeks to bring stakeholders new resources to help develop their ethics literacy and engage in a broader dialogue about social change while incorporating real-world, local, community-based ethical issues into teaching across the STEM curriculum. The authors characterize the REACH project components and describe student and instructor feedback about the learning process.

References

Berne, R. W., and J. Schummer. 2005. Teaching societal and ethical implications of nanotechnology to engineering students through science fiction. *Bulletin of Science, Technology & Society* 25(6): 459–468.

Bielefeldt, A.R., N. Canny, C. Swan, and D.W. Knight. 2016. Contributions of learning through service to the ethics education of engineering students. *International Journal for Service Learning in Engineering* 11(2). https://doi.org/10.24908/ijsle.v11i2.6392

Burton, E., J. Goldsmith, and N. Mattei. 2018. How to teach computer ethics through science fiction. *Communications of the. ACM* 61 (8): 54–64. https://doi.org/10.1145/3154485.

Smith, C., and J. Watson. 2020. From streams to streaming: a critique of the influence of STEM on students' imagination for a sustainable future. *Journal of Applied Teaching and Learning* 3(Special Issue): 21–29. https://eprints.utas.edu.au/36825/. Accessed 21 Mar 2023

Snieder, R., and Q. Zhu. 2020. Connecting to the Heart: Teaching Value-Based Professional Ethics. *Science and Engineering Ethics* 26: 2235–2254. https://doi.org/10.1007/s11948-020-00216-2

Zhu, Q., and S. Woodson. 2020. Educating self-relective engineers: Ethics autobiography as a tool for moral pedagogy in engineering. *Teaching Ethics* 20 (1–2): 31–46. https://doi.org/10.5840/tej20214190

Chapter 15
Inclusivity in the Education of Scientific Imagination

Michael T. Stuart and Hannah Sargeant

Abstract Scientists imagine constantly. They do this when generating research problems, designing experiments, interpreting data, troubleshooting, drafting papers and presentations, and giving feedback. But when and how do scientists learn how to use imagination? Across 6 years of ethnographic research, it has been found that advanced career scientists feel comfortable using and discussing imagination, while graduate and undergraduate students of science often do not. In addition, members of marginalized and vulnerable groups tend to express negative views about the strength of their imaginations and the general usefulness of imagination in science. After introducing these findings and discussing the typical relationship between a scientist and their imagination across a career, we argue that reducing the number or power of active imaginations in science is epistemically counterproductive. We suggest several ways to bring imagination back into science in a more inclusive way, especially through courses on imagination for scientists, good role models, and exemplar-based learning.

Keywords Imagination · Creativity · Science education · Inclusive education · Virtue epistemology · Ethics of science education

M. T. Stuart (✉)
Department of Philosophy, University of York, York, UK
e-mail: mike.stuart@york.ac.uk

H. Sargeant
Department of Philosophy, University of York, York, UK

Aerospace Engineering, University of Leicester, Leicester, UK
e-mail: hs564@le.ac.uk

E. Hildt et al. (eds.), *Building Inclusive Ethical Cultures in STEM*, The International Library of Ethics, Law and Technology 42,
https://doi.org/10.1007/978-3-031-51560-6_15

15.1 Introduction

Educators are supposed to provide people with skills that help them in life. Creativity is such a skill. How can educators increase it? Philosophers typically portray creativity as the ability to produce ideas that are both *new* and *valuable*. How do you teach someone to produce a new, valuable idea? For a start, they will need some relevant background knowledge, experience, and especially *imagination* (Gaut 2003; Hills and Bird 2019; Stokes 2014; Stuart 2020). Indeed, the connection between imagination and creativity is so tight that scientists often use them as synonyms (Sánchez-Dorado 2020; Stuart 2019c).

So, if we want to improve creativity, we can start by improving imagination. But how do we do that? The answer depends on what imagination is. Some philosophers argue that imagination is a mental attitude that we take toward some content (e.g., Currie et al. 2002; Arcangeli 2017; Nichols 2006; Kind and Kung 2016). For example, we might believe, doubt, desire, or imagine that we have won the lottery. Belief, doubt, desire, and imagination are different mental attitudes we can take to the same content. They are different, at least because they have different functions: believing we have won the lottery might cause us to jump and down, desiring to win the lottery might cause us to buy a ticket, and imagining that we have won the lottery might only cause a smile. However, if imagination is just one kind of mental attitude that we can have toward some content, it is not clear how it could be improved: we either take that attitude towards some content, or we don't. A promising new view is that imagination should be understood as an *ability*. This is a useful view for a discussion about education because abilities can be improved (Kind 2020, 2022; Stuart 2019b). Insofar as educators do (and should) aim to develop scientific creativity, then they do (and should) aim to develop the *ability* of scientific imagination. "Attending to the workings of imagination is not a soft option" (Midgley 1992, 24).

It was noticed already more than a century ago (Ribot 1906) that there is very little literature on imagination in science education. This continues to be true today (though see Hadzigeorgiou 2016 and Eijck and Roth 2012). Drawing on qualitative observations performed in five research laboratories (all of which contained graduate students, undergraduate students, or both) and interviews with 13 scientists, we claim that while scientists recognize a need to foster imagination in their students, no standard recognized methods yet exist for teaching the skill of imagination (Stuart 2022a).[1] In general, scientists do not talk about imagination. Here are a few indicative quotations from established scientists: "It's not something that people talk about" (interview, PI, 19/06/2019). "Everybody imagines, but no one really talks about it, or how to do it better, or when to spend time on it" (interview, postdoc, 22/04/2019). "I think the only time when I have this discussion is when PhDs

[1] While this paper builds on the results of Stuart (2019c, 2022a), all quotations given are presented here for the first time unless otherwise indicated. For more information on the sociological methods used to collect and analyze these quotations and observations, see (Stuart 2019c).

transition towards postdocs, or postdocs transition to a point where they need to get their own funding. So, where they have to come up with their own research ideas. And that's when I had this discussion where people were scared that they wouldn't have the imagination to come up with these ideas. So I think that's the only point in time" (interview, PI, 19/06/2019).

Of course, scientists do eventually learn to use imagination. They must. But how they do this greatly depends on their individual preferences, abilities, and values, as well as the research culture of their field, which determines the kinds of problems that need to be solved and the methods used to solve them. Also important are the experiences, preferences, abilities, values, and teaching styles of their educators. These can vary dramatically, such that even in a single department, there can be very large differences between attitudes that principal investigators (PIs) take towards imagination-use, and these differences tend to be reflected in pedagogical practices (Stuart 2019c, 2022a).

While imagination is not taught explicitly in science classrooms or laboratories, it is still important to identify how it is taught *implicitly* across the different sciences, educational stages, and geographical locations. Meanwhile, raising awareness about this vital part of science education can prompt educators to think about it more directly, which can lead to the identification and dissemination of successful strategies. This will take time. However, even with the little information we have, we can already show that the existing processes of teaching imagination (whatever they are) are not maximally inclusive.

Why? Despite all the differences in attitudes and approaches to imagination across the labs observed, one thing seems to hold everywhere: whether a scientist values imagination and whether they feel confident in using their own imagination, depends not only on career stage, but also on whether someone is a member of a traditionally marginalized group. People from such groups (whether based on race, gender, ability, etc.) tend to feel that imagination is less important in science, and they tend to have less faith in the strength of their own imagination as compared to non-marginalized people at the same career stage (Stuart 2019a). Statements like the following are common.

> I was just thinking how there is this absolute requirement to be imaginative, to get paid, because you have to come up with a new idea to get the funding, to have a job. And that's really hard because I don't feel like I'm an imaginative person, I tell everyone I'm not very creative, and I compare the two words quite a lot. So yeah, I find it really hard to come up with new ideas because it's something I've never had to think about before...You have these constraints: you have to be imaginative to do something new, but it has to be, whatever the idea is, it has to be able to be studied within a restricted timeframe, and the restricted amount of funding. So you have to come up with something creative that ticks all of those boxes, and that's really hard! I have no idea how to do that! (interview, Ph.D. student, 14/01/2020)

This was a female Ph.D. student whose imagination had already brought her great scientific success. This kind of sentiment appeared regularly for members of marginalized groups, and it was stronger in those who found themselves at the intersection of the more vulnerable end of the sex, gender, race, geographic, age, and career stage continuums. Those at the other end of those continuums often expressed the opposite sort of view: their own imagination was powerful and played a central role in their identity as a scientist, and was partially responsible for whatever success they had already achieved. It is not the case that people in marginalized groups *identify themselves* as having different views toward their own imaginations. The point is that sociological investigation reveals that such people tend to have systematically different views. This is something that can only be noticed with a sufficient number of data points.

We do not want to claim that people at the confident end of the spectrum are incorrect about the power of their own imaginations or that any of them specifically are to blame for this state of affairs. Perhaps, given the same privilege, all scientists would agree that imagination is important for science. No implicit or explicit biases were observed in privileged scientists acting against less privileged scientists, and indeed many of the labs were quite diverse in terms of culture, race, gender, sexual orientation, ability, age, etc. So we do not locate the problem here, either. Instead, we hypothesize that it is a systemic educational issue, and so in Sect. 4, we consider systemic solutions.

Before moving on, we should mention that this chapter's topic lies at the intersection of many different literatures, including education studies, philosophy of science, ethics, gender and race studies, and cognitive science. Particularly important are the feminist (Arámbula-Greenfield 1995; Barton 1997, 1998; Hussénius et al. 2013; Hussénius 2014) and anti-racist literatures on science education (e.g., Anderson and Herr 2007; Dennick 1992; Gill and Levidow 1987; Hines 2007; Law 2017; Kishimoto 2018; and the contributions to the present volume), as well as the discourse on imagination in non-scientific educational contexts (e.g., Rorty 2009; Hatt 2022; Wright 2021; Dewey 1980; Murdoch 1994, 2001; Greene 2000; Nussbaum 1998). A complete understanding of the issues we are going to raise requires insights from all these fields, which we will discuss as we go, as best we can. However, neither of the authors of the present paper are scholars in education studies. One is a philosopher of science interested in scientific imagination, and the other is a scientist who shares an interest in imagination. As we both are passionate about education and the ethics of science, we decided to combine our perspectives in a spirit of interdisciplinarity in a way that we hope is useful, making no claims to completeness or general authority.

15.2 No Imagination Allowed

Scientists approve of the use of imagination in contexts where the usual methods have already failed (or would fail) to provide a solution to a maximally specific problem (Stuart 2019b). More generally, imagination is crucial in periods of uncertainty, whether that uncertainty pertains to a contradiction in the background literature, how to design or interpret an experiment, how to write up a result, how to reply to reviewers, or how to give a presentation. Summarizing interview data from 63 scientists, Schickore and Hangel (2019) show that scientists are often uncertain about these kinds of things. In other words, scientists very frequently find themselves in contexts where imagination is helpful for resolving specific problems.

Education should prepare future scientists for this by preparing their imaginations. This is not currently at the top of the science education to-do list, which is understandable: scientists must also learn the techniques and theory of their domain before imagining how to transcend these. So, we should expect that imagination will not be the main focus in early periods of science education. Still, scientists interviewed agree that the majority of their experiences pursuing undergraduate degrees in science or engineering sharply discouraged the use of imagination (Stuart 2019c). And while we might think that imagination is eventually taught or nurtured in graduate school, this is only sometimes true, and when it is, it can be too late. The endurance test that is science education does not perfectly select for good future scientists. It pushes away many who might have been great. Consider the following reflections of famous physicists David Peat and David Bohm.

David Peat:

> As far back as I can remember, I was always interested in the universe. I can still remember standing under a street lamp one evening—I must have been eight or nine—and looking up into the sky and wondering if the light went on forever and ever, and what it meant for something to go on forever and ever, and if the universe ever came to an end…These sorts of ideas continued right through school, along with a feeling of the interconnectedness of everything. It was almost as if the entire universe were a living entity. But of course, when I got down to the serious business of studying science at university, all this changed. I felt that the deepest questions…were never properly answered…Instead, we were all encouraged to focus on getting concrete results that could be used in published papers and to work on problems that were "scientifically acceptable." (Bohm and Peat 1987, pp. ix–x)

David Bohm:

> I, too, felt that kind of wonderment and awe in my early days, along with an intense wish to understand everything, not only in detail but also in its wholeness.
>
> However, in graduate school...I found that there was a tremendous emphasis on competition and that this interfered with such free discussions. *There was a great deal of pressure to concentrate on learning formal techniques* for getting results. It seemed that there was little room for the desire to understand in the broad sense that I had in mind…*Although I was quite capable of mastering these mathematical techniques, I did not feel that it was worth going on with*, not without a deeper philosophical ground and the spirit of common inquiry. You see, it is these very things that provide the interest and motivation for using mathematical techniques to study the nature of reality. (Bohm and Peat 1987, pp. xi–xiii, emphasis added)

Philosophers of science have commented on this aspect of science education as well. Paul Feyerabend argued that science education "leads to a deterioration of intellectual capabilities, of the power of the imagination. It destroys the most precious gift of the young, their tremendous power of imagination" (Feyerabend 1975, 96–7). He adds,

> Teachers using grades and the fear of failure mould the brains of the young until they have lost every ounce of imagination they might once have possessed. This is a disastrous situation, and one not easily mended…Agreement with science, decision to work in accordance with the canons of science should be the result of examination and choice, and not of a particular way of bringing up children. (160–162)

This situation has also drawn the attention of scholars in education studies, for example, the Imaginative Education Research Group at Simon Fraser University (2001–2015), which included scholars Kieran Egan (Egan and Madej 2010; Egan et al. 2015), Gillian Judson (see, e.g., Judson 2015) and Yannis Hadzigeorgiou (see, e.g., his 2016). Hadzigeorgiou writes, "Teachers often fail to use students' imaginations and creativity in their science classrooms, despite the fact that the 'doing' of science requires much imagination" (2016, viii).

The current way of doing things has negative epistemic effects.[2] Most obviously, imagination and creativity are required to solve scientific problems. Insofar as solving problems is part of scientific progress (Shan 2019, 2022), numbing imagination is regressive for science. Equally significantly, it can reduce the diversity of voices in science by driving away those who strongly value imagination. This is counterproductive because science requires diversity if it wants to make any claim to objectivity. Losing people who are both imaginative and marginalized is worse, because it will often be the imagination of marginalized perspectives that is most valuable in solving difficult problems.

To better understand the problem, in the next section, we shift into the first person and present the story of Hannah Sargeant, one of the co-authors of this paper. Dr. Sargeant is a planetary scientist who works on space instrumentation. The story of any scientist will be winding and personal. Still, the specific relationship between Dr. Sargeant and her imagination reflects the typical experience of many scientists. We present it here to focus the discussion, transition into a more careful exegesis, and motivate some positive proposals. Dr. Sargeant will not focus on her experiences with marginalization but, more generally, lay out the evolving relationship with imagination over a career.

15.3 A "Typical" Relationship with Imagination

15.3.1 Reflections of a Space Scientist

In my early years of school education, when I didn't fully grasp what science was, I was primarily interested in understanding how everything and anything works. This curiosity and sense of discovery are generally nurtured in UK primary education. One other potentially contributing factor to this way of learning is the fact that

[2] There are also negative *ethical* effects, e.g., by introducing unnecessary suffering into the life of scientists. While no less important, we will focus on the negative epistemic effects.

primary education teachers often lack confidence when teaching scientific theories and facts. However, there is a lot of themed working with the overlap of subjects in primary education which means that the humanities and the arts are integrated into other subjects, including science. This way of learning can nurture a sense of curiosity and imagination in students. Imagination is also nurtured when discussing the types of jobs that people have. The idea of becoming an astronaut and being among the stars is celebrated, and you are encouraged to imagine what such an experience would be like. However, as I progressed through to high school, and especially with examinations becoming a more prominent feature in school life, the role of imagination in science classes became diminished. In high school, I learned equations and theories with extremely prescribed and constrained problems to apply them to. Experimental work had a focus on developing lab skills and repeating results to confirm known theories. We would never carry out experiments without expecting a certain result.

Another key point is that imagination is never mentioned or knowingly encouraged in the scientific process. For a long time, I thought that curiosity and imagination were for the arts alone. As someone who was not successful in the arts, I, therefore, believed that I did not have imagination.

As I continued on my path to understanding how the physical world operates, I began my physics undergraduate degree. Here we learned more complex theories and formulas with trickier questions to apply them to. At this point, we were being taught by professional scientists; however, the reality of a professional scientist was not communicated to us, and the problems we were solving were still very prescribed with known solutions. Meanwhile, unbeknownst to us, our teachers were conducting research projects and using imagination to enable scientific discovery and find novel solutions to problems. As an undergraduate, I conducted two relatively small research projects, which were daunting at the time as I would always need to figure out where to start. This knocked my confidence and made me feel like I didn't have what it takes to be a "real" scientist. Solving novel problems was a slow and difficult process, and I often ended up at a dead end with no direction. At this stage, I would ask a professor for help, but that would involve them offering up a solution, and the penny still hadn't dropped that it was imagination that I needed to help me to solve the problem myself. If someone had asked me how I used imagination, I would have said that I didn't. There were no discussions about imagination as part of the scientific process.

After completing my physics degree, I trained as a high school teacher. For the first time, I was asked to make observations and record my opinions. As a "scientist," I thought that opinions were useless and that only facts were relevant. I was left frustrated, losing marks because I didn't develop my own opinions; I wasn't thinking for myself. Looking back, I see how naïve I was and that my education to date had simply been the "rinse and repeat" of applying scientific theories to problems. Imagination and critical thinking had been omitted and devalued.

It was only once I began my masters in Space Exploration Systems that I began to really develop these critical scientific skills. The reality of the scientific process was slowly revealing itself. I was beginning to learn about some of the most cutting-edge technologies and research areas, which importantly also meant I was hearing "we don't know" in response to my questions. I was finally realising what a scientist does; they sit on the edge of the known and unknown, and they are not just solving problems in which the answer already exists, which was my experience as a scientist-in-training. As part of my studies, we were beginning to look at applications of the scientific theories we had been taught, even theoretically. The problems we were solving were also a lot more complex and open-ended. For example, we completed a large group project looking at the design of the infrastructure for the development of a human-tended Mars base. We could apply the theories of astrodynamics and rocket engine design to the development of a series of space craft and routes and launch windows to achieve this goal. There was no single correct answer, so we were interested in thinking outside the box and innovating to create the most efficient and affordable solution. Instead of dealing with constrained problems (often described as "spherical chicken in a vacuum" problems, where many simplifications/assumptions are made such that the problem is completely unrealistic), we were considering many more variables and real-life scenarios. We had a lot more independence and were forced to think imaginatively. This was challenging without any specific training, in the sense that it was mostly just trial and error and gaining confidence.

The transition to my Ph.D. studies involved a more significant shift to real-world problems and working on actual mission payload design and operation. The pressure was, therefore, significantly higher than anything I'd worked on before. The goal of a Ph.D. is to make a novel contribution to knowledge, and so most of the time, I was trying to solve problems that had never been looked at. Imagination was critical to my Ph.D. studies as I sought the right tools to solve the problems at hand. Sometimes this meant looking at similar problems and combining the methods that others had used, or looking at problems from a different perspective and trying to understand the physical processes occurring at the micro and macro level to improve our understanding of the problem. Generally, the problem-solving process involved a lot of failure. For example, I wanted to perform a reaction in a closed system and use the change in gas pressure to measure reaction rate. However, there appeared to be a loss of gas without any identifiable leaks. It was only when I considered the behaviour of gas molecules in a low-pressure system did I realise that the molecules could be condensing, and so, not in the gas phase at all. To get to this point, I needed the scientific understanding from my undergraduate training and the practical problem-solving expertise of lab mates and supervisors to rule out other causes. After my Ph.D. I worked as a postdoctoral researcher, and at first, I wanted to work on prescribed problems, as I still didn't feel imaginative enough to come up with new research ideas on my own. As I worked on different problems, new ideas would slowly come. The generation of ideas is accelerated with every new skill I learn, as it is in the overlap of skills/expertise where new approaches to problems are identified. With a couple of years of postdoctoral experience, I am no longer intimidated by the idea of generating my own research themes; in fact, I prefer it to conducting the research plans of others. I finally feel confident in the use of imagination in my work. Interestingly, only after simply discussing with the lead author of this work about the use of imagination in science did I realise how important it is. I now consciously consider how I can utilise imagination in my work and how we can integrate this into scientific education.

15.3.2 Imagined Careers

There are several important points to note about this evolving relationship with imagination. First, imagination is celebrated in early education until the introduction of big standardized tests for entering university all but dissipate that enthusiasm, which is damaged further at the university level by competition and assignments meant to "weed out" certain kinds of student. From the student's perspective, interacting with professors is difficult and often embarrassing, and such interactions do not typically encourage more imagination. Finally, it is nearly impossible to guess what a life of science amounts to until at least reaching graduate school. Most of these story beats will ring true to anyone with an undergraduate science training, especially those not from privileged backgrounds.

Luckily, Dr. Sargeant was able to persist and eventually re-ignited some of her earlier curiosity and imagination. Combined with hearing "we don't know" more and more often in graduate school, she began to use her imagination more and more to find, and then solve, cutting-edge problems. This kind of work became exciting and fulfilling, but the feeling that her imagination was lacking persisted into postdoctoral research when it became crucial to come up with new projects of her own. Dr. Sargeant now feels more confident in the power of her imagination.

One final thing to notice is how the relationship with scientific imagination changes over a typical scientist's career. At first, students imagine what science is about and what it would be like to do science, and they ask imaginative questions about the world. Then, imagination gets sidelined in favour of developing skills necessary to pass standardized tests, solve problem sets, reproduce prescribed experimental results, and find solutions to problems that are "small" enough to solve on one's own. Then, as the projects get bigger and more cutting edge, there is a shift to imagining *together*. Undergraduate education is currently better at weeding out imaginative people than it is at preparing students to use their imagination in finding solutions to open-ended cutting-edge problems, or to imagine together with others.

15.4 Improving Imagination Education

How might educators support the development of scientific imagination? One very basic but crucial step is to address the fact that students come to believe (typically during high-school and undergraduate education) that imagination is not essential for the practice of science. Students do not know what the daily work of a scientist is like, and therefore do not know what role imagination plays, if any, in that work.

To remedy this, we suggest that at the critical juncture when science education pivots from inspiring students to a focus on theory and technique, students must be given a chance to see, first-hand, if possible, what the daily practice of science is like, where that practice includes the use of imagination. Through exposure to this kind of experience, students can recognize the crucial role of imagination for professionals in their art, which helps keep them motivated. Allowing students to appreciate working scientists using imagination in their daily work would enable students to reconceptualize the coming years of theory and technique for what they are: necessary steps on the path to something almost completely different. Professional scientists frequently return to their techniques and theory, but with respect to the role of imagination in their work, their professional life is entirely different from their life as a student. Students of music or art understand this, even at an early stage, because they have exemplars to engage with almost any time they please, via their teachers or online videos. They know that imagination will be valued, because they see and hear it exercised by professionals in their discipline. Students of science typically do not have such exemplars. And this can cause imaginative individuals to lose motivation insofar as they are prevented from recognizing that they are in a discipline that will (eventually) value their imagination, even if, at the moment, their experience in the classroom suggests otherwise.

In sum, educators should do their best to destigmatize discussions of imagination in science at all stages. The typical trajectory of imagination should be explained to students interested in pursuing science, and students should be made aware that their imagination will eventually be valued. Also, more precise language should be developed and employed to discuss the imagination, including the different kinds of imagination and ways that it can be applied. For example, it might be helpful for students to learn that imagination as a tendency to fantasize or think in a completely unrestrained way is different from imagination as an ability to produce pursuit-worthy hypotheses.

One way to achieve these goals is by designing and providing classes specifically on the imagination at the undergraduate level aimed at science students, as trialed successfully by Chiodo et al. (2020) for engineering students at the Politecnico di Milano. Among other things, this course required students to reflect on the role of imagination in science, and it was highly valued by students (for a different kind of initiative, see Brown 2020). It might be best for these courses to involve philosophers of science at both the planning and implementation stages (Green et al. 2021; Jaksland 2021; de Regt and Koster 2021; Lusk 2022).

However, it might not always be possible or desirable to dedicate an entire course to scientific imagination. And, for some students, such a course might come too early or too late. Still, it should be possible to have a discussion about imagination in science at any stage.

What else can be done? We now present some original data and analysis drawing from interviews that one of us (Stuart) performed with scientists at the PI level over the past 6 years. Specifically, the following list of strategies has been extracted, where each item was suggested in the context of asking how to help struggling students "imagine better," where the definition of "better" was left up to the scientist. Many of these are recognizable as general pedagogical principles, which would be useful in many different contexts but may or may not speak to inclusivity. As we will see below, a number of general themes underlie these suggestions that we can explore in the context of educating scientific imagination in an inclusive way.

1. Use open-ended prompts. These might include metaphors, thought experiments, visualizations, etc.
2. Increase student self-confidence.
3. Provide good role models.
4. Give students responsibility.
5. Encourage a critical attitude.
6. Encourage discussion, both inside and outside of the lab.
7. Don't be dogmatic.
8. Avoid competitive environments.
9. Encourage everyone to admit what they don't know.

These suggestions fit well with a virtue theoretic account of scientific imagination (Stuart 2022a). In general, virtue theory focuses on the properties of a person as a whole, as opposed to their individual actions or the consequences of their actions. Roughly, virtues are character traits that make someone excellent. These might be traits given to the individual by genetic lottery, like good eyesight, or they might be traits that the person worked hard to develop, like humility, courage, and sensitivity to evidence. A good imagination could be either kind of virtue, so we won't discuss this distinction further. (For a discussion of imagination that separates it into two kinds which correspond to these two kinds of virtue, see Stuart 2019b, and for a more general argument that cognitive processes and virtue theory can be connected in this way, see Ohlhorst 2022).

Virtue theory claims that a good scientist possesses the correct overall balance of traits. For example, a good scientist is neither too skeptical nor too open-minded. This is the right way to be, even if many of actions turn out to be "incorrect" in the sense that they had suboptimal consequences. What is essential for our purposes is that when scientists talk about imagination in pedagogical contexts, they naturally adopt the language of virtue theory (Stuart 2022a), and this seems appropriate given that virtue theory has always had a close connection to education, and yields implementable strategies (Bezuidenhout 2017; Nersessian 2022; Orona et al. 2023), while it can be difficult to see how we might implement competing frameworks capable of defining scientific "goodness" (like deontology and consequentialism) in an educational context (Stuart 2022a).

Applying virtue theory to scientific imagination captures items (1–9) in the above list. Good teachers recognize that virtues cannot be taught but *can be learned*. For example, there are no words an educator can say or actions they can perform that would *make* someone brave. Still, if a person wants to become brave, they can internalize words of wisdom and emulate bravery seen in others to develop their bravery, and in this way, become brave. Equally, we cannot *make* someone a good imaginer with uttered words or demonstrative actions. But, for students who want to be imaginative, we can *set them tasks that require imagination* (this encompasses the above list items 1, 4, 7), we can point to (or be) *good role models* (list items 3, 5, 6, 7, 9), and we can *support the virtues that enable imaginativeness* (items 1, 2, 4, 5, 6, 7, 8, 9).

We will discuss all three of these in some detail, paying attention to whether and how they speak to the issue of inclusivity.

15.4.1 Prompting Imagination with Tools

Lev Semenovich Vygotsky convinced a generation of scholars about the importance of imagination for education, (Vygotsky 1967; for an important precursor see Ribot 1906). Much of the literature we see today focuses on the imagination (very loosely defined) of children (Caiman and Lundegård 2018; Fleer 2015; Egan 1990, 1998, 2005; Bascandziev and Harris 2020; Skolnick Weisberg 2020). In Vygotsky's work, we can see the influence of a Kantian sense of imagination as a faculty required for all meaningful thought, but also the Lockean-Humean sense of imagination as the ability to combine ideas together. Vygotsky noted how imagination changes and develops: at first, a child has only a little experience to draw on, which is a limiting factor. But as they accumulate experiences, they have more to draw on, and their imaginings can grow in complexity. At first, they cannot control their imaginings very well, but as they develop, they learn to do this. Finally, young children have difficulty separating what is imagined from what is believed to be true, but they eventually learn to hold their imaginings at arm's length. Vygotsky also drew attention to the fact that imagination is closely connected with play, language, and thought, and that the nature of this connection also changes over time.

The insights that Vygotsky's work inspired can be extended beyond children and adolescents to the university context. For example, as sophistication with language increases, students speak less about play and more about how they *would* play if they were to play (Gajdamaschko 2005). This coheres nicely with the view common in philosophy of science that models, metaphors, thought experiments, and theories can be thought of as sophisticated games of make-believe, which are not so much played as discussed (Friend 2020; Frigg and Nguyen 2020). It also coheres with the view in science education studies that students shift in their imagination from playing a justificatory role to a merely illustrative or communicative role as they progress into higher education (Özdemir 2009). But rather than portraying models and metaphors as acts of play or games of make-believe, Vygotsky portrayed these devices as *tools* that could assist imagination (along with language, algebra, art, diagrams, maps, blueprints, and so on; see Leont'ev 1997, 22). These tools can empower the imagination, and we should use them to help our students. Diagrams, thought experiments, narratives, models, and so on, work (when they do, see

Stephens and Clement 2012; Bascandziev and Harris 2020; Bancong and Song 2020; McCrudden et al. 2011; Coll 2006) because they require students to exert their imaginative effort. Insofar as the imagination is exercised and practiced, and insofar as feedback is possible, imagination can be trained and improved via these means (Egan 1990, 1998, 2005; Kind 2020; Stuart 2018, 2022b; Hadzigeorgiou 2016).

In sum, one option for improving imagination is carefully using tools that require open-ended imagining. Might this address the lack of inclusivity? Perhaps not. It is possible to develop better tools for marginalized groups, but it is unclear whether or how this could work in practice (Kauffman et al. 2022). For example, there might not be any specific metaphors, thought experiments, or diagrams that work well for only one marginalized group or intersection of groups. And defending the specialized use of tools directed toward certain groups might require making essentializing assumptions about the cognitive, social, cultural, or physical properties of members of those groups, as such assumptions are usually empirically, morally, and politically unjustified given inter-group variation. Of course, we should continue to develop the best tools for imagination education possible and all useful tools should be made available to everyone. However, we do not think the use of educational tools for increasing the power of imagination can on its own address inclusivity issues.

15.4.2 Role Models

As mentioned above, one way to increase the skill of scientific imagination (thought of as a scientific virtue) is by exposure to good role models. For someone who wants to develop some virtue (like courage, open-mindedness, or compassion), exposure to someone who is courageous, open-minded, or compassionate shows them how such a person exercises that virtue in practice. The student copies the behaviour of the role model by acting in similar ways in similar situations until that virtue becomes internalized in themselves such that they know when and how to apply it in general. The same thing can work with imagination. Commonsensically, students trained by a PI that is more vocal about imagination will feel more comfortable talking about and using imagination than students who learn under a PI who is less vocal about imagination.

One problem for developing the imagination is that many students do not have exposure to the right kind of role models until graduate school. They might see other students as role model students and professors as role model teachers, but they do not get to see role model scientists who are actually *doing* science. And this is especially true for *imagination* role models as the imagination of a professional scientist is almost impossible to witness and appreciate without being a professional scientist oneself, because until this point it is rare to encounter other scientists who are actually imagining.

Where might good imagination role models come from? It has been proposed that we can increase student exposure to good role models by telling stories about exemplary scientists (Eijck and Roth 2012). This suggestion has some potential. However, it is easy for a science student, especially one who is marginalized and not fully confident, to read stories about people like Einstein, Feynman, Darwin, etc., and feel that they are not (and will not be) like them. This is at least partially because the kinds of heroes we typically find represented in such stories are Anglo-American or European men from privileged backgrounds.

An exciting number of resources are now available that tell the stories of scientists who represent a more diverse set of people, like chemists Angie Turner King, Dorothy Hodgkin, and Marie Daly, mathematicians Katherine Johnson and Maryam Mirzakhani, astronaut Mae Jemison, astrophysicist Jocelyn Bell Burnell, and many others (see e.g., Bolden 2020; Johnson 2019; Moss 2020; Johnson 2020; Brown 2011; Ignotofsky 2016; Prescod-Weinstein 2021). Exposure to more representative role models would be useful for everyone. We hope these will be added to science curricula and eventually form part of the public understanding of science. However, storytelling still might not be the best way to expose students to good role models, mainly because we think that such exposure should be "in person" as much as possible, as direct contact with role models gives students the ability to receive real-time feedback on their imaginings.

One way to give students in-person access to role models is through joint research projects with professional scientists (see, e.g., Chng and Mei 2020; Nersessian and Newstetter 2014; Richter and Paretti 2009). This practice can provide students with role models that are different from those they merely read about or professors whose daily work is inaccessible to them. Insofar as undergraduates come to be exposed to the uncertainties inherent in cutting-edge scientific work, this kind of exercise can help to show students *that* imagination can be important in the work of science. But this will only be effective to the extent that students (a) get to witness professional scientists in moments of uncertainty who resolve that uncertainty using their imagination and make this clear to the student, and (b) learn that such imaginative problem-solving occasions are common in science. With these conditions fulfilled, this kind of problem-based participatory learning could be a useful strategy. Ideally, the role model should already be educated on how to talk about imagination.

In summary, role models have the potential to ameliorate the inclusivity issues highlighted above, at least partially. Students from marginalized backgrounds could learn earlier in their careers to see themselves as insiders, not outsiders, and learn that imagination matters for doing science. However, storytelling and joint research projects can make things worse by knocking down marginalized students' confidence further, especially when the role models are not selected inclusively. It might take years to overcome and reverse the effects that negative experiences with working scientists can have. And these kinds of initiatives might take place at the wrong time: storytelling typically comes early in education (to spur students to try science), but it is not there to help students who have become disenchanted by the competitive, rote, test-based work in upper secondary and lower undergraduate years. Joint research projects typically take place late in undergraduate education, by which point science education has already alienated many potential scientists.

15.4.3 Supporting Virtues

Over the past 15 years, Nancy Nersessian and her colleagues have performed ethnographic research on several different labs (tissue engineering, neural engineering, computational systems biology, and experimental systems biology). One of the outcomes of this work is a new award-winning educational program that focuses on producing virtuous biomedical scientists (see e.g., Nersessian and Newstetter 2014; Nersessian 2022). Interdisciplinary researchers require several virtues, including cognitive flexibility (to understand problems from multiple perspectives), methodological versatility (to be competent with several different methods), resilience (to deal with constant failure and uncertainty), interactional expertise (to navigate across disciplinary boundaries), and epistemic awareness (to appreciate the norms and values operating in different fields) (Nersessian 2022, 296–298). We have argued that a good imagination is also a virtue that scientists should possess. This virtue is among the most fundamental, as it is helpful or necessary for many of the other virtues that scientists should possess, including the ones just mentioned.

Imaginativeness (like all other virtues) should aim to find a "golden mean" between two extremes. That is, between, on the one extreme, having your head always in the clouds, and on the other, being stuck inside a box. Different PIs have different strategies for helping students find the right balance. Many embrace a negative strategy: "I often just leave students on their own. I'm always very happy to talk to them…but I don't like so much to tell students what to do. Because I think, too much of that stifles their imagination" (interview, 22/05/2019). Another PI in a different field agreed: "I lead them with problems to solve. And I don't try to solve the problem for them" (interview, 08/12/2018). This agrees with a fundamental assumption of most accounts of virtue theory: virtues must be developed through the effort of the student, not only the teacher. To hone the virtue of imaginativeness, PIs might have various (if implicit) strategies that include using open-ended prompts and tools (as discussed in Sect. 4.1) or being a role model for the students (as mentioned in Sect. 4.2). But there is another essential thing that educators can do, and this is to develop the virtues that support imaginativeness, which enable students to perform the actions required to develop imagination.

To do this, educators can discourage competition in the lab, increase a skeptical attitude toward experimental results and theoretical dogma, and foster a supportive environment in which people can feel safe to fail and admit that they don't know something, and to joke around. These kinds of changes support scientific virtues like courageousness, kindness, honesty, resilience, and curiosity.[3] We think there is one virtue that is required for many of these, and that is self-confidence (sometimes called intellectual courage). When asked about imagination, the words of marginalized scientists unequivocally point toward a felt lack of self-confidence. Research on the effects of marginalization shows why this is to be expected. Someone is marginalized when they are on the margins of a culture, social group, or power structure (Billson 1988). Whether this is explicit to them or not, marginalized people tend to have less control over the determining factors of their lives, fewer available resources, and fewer opportunities. This naturally leads to lower confidence and self-esteem (Burton and Kagan 2003). Always being perceived as "other" or different, and always wondering about tokenism, bias, and stereotyping "is suffocating" (Nugent et al. 2016). It can produce an amplified and exclusive focus on one's negative traits or failures, as well as a greater felt need to justify oneself as belonging.

In the case of science education, low self-confidence has at least two negative consequences for the development of imagination. First, it can make a student feel that insofar as imagination is required in science, they are not going to be someone who will be very imaginative, which might also make them less prone to try. As Lin et al. (2015) show, higher extroversion correlates with using imagination to come up with new ideas, and higher motivation correlates with an increased ability to use imagination to draw connections between concepts and achieve goals. The combination of motivation and a safe social climate together strongly predicts whether imagination will be used at all. As marginalization is an aspect of social climate that can reduce extroversion and motivation, the development of imagination will suffer in social climates that include marginalizing factors (which is, unfortunately, most social climates). Second, a marginalized student will feel a greater need to prove that they are a good scientist who belongs to the social group as much as their colleagues do. Insofar as their image of a good scientist does not include using imagination but instead things like mathematical and technical competence, they will downplay any reliance on the imagination in favour of mathematical and technical skills (Stuart 2019a, c).

Finally, confidence typically drops when entering a new career stage, e.g., from undergraduate to graduate school or graduate school to postdoctoral research. And this drop will disproportionately affect marginalized students. Confidence typically rises when accumulating experience in solving problems. But this will be less pronounced for marginalized students. How can we raise confidence in marginalized students overall? Common suggestions include increasing the representation of

[3] We thank Elisabeth Hildt for prompting us to think more about curiosity, especially in light of the discussion in Sect. 3.1. There is precious little work in philosophy of science on curiosity (Inan 2017; Inan et al. 2018; Papastephanou 2019; Miščević 2020). The way we see the connection between creativity, imagination, and curiosity is as follows. Taking up a virtue-theoretic stance, creativity can be understood as the disposition to use (and be motivated to use) imagination to generate new and valuable ideas and to see those ideas through (this is inspired by, but distinct from, the account presented in Hills and Bird 2019). That motivation to find new ideas and see them through is the effect of curiosity, which might be portrayed as a desire for understanding (Miščević 2020).

marginalized people among faculties and in syllabi, changing funding structures that keep marginalized people out of academia, eliminating targeted surveillance of marginalized students, addressing inequalities resulting from school districting, increasing access to the internet, and improving teacher preparation. Reasons have been given to support these measures elsewhere, which we will not rehearse here. Insofar as they are effective, they would reduce marginalization and the negative effects that marginalization has on self-confidence.

15.5 Conclusion

Imagination is necessary for science, and a good imagination makes good science better. Current trends in scientific pedagogy and elsewhere conspire to hide the role of imagination in science from students, and also to make the education of imagination less inclusive than it should be.

In this chapter, we have tried to provide a map of the changing relationship scientists have with their imaginations over the course of a career, and canvassed several ways to improve the current situation. These include: providing courses on imagination to scientists that could give students a better idea of the role of imagination in the daily work of professional scientists, and new language to discuss scientific imagination; using open-ended tools that require students to use their imaginations, including thought experiments, models, and metaphors; exposing students to role models, whether through storytelling or (better) in person via joint research initiatives; and finally, supporting the virtues that are needed to have confidence in one's ability to use imagination, by discouraging competition, encouraging healthy skepticism, and fostering an environment where people feel safe to fail and play. Perhaps the most critical positive suggestion, which we also think is relatively easy to put into practice, is setting up an exemplar-based learning system where students get to occasionally see the daily work of scientists, and the use of imagination in particular, which is important insofar as it would help imaginative students stay motivated by understanding that the theory and technique they are learning are actually stepping stones to a future time at which their imagination will be highly valued.

There are many more topics we could discuss regarding imagination and scientific education. We will mention just three. Almost every scientific discipline is now thoroughly interdisciplinary. Are there special challenges when teaching students how to imagine in an interdisciplinary way? Second, there is a close connection between (certain kinds of) imagination and emotion (Arcangeli 2017). Emotion is also essential in science (Kozlov 2023), though it is not as well discussed as it should be in the education literature (Weil 2002). Can this connection be used in a positive way to assist in the development of scientific imagination in students? Finally, scientists are now using tools like artificial intelligence to help them extend the power of their imagination. How might this change what is required from the imagination of future scientists, and how can we prepare them for this in advance?

We think there are bound to be connections between all three issues and the inclusivity of the education of scientific imagination. Thus, there is still very much to be done. We hope the imaginations of scholars will be up to the task.

References

Anderson, Gary, and Kathryn Herr. 2007. Anti-racist teaching. In *Encyclopedia of activism and social justice*, 173–175. Thousand Oaks: Sage. https://doi.org/10.4135/9781412956215.

Arámbula-Greenfield, Teresa. 1995. Teaching science within a feminist pedagogical framework. *Feminist Teacher* 9 (3): 110–115.

Arcangeli, Margherita. 2017. *Supposition and the imaginative realm: A philosophical inquiry*. London: Routledge. https://www.routledge.com/Supposition-and-the-Imaginative-Realm-A-Philosophical-Inquiry/Arcangeli/p/book/9781138223042.

Bancong, Hartono, and Jinwoong Song. 2020, June. Exploring how students construct collaborative thought experiments during physics problem-solving activities. *Science & Education*. https://doi.org/10.1007/s11191-020-00129-3.

Barton, Angela Calabrese. 1997. Liberatory science education: Weaving connections between feminist theory and science education. *Curriculum Inquiry* 27 (2): 141–163.

———. 1998. *Feminist science education*. 1998th ed. New York: Teachers College Press.

Bascandziev, Igor, and Paul L. Harris. 2020. Can children benefit from thought experiments? In *The scientific imagination*, ed. Arnon Levy and Peter Godfrey-Smith, 262–279. Oxford University Press. https://oxford.universitypressscholarship.com/view/10.1093/oso/9780190212308.001.0001/oso-9780190212308-chapter-12.

Bezuidenhout, Louise. 2017. The relational responsibilities of scientists: (Re) considering science as a practice. *Research Ethics* 13 (2): 65.

Billson, Janet. 1988. No owner of soil: The concept of marginality revisited on its sixtieth birthday. *International Review of Modern Sociology* 18 (2): 183–204.

Bohm, David, and David Peat. 1987. *Science, order and creativity*. London: Routledge.

Bolden, Tonya. 2020. *Changing the equation: 50+ US Black women in STEM*. Abrams. https://www.slj.com/review/changing-the-equation-50-us-black-women-in-stem.

Brown, Jeannette. 2011. *African American women chemists*. Illustrated ed. New York: Oxford University Press.

Brown, Matthew. 2020. *Science and moral imagination*. University of Pittsburgh Press. https://upittpress.org/books/9780822946267/.

Burton, Mark, and Carolyn Kagan. 2003. Community psychology: Why this gap in Britain. *History and Philosophy of Psychology* 4 (January): 10–23.

Caiman, Cecilia, and Iann Lundegård. 2018. Young children's imagination in science education and education for sustainability. *Cultural Studies of Science Education* 13 (3): 687–705. https://doi.org/10.1007/s11422-017-9811-7.

Chiodo, Simona, V. Schiaffonati, M. Arcangeli, and E. Bernadó-Mansilla. 2020. Beyond creativity: Reflection and practice of imagination with students of technical universities. In *2020 IEEE* Global Engineering Education Conference (EDUCON), 1225–1229. https://doi.org/10.1109/EDUCON45650.2020.9125158.

Chng, Huang Hoon, and Wu Siew Mei. 2020. Undergraduate research in the university curriculum: An institutional perspective. In *International perspectives on undergraduate research: Policy and practice*, ed. Nancy H. Hensel and Patrick Blessinger, 57–77. Cham: Springer. https://doi.org/10.1007/978-3-030-53559-9_4.

Coll, Richard K. 2006. The role of models, mental models and analogies in chemistry teaching. In *Metaphor and analogy in science education*, Science & Technology Education Library, ed. Peter J. Aubusson, Allan G. Harrison, and Stephen M. Ritchie, 65–77. Dordrecht: Springer. https://doi.org/10.1007/1-4020-3830-5_6.

Currie, Gregory, Professor of Philosophy Gregory Currie, and I. Ravenscroft. 2002. *Recreative minds: Imagination in philosophy and psychology*. Oxford: Clarendon Press.

de Regt, Henk W., and Edwin Koster. 2021. The hows and whys of philosophy of science teaching: A comparative analysis. *European Journal for Philosophy of Science* 11 (4): 106. https://doi.org/10.1007/s13194-021-00421-1.

Dennick, Reg. 1992. *Multicultural and antiracist science education: Theory and practice*. University of Nottingham.

Dewey, John. 1980. *Art as experience*. New York: Perigee Books. http://archive.org/details/deweyjohnartasanexperience.

Egan, Kieran. 1990. *Romantic understanding: The development of rationality and imagination, ages 8–15*. New York: Routledge.

———. 1998. *The educated mind: How cognitive tools shape our understanding*. Chicago: University of Chicago Press. https://press.uchicago.edu/ucp/books/book/chicago/E/bo5542156.html.

———. 2005. *An imaginative approach to teaching*. San Francisco: Jossey-Bass, An Imprint of Wiley.

Egan, Kieran, and Krystina Madej. 2010. *Engaging imagination and developing creativity in education*. Newcastle upon Tyne: Cambridge Scholars.

Egan, Kieran, Gillian Judson, and Krystina Madej. 2015. *Engaging imagination and developing creativity in education*. Newcastle upon Tyne: Cambridge Scholars.

Feyerabend, Paul. 1975. *Against method: Outline of an anarchistic theory of knowledge*. London/New York: Verso.

Fleer, Marilyn. 2015. Imagination and its contributions to learning in science. In *A cultural-historical study of children learning science: Foregrounding affective imagination in play-based settings*, Cultural studies of science education, ed. Marilyn Fleer and Niklas Pramling, 39–57. Dordrecht: Springer. https://doi.org/10.1007/978-94-017-9370-4_3.

Friend, Stacie. 2020. The fictional character of scientific models. In *The scientific imagination*, ed. Arnon Levy and Peter Godfrey-Smith, 102–127. Oxford: Oxford University Press. https://oxford.universitypressscholarship.com/view/10.1093/oso/9780190212308.001.0001/oso-9780190212308-chapter-5.

Frigg, Roman, and James Nguyen. 2020. *Modelling nature: An opinionated introduction to scientific representation*, Synthese library. Springer. https://doi.org/10.1007/978-3-030-45153-0.

Gajdamaschko, Natalia. 2005. Vygotsky on imagination: Why an understanding of the imagination is an important issue for school teachers. *Teaching Education* 16 (1): 13–22. https://doi.org/10.1080/1047621052000341581.

Gaut, Berys. 2003. Creativity and imagination. In *The creation of art: New essays in philosophical aesthetics*, ed. Berys Gaut and Paisley Livingston, 148–173. Cambridge: Cambridge University Press. https://risweb.st-andrews.ac.uk/portal/en/researchoutput/creativity-and-imagination(0769616d-8027-4b94-8473-315ecc2bb954).html.

Gill, Dawn, and L. Levidow, eds. 1987. *Anti-racist science teaching*. London: Free Association Books.

Green, Sara, Hanne Andersen, Kristian Danielsen, Claus Emmeche, Christian Joas, Mikkel Willum Johansen, Caio Nagayoshi, Joeri Witteveen, and Henrik Kragh Sørensen. 2021. Adapting practice-based philosophy of science to teaching of science students. *European Journal for Philosophy of Science* 11 (3): 75. https://doi.org/10.1007/s13194-021-00393-2.

Greene, Maxine. 2000. *Releasing the imagination: Essays on education, the arts, and social change*. 1st ed. San Francisco: Jossey-Bass.

Hadzigeorgiou, Yannis. 2016. *Imaginative science education: The central role of imagination in science education*. Cham: Springer.

Hatt, Blaine E., ed. 2022. *Crushing ICE: Short-on-theory, long-on-practical approaches to imagination creativity education*. Altona: Friesen Press.

Hills, Alison, and Alexander Bird. 2019. Against creativity. *Philosophy and Phenomenological Research* 99 (3): 694–713. https://doi.org/10.1111/phpr.12511.

Hines, S. Maxwell, ed. 2007. *Multicultural science education: Theory, practice, and promise*. Peter Lang. https://www.peterlang.com/document/1103911.

Hussénius, Anita. 2014. Science education for all, some or just a few? Feminist and gender perspectives on science education: A special issue. *Cultural Studies of Science Education* 9 (2): 255–262. https://doi.org/10.1007/s11422-013-9561-0.

Hussénius, Anita, Kristina Andersson, Annica Gullberg, and Kathryn Scantlebury. 2013. Ignoring half the sky: A feminist critique of science education's knowledge society. In *Science education for diversity: Theory and practice*, Cultural studies of science education, ed. Nasser Mansour and Rupert Wegerif, vol. 8, 301–315. Dordrecht: Springer. http://urn.kb.se/resolve?urn=urn:nbn:se:hig:diva-9969.

Ignotofsky, Rachel. 2016. *Women in science: 50 fearless pioneers who changed the world*. NO-VALUE ed. New York: Ten Speed Press.

Inan, Ilhan. 2017. *The philosophy of curiosity*. 1st ed. New York/London: Routledge.

Inan, Ilhan, Lani Watson, and Dennis Whitcomb, eds. 2018. *The moral psychology of curiosity*, Moral psychology of the emotions. Vol. 8. Illustrated ed. London/New York: Rowman & Littlefield Publishers.

Jaksland, Rasmus. 2021. Teaching scientific creativity through philosophy of science. *European Journal for Philosophy of Science* 11 (4): 110. https://doi.org/10.1007/s13194-021-00427-9.

Johnson, Katherine. 2019. *Reaching for the moon: The autobiography of NASA mathematician Katherine Johnson*. S&S Atheneum. https://www.slj.com/review/reaching-for-the-moon-the-autobiography-of-nasa-mathematician-katherine-johnson.

Johnson, Jen Cullerton. 2020. *The story of environmentalist Wangari Maathai*. Reprint ed. New York: Lee & Low Books.

Judson, Gillian. 2015. *Engaging imagination in ecological education: Practical strategies for teaching*. Vancouver: Pacific Educational Press. ISBN: 978-1-926966-75-5.

Kauffman, James M., Dimitris Anastasiou, Garry Hornby, Joao Lopes, Mack D. Burke, Marion Felder, Bernd Ahrbeck, and Andrew Wiley. 2022. Imagining and reimagining the future of special and inclusive education. *Education Sciences* 12 (12): 903. https://doi.org/10.3390/educsci12120903.

Kind, Amy. 2020. The skill of imagination. In *Routledge handbook of skill and expertise*, ed. Ellen Fridland and Carlotta Pavese, 335–346. London: Routledge.

———. 2022. *Imagination and creative thinking. Elements in philosophy of mind*. Cambridge: Cambridge University Press. https://doi.org/10.1017/9781108973335.

Kind, Amy, and Peter Kung, eds. 2016. *Knowledge through imagination*. Oxford/New York: Oxford University Press.

Kishimoto, Kyoko. 2018. Anti-racist pedagogy: From faculty's self-reflection to organizing within and beyond the classroom. *Race Ethnicity and Education* 21 (4): 540–554. https://doi.org/10.1080/13613324.2016.1248824.

Kozlov, Anatolii. 2023. Emotions in scientific practice. *Interdisciplinary Science Reviews* 48 (2): 329–48. https://doi.org/10.1080/03080188.2023.2193073.

Law, Ian. 2017. Building the anti-racist university, action and new agendas. *Race Ethnicity and Education* 20 (3): 332–343. https://doi.org/10.1080/13613324.2016.1260232.

Leont'ev, A.N. 1997. On Vygotsky's creative development. In *The collected works of L. S. Vygotsky: Problems of the theory and history of psychology*, Cognition and language, ed. Robert W. Rieber and Jeffrey Wollock, 9–32. Boston: Springer. https://doi.org/10.1007/978-1-4615-5893-4_2.

Lin, Ju-Sen, Wen-Shan Chang, and Chaoyun Liang. 2015. The imagination constructs with science students: Interplay among social climate, intrinsic motivation, and personality traits. *Imagination, Cognition and Personality* 34 (4): 340–359. https://doi.org/10.1177/0276236615572589.

Lusk, Greg. 2022. Is HPS a valuable component of a STEM education? An empirical study of student interest in HPS courses within an undergraduate science curriculum. *European Journal for Philosophy of Science* 12 (1): 19. https://doi.org/10.1007/s13194-021-00433-x.

McCrudden, Matthew T., Joseph P. Magliano, and Gregory Schraw. 2011. The effect of diagrams on online reading processes and memory. *Discourse Processes* 48 (2): 69–92. https://doi.org/10.1080/01638531003694561.

Midgley, Mary. 1992. *Science as salvation: A modern myth and its meaning*. 1st ed. London: Routledge.

Miščević, Nenad. 2020. *Curiosity as an epistemic virtue*, Palgrave innovations in philosophy. Cham: Springer. https://link.springer.com/book/10.1007/978-3-030-57103-0.

Moss, Caroline. 2020. *Work it, girl: Mae Jemison*. Quarto/Frances Lincoln. https://www.slj.com/review/work-it-girl-mae-jemison.

Murdoch, Iris. 1994. *Metaphysics as a guide to morals*. Reprint ed. London: Penguin Books.

———. 2001. *The sovereignty of good*. 2nd ed. London: Routledge.

Nersessian, Nancy J. 2022. *Interdisciplinarity in the making: Models and methods in frontier science*. Cambridge, MA: MIT Press. https://doi.org/10.7551/mitpress/14667.001.0001.

Nersessian, Nancy J., and Wendy C. Newstetter. 2014. Interdisciplinarity in engineering research and learning. In *Cambridge handbook of engineering education research*, ed. Aditya Johri and Barbara M. Olds, 713–730. Cambridge: Cambridge University Press. https://doi.org/10.1017/CBO9781139013451.043.

Nichols, Shaun. 2006. *The architecture of the imagination: New essays on pretence, possibility, and fiction*. Oxford: Oxford University Press.

Nugent, Julie S., Alixandra Pollack, and Dnika J. Travis. 2016. *The day-to-day experiences of workplace inclusion and exclusion*. New York: Catalyst. https://www.catalyst.org/research/the-day-to-day-experiences-of-workplace-inclusion-and-exclusion/.

Nussbaum, Martha C. 1998. *Cultivating humanity: A classical defense of reform in liberal education*. Cambridge, MA: Harvard University Press.

Ohlhorst, Jakob. 2022. Dual processes, dual virtues. *Philosophical Studies* 179 (7): 2237–2257. https://doi.org/10.1007/s11098-021-01761-7.

Orona, Gabe Avakian, Duncan Pritchard, Richard Arum, Jacqueline Eccles, Quoc-Viet Dang, David Copp, Daniel Alexander Herrmann, Bruce Rushing, and Steffen Zitzmann. 2023. Epistemic virtue in higher education: Testing the mechanisms of intellectual character development. *Current Psychology*, July. https://doi.org/10.1007/s12144-023-05005-1.

Özdemir, Ömer Faruk. 2009. Avoidance from thought experiments: Fear of misconception. *International Journal of Science Education* 31: 1–20.

Papastephanou, Marianna. 2019. *Toward new philosophical explorations of the epistemic desire to know: Just curious about curiosity*. 1st ed. Cambridge: Cambridge Scholars Publishing.

Prescod-Weinstein, Chanda. 2021. *The disordered cosmos: A journey into dark matter, spacetime, and dreams deferred*. 1st ed. New York: Bold Type Books.

Ribot, Théodule. 1906. *Essay on the creative imagination*. Translated by Albert H. N. Baron. London: Open Court. https://www.gutenberg.org/ebooks/26430/pg26430-images.html.utf8.

Richter, David M., and Marie C. Paretti. 2009. Identifying barriers to and outcomes of interdisciplinarity in the engineering classroom. *European Journal of Engineering Education* 34 (1): 29–45. https://doi.org/10.1080/03043790802710185.

Rorty, Amelie. 2009. Educating the practical imagination: A prolegomena. In *The Oxford handbook of philosophy of education*, ed. Harvey Siegel. https://doi.org/10.1093/oxfordhb/9780195312881.003.0012.

Sánchez-Dorado, Julia. 2020. Novel & worthy: Creativity as a thick epistemic concept. *European Journal for Philosophy of Science* 10 (3): 1–23. https://doi.org/10.1007/s13194-020-00303-y.

Schickore, Jutta, and Nora Hangel. 2019. 'It might be this, it should be that…' uncertainty and doubt in day-to-day research practice. *European Journal for Philosophy of Science* 9 (2): 31. https://doi.org/10.1007/s13194-019-0253-9.

Shan, Yafeng. 2019. A new functional approach to scientific progress. *Philosophy of Science* 86 (4): 739–758. https://doi.org/10.1086/704980.

———, ed. 2022. *New philosophical perspectives on scientific progress |Yafeng Shan|*. New York: Routledge. https://doi.org/10.4324/9781003165859.

Skolnick Weisberg, Deena. 2020. Is imagination constrained enough for science? In *The scientific imagination*, ed. Arnon Levy and Peter Godfrey-Smith, 250–261. Oxford University Press. https://oxford.universitypressscholarship.com/view/10.1093/oso/9780190212308.001.0001/oso-9780190212308-chapter-11.

Stephens, Lynn A., and John J. Clement. 2012. The role of thought experiments in science and science learning. In *Second international handbook of science education*, ed. B. Fraser, K. Tobin, and C. McRobbie, 157–175. Dordrecht: Springer.

Stokes, Dustin. 2014. The role of imagination in creativity. In *The philosophy of creativity: New essays*, ed. Elliot Samuel Paul and Scott Barry Kaufman, 157–184. Oxford: Oxford University Press.

Stuart, Michael T. 2018. How thought experiments increase understanding. In *The Routledge companion to thought experiments*. Routledge. https://doi.org/10.4324/9781315175027.ch30.

———. 2019a. Ethics of scientific imagination: Who gets to use imagination in science? *The Junkyard* (blog). https://junkyardofthemind.com/blog/2019/9/27/ethics-of-scientific-imagination-who-gets-to-use-imagination-in-science

———. 2019b. Towards a dual process epistemology of imagination. *Synthese*. https://doi.org/10.1007/s11229-019-02116-w.

———. 2019c. Everyday scientific imagination: A qualitative study of the uses, norms, and pedagogy of imagination in science. *Science & Education* 28 (6–7): 711–730. https://doi.org/10.1007/s11191-019-00067-9.

———. 2020. The productive anarchy of scientific imagination. *Philosophy of Science* 87 (5): 968–978. https://doi.org/10.1086/710629.

———. 2022a, May. Scientists are epistemic consequentialists about imagination. *Philosophy of Science*: 1–22. https://doi.org/10.1017/psa.2022.31.

———. 2022b. Sharpening the tools of imagination. *Synthese* 200 (6): 451. https://doi.org/10.1007/s11229-022-03939-w.

van Eijck, Michiel, and Wolff-Michael Roth. 2012. *Imagination of science in education: From epics to novelization*. Dordrecht: Springer.

von Wright, Moira. 2021. Imagination and education. In *Oxford research encyclopedia of education*. https://doi.org/10.1093/acrefore/9780190264093.013.1487.

Vygotsky, Lev Semenovich. 1967. Imagination and creativity in childhood. *Journal of Russian & East European Psychology* 42 (1): 7–97. https://doi.org/10.1080/10610405.2004.11059210.

Weil, Simone. 2002. *Gravity and grace*. 1st ed. London/New York: Routledge.

Chapter 16
Tinkering with Technology: How Experiential Engineering Ethics Pedagogy Can Accommodate Neurodivergent Students and Expose Ableist Assumptions

Janna van Grunsven, Trijsje Franssen, Andrea Gammon, and Lavinia Marin

Abstract The guiding premise of this chapter is that we, as teachers in higher education, must consider how the content and form of our teaching can foster inclusivity through a responsiveness to neurodiverse learning styles. A narrow pedagogical focus on lectures, textual engagement, and essay-writing threatens to exclude neurodivergent students whose ways of learning and making sense of the world may not be best supported through these traditional forms of pedagogy. As we discuss in this chapter, we, as engineering ethics educators, designed and implemented a new engineering ethics exercise with which we aimed to promote inclusivity *at the levels of form and content*. At the *content* level, students were invited to critically engage with inclusivity-undermining ableist assumptions in technology development. This took shape, at the *form* level, through a hands-on 'material tinkering' workshop in which students collaboratively and creatively altered (or 'hacked') artifacts used in contexts of disability and healthcare, so as to operationalize values of inclusivity and accessibility. Our hunch was that this hands-on tinkering workshop would simultaneously encourage a meaningful way of engagement with these ethical issues and values, while also enacting a more inclusive learning environment by enriching the range of pedagogical activities and learning formats available to our students.

As we aim to show in this chapter, we believe this hunch largely panned out – though there are clear areas for future improvement pertaining to the pilot exercise itself and the research we conducted on the exercise. We begin by offering a description of our tinkering exercise. We discuss the exercise's source of inspiration (Sect. 16.2.1) and its implementation (Sect. 16.2.2), which is visually captured via

This research was supported by the 4TU Centre for Engineering Education, grant number: TBM_ERE_2021_02_4TU.CEE.TUD

J. van Grunsven (✉) · T. Franssen · A. Gammon · L. Marin
Delft University of Technology, Delft, Netherlands
e-mail: J.B.vanGrunsven@tudelft.nl

E. Hildt et al. (eds.), *Building Inclusive Ethical Cultures in STEM*, The International Library of Ethics, Law and Technology 42,
https://doi.org/10.1007/978-3-031-51560-6_16

photographic documentation. We then discuss (Sect. 16.3) how we utilized a triangulated research method to assess the pedagogical value of the exercise. After we discuss our findings, we conclude by identifying areas for future improvement (Sect. 16.4).

Keywords Neurodiversity · Ableism · Engineering ethics education · Tinkering · Inclusivity

16.1 Introduction

The Ethics and Philosophy of Technology department at Delft University of Technology has a long history of teaching engineering ethics to its large body of engineering students. This history reflects a commitment to combining traditional pedagogical approaches (lecturing, reading ethical theory, writing essays) with less traditional exercises that call on students to engage with ethical issues in a more embodied interactive manner (Doorn and Kroesen 2013; Van Grunsven et al. 2021). Our guiding assumption has been that active learning through interactive embodied exercises, such as role-play, makes the ethical issues at stake in engineering contexts more experiential to engineering students and that this helps foster important ethical competencies such as moral sensitivity, imagination, and reflection.

While examining this assumption in a 4-year research project (link to the project), we have become increasingly preoccupied with the idea that the embedding of non-traditional embodied interactive exercises is warranted not only from a pedagogical perspective but also from a perspective of social justice and inclusivity. On a conservative estimate, 10% of TU Delft's student population studies with a disability. Since this number is based on students who self-identify and voluntarily report as disabled, it may, in fact, be more reasonable to assume that up to 30% of students in higher education study with a disability.[1] With ADHD, autism, and dyslexia making up a large portion of these disabilities, this means that our student body is emphatically neurodiverse. Indeed, "Evidence … shows that in engineering degrees neurodiverse students are overrepresented" (Saunders-Smits and van den Bogaard 2019). Neurodiversity refers to the idea that people experience, understand, and interact with the world in many different ways and that those differences ought to be *valued* rather than labeled as *deficient* deviations from an assumed norm of typicality (Cf. Van Grunsven 2020).[2]

[1] See Expertisecentrum Handicap + Studie (Dutch Expertise Centre on Studying with a Disability) https://www.ecio.nl/wp-content/uploads/sites/2/2019/09/70jaarhandicapstudie-min.pdf, p.35 Accessed April 20th 2022.

[2] See Chapman (2020) for a discussion of why the concept of 'neurodiversity' is best understood as a 'moving target'.

Like many other universities across the globe, TU Delft has signed the United Nations Convention on the Rights of Persons with Disabilities, thereby committing itself to actively valuing neurodiversity and promoting an inclusive learning environment. However, such an explicit commitment is, of course, just a first step in the realization of more inclusive, equitable education. Among other things, we as teachers in higher education must consider how the content and form of our teaching can foster inclusivity by being responsive to neurodiverse learning styles. The question is how we can develop "pedagogy … that addresses multiple ways of thinking?" (The National Association for Multicultural Education 2021). How can we "make education maximally accessible," providing "different ways for students to gain knowledge and formulate what they know" (Shmulsky et al. 2021b)?

A narrow pedagogical focus on lectures, textual engagement, and essay-writing threatens to exclude neurodivergent students whose ways of learning and making sense of the world may not be best supported through these traditional forms of pedagogy (Gardner 2000; Armstrong 2009). Research suggests that many dyslexic and a significant portion of autistic students are more likely to thrive in educational settings that encourage the use of visual–spatial talents (Cf. Davis 1997; Grandin 2009, 2023) Similarly, many students with ADHD would seem to benefit from pedagogy that requires "creative divergent thinking" which is "the ability to generate multiple ideas or solutions to a problem;" for instance identifying unexpected new uses for everyday use objects (White and Shah 2006). Against this backdrop, we designed and implemented a new engineering ethics exercise with which we aimed to promote accessibility and inclusivity *at the levels of form and content.*[3] At the *content* level, students were invited to critically engage with inclusivity-undermining ableist assumptions in technology development. This took shape, at the *formal* level, through a hands-on 'material tinkering' workshop in which students collaboratively and creatively altered (or 'hacked') artifacts used in contexts of disability and healthcare, so as to operationalize values of inclusivity and accessibility. Our hunch was that this hands-on tinkering workshop would simultaneously encourage a meaningful way of engagement with these ethical issues and values, while also enacting a more inclusive learning environment, enriching the range of pedagogical activities and learning formats available to our students.

As we aim to show in this chapter, we believe this hunch largely panned out – though there are clear areas for future improvement pertaining to the pilot exercise itself and the research we conducted on the exercise. We begin by offering a description of our tinkering exercise. We discuss the exercise's source of inspiration (Sect.

[3] There are discussions within critical disability studies and crip technoscience about the difference between and the limits of the concepts of 'inclusivity' and 'accessibility.' Both concepts, it is argued, can have assimilatory undertones. Though we believe these conceptual disputes matter (and can have practical implications), it is beyond the scope of this chapter to delve into them here. For our current purposes, we treat inclusivity and accessibility as broadly the same and we understand them as concepts that capture a need to recognize, value and accommodate the various forms of diversity among human beings. For a critical discussion of the concept 'accessibility' see (Hamraie and Fritsch 2019). For a discussion of the difference between the concepts of 'inclusivity' and 'accessibility' see (Van Grunsven and IJsselsteijn 2023).

16.2.1) and its implementation (Sect. 16.2.2), which is visually captured via photographic documentation. We then discuss (Sect. 16.3) how we utilized a triangulated research method to assess the pedagogical value of the exercise. Drawing on data gathered through (Sect. 16.1) a survey, (Sect. 16.2) open-ended interviews, and (Sect. 16.3) ethnographic and instructor observations, two of the questions that we aimed to answer through triangulation were:

Research question 1 [RQ1]: Did our collaborative tinkering exercise offer an alternative form of engineering ethics pedagogy, capable of contributing to a more inclusive learning environment?

Research question 2 [RQ2]: Did our collaborative tinkering exercise stimulate moral sensitivity regarding issues of ableism, inclusivity, and accessibility in contexts of technology development?[4]

After we discuss our findings, which largely affirmed RQ1 and partially confirmed RQ2, we conclude by identifying areas for future improvement (Sect. 16.4).

16.2 The Exercise

16.2.1 Inspiration Behind the Exercise

The source of inspiration behind the tinkering exercise was a TED talk by artist and disability rights activist Sue Austin. In the talk entitled "Deep sea diving …. in a Wheelchair," Austin powerfully captures her multi-layered experience of becoming a wheelchair user, or, as she prefers, a 'powerchair' user. Austin recollects how, on the one hand, the chair was instantly empowering, a source of joy. After an extended period of illness, the artifact expanded Austin's access to the world in a spatial and bodily sense, allowing her to race down the streets and feel the wind blowing through her hair. However, in a social sense, she felt instantly excluded, as others seemed to see her primarily in terms of loss and deficiency. To challenge this image, reclaim her visibility in social space, and articulate the empowering joy-providing experiences that her wheelchair had brought her, Austin began transforming the artifact into a deep-sea diving device, making video recordings of herself floating along the ocean's corals. As she explains in her TED talk, when people watch her videos, they are "seeing an object they have no frame of reference for" such that "they have to think in a completely new way."

> For me, this means that they are seeing *the value of difference*, the joy it brings, when instead of focusing on loss or limitation, we see and discover the power and joy of seeing the world from exciting new perspectives. For me the wheelchair becomes a vehicle of transformation. … Because nobody's seen or heard of an underwater wheelchair before …

[4]Another question that we raised, and that we discuss in a different paper that is currently under review, is to whatextent and in what ways the exercise enlivened the moral imagination of our students.

> creating this spectacle is about creating new ways of seeing, being and knowing. (Austin 2012, our italics)

A Similar attempt to alter pernicious yet commonplace ways of seeing disabled people is offered by non-speaking autistic blogger Mel Baggs. In their widely viewed short video "In My Language,"[5] Baggs challenges viewers' assumptions about the idiosyncratic ways in which many autistic people behave and engage with their environment. Where these behaviors and engagements are often dismissed as pathological and problematic, Baggs invites us to see them as deeply communicative and meaningful. In doing so, they confront us with the pervasiveness of ableism. As many disability rights activists and scholars have shown, ableism is a pernicious value-system that gets materialized into the world through a wide range of technological artifacts (Shew 2020). From lecterns expressing norms about stature and traffic signals timed for fast-moving pedestrians to communication devices designed in accordance with neurotypical communication norms, the world is built with a certain kind of body-mind in mind (Hendren 2020; Van Grunsven and Roeser 2022). At the same time, Austin's artwork wagers that it is also through the tweaking, tinkering with, and disrupting of technological artifacts that entrenched ableist ways of seeing and imagining disabled people can be called into question.

At the theoretical level, this idea is emphatically put forth in the field of *Crip Technoscience* (Hamraie and Fritsch 2019). Crip Technoscience situates itself as an emancipatory alternative to what it calls "disability technoscience." Disability technoscience frames the lives of disabled people as marked by loss and deficiency, which ought to then be overcome via technology. Operating from within this perspective, well-intending (usually 'able-bodied') engineers tend to view themselves as self-proclaimed problem-solvers offering technological interventions to disabled people, who, in turn, are framed as the passive non-agential recipients of (allegedly much-needed) support (Hamraie and Fritsch 2019; Shew 2020). Crip Technoscience resists this perspective on disabled people and their relation to technology. It draws attention to the numerous ways in which disabled people have always actively hacked and tinkered with their material-technological environment, claiming access to the world as skilled, knowledgeable agents of world-making, instead of waiting to be invited as the mere beneficiaries of technological assistance.

Viewing disabled people as world-making agents of change and as crucial experience experts is also part of the *Warm Technology* framework. This perspective has recently emerged in the field of human-computer interaction, applied within the context of Alzheimer's disease (IJsselsteijn et al. 2020). Much like Sue Austin and the representatives of Crip Technoscience, Warm Technologists resist the typical emphasis on deficiency and loss that so often guides engineering projects in healthcare technology: "Warm Technology is born from an emancipatory view of living with dementia. It is to de-emphasize disease and deficiency, and instead focus on the unique identity of the person, on the myriad of ways in which the person inhabits their world as a place of familiarity"(Van Grunsven and IJsselstein 2023). It aims to

[5] https://www.youtube.com/watch?v=JnylM1hI2jc.

design technology capable of "affirming old age – enabling people to remain open and attached to the world and to other people" (IJsselstein et al. 2020, 33).

These emancipatory perspectives on disability, technology, and inclusivity were offered to the students in a variety of formats. Following the inclusive principles of universal design for learning, students had access to video materials, written academic articles, personal testimonials, as well as in-person and pre-recorded subtitled video lectures.[6] This set up the theoretical backdrop against which our tinkering workshop took place.

16.2.2 Implementation

Austin's work exposed us, as engineering ethics educators, to the possibility that hands-on tinkering with artifacts can stimulate critical reflection on ableist biases, opening up an experiential engagement with the ways in which ethical values such as inclusivity and accessibility can be promoted (or thwarted) through material design choices. This prompted us to develop our hands-on tinkering workshop, during which students would work together in small groups to transform artifacts used in disability, illness, and rehabilitation contexts. These transformations had to be value-oriented. That is, students would have to consider how concrete material changes to the artifact could expose ableist assumptions and/or improve (or undermine) the values of accessibility and inclusivity for relevant stakeholders.

We received funding from our department to purchase scrap materials used for tinkering (see Image 16.1) and 15 artifacts, including a tricycle walker, a dressing stick, a foldable walking cane, hearing aids, and a picture memory phone designed for people with dementia (see Image 16.2). We should note that students were not restricted to these purchased artifacts. Using a suggestion from *Student Onbeperkt* – TU Delft's student-run organization for students with a disability – we

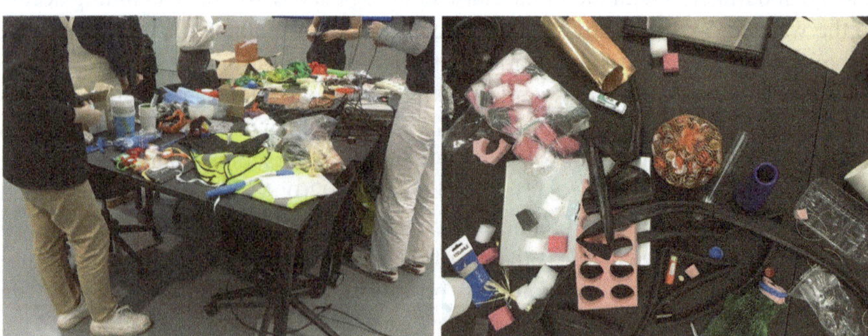

Image 16.1 An overview of tinkering material

[6] https://udlguidelines.cast.org/.

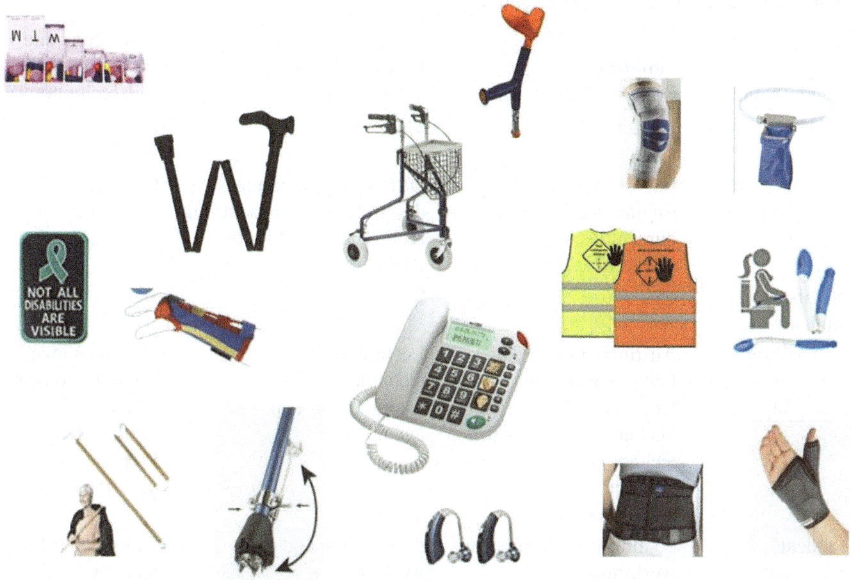

Image 16.2 An overview of the artifacts used in the exercise

gave students the option to select a personal artifact, thereby aligning ourselves with the view "that students and their life histories and experiences should be placed at the center of the teaching and learning process" (The National Association for Multicultural Education). Interestingly, none of the approximately 200 students who participated in the exercise opted for a personal artifact. We revisit this point in Sect. 16.4.1.

In collaboration with several other members of our department, we embedded the exercise in three different courses in the fall semester of 2021.[7] These courses were: the ethics of healthcare technologies, introduction to responsible innovation, and philosophy and history of science and technology. While they differ in a number of respects (see Table 16.1), these courses share an emphasis on ethical issues in engineering and design contexts. Some of the instructors involved in these courses explicitly presented the workshop to the students as an effort to value neurodiversity in our pedagogy. Such explication can signal to students who identify as neurodiverse that "they belong," while normalizing the idea that learning and knowledge-acquisition takes on many different shapes (Shmulsky et al. 2021a). As a form of consciousness-raising, the latter can be particularly important for students who do

[7] During the early developmental phase of the exercise, Samantha Copeland made significant contributions. In ordering the artifacts and tinkering materials as well as teaching the exercise, Cristina Richie was involved.

Table 16.1 Description of the three courses

Course name	Introduction to Responsible Innovation	Ethics of Healthcare Technologies	Philosophy and History of Science and Technology
Level	Bachelor (Second year)	Master	Bachelor (Second year)
Student background	Interdisciplinary: Engineering, design, humanities & (social) sciences)	Different engineering & design backgrounds (strong representation of biomedical engineering)	Clinical technology
Number of students	65	24	105
Preparation before the workshop	Austin's video Lecture & literature on Crip Technoscience & warm technology Artifact selection Preparatory questions	Lecture on crip technologies; Literature on crip technology	Austin's video Literature on Crip technology Artifact selection Preparatory questions
Student presentations on the workshop	10 min, 1 week after the workshop	5 min at the end of workshop	5 min at the end of workshop

not yet recognize their own mind as neurodivergent, but who would benefit from such acknowledgment.[8]

Within each of these three courses, a three-hour workshop took place, during which students worked in groups of 4–6 to tinker with their chosen artifact.[9] In the two BSc courses, the groups already selected their artifact several weeks prior to the workshop. This gave them time to brainstorm and to meet the requirement of consulting relevant stakeholders by reading testimonials on blogs, talking to friends and family who might count as a stakeholder, watching documentaries, etc. The extent to which students utilized this opportunity differed widely between groups, but many groups did little stakeholder research beforehand. We discuss this further, and why it is particularly problematic in the context of this exercise, in Sects. 16.3.2 and 16.4.1.

The workshop was divided into two 'rounds.' **Round one**: After examining their selected artifact by touching it, walking around it, discussing it, and in some instances using it in different indoor and outdoor settings – each student group started with a first 'redesign' or iteration of the artifact. A walking cane, a hearing device for children, and a stoma were aestheticized, transforming the 'medical look' of these devices into a more eye-pleasing one; a picture memory phone was visually simplified and enriched with tactile elements; a walker was motorized to facilitate

[8] See also https://www.nicole-brown.co.uk/invisible-disabilities-academia/.

[9] The workshop held in the ethics of health care technologies course was restricted to 90 min due to scheduling constraints.

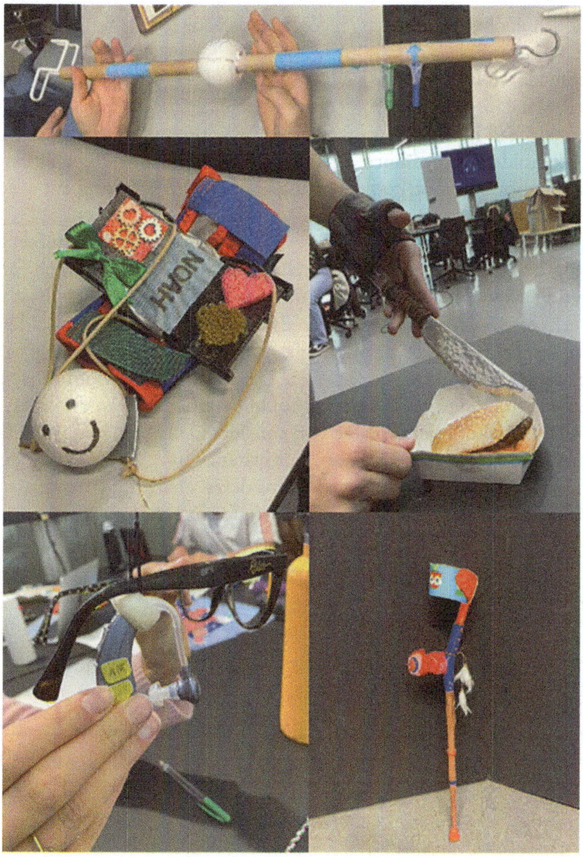

Image 16.3 sample of artifacts after second iteration

up-hill mobility, etc. This process took 45 min and was photographically recorded by the students (Image 16.3).

After this initial design round, student groups were paired up and asked to observe and constructively comment on each other's redesign. To encourage targeted feedback, a selection of constructive questions was provided to the BSc students in advance (see Image 16.4).

Round 2 In the second round, students made another iteration of the same artifact, taking the other group's feedback into account. At the end, they took additional pictures of their artifact, so that later iterations could be compared and reflected upon in ensuing in-class presentations.

During the workshop, the instructors walked around to observe and ask open questions to stimulate discussion and creativity. To avoid steering students in a certain direction and to leave sufficient space for them to come up with their own ideas, we deliberately chose not to provide feedback that would have included concrete

Feedback questions/recommendations (pick one)

1) Have you considered incorporating another value in your design as well, namely value X, for instance by making change Y?
2) Have you considered how your artefact could be improved to accommodate a very particular stakeholder (for instance, a parent who may have young children, a teenager who cares a lot about looking cool, an aging person who is feeling cut off from their social environment). Here is one way you could maybe do that ________
3) Have you thought about adding an additional functionality to the artefact (such as a snow-plowing or deep see diving functionality to a wheelchair)? Which functionality would you add and what purpose would this serve (for instance, would it serve a practical purpose? Perhaps a practical service catered towards a specific stakeholder? Or would it serve a more critical reflection-promoting purpose, encouraging people to reflect on their own ableist assumptions about the artefact? Or would it serve both those purposes at once?). We were thinking functionality X could be interesting to consider, because ___
4) In line with Crip Technoscience, have you considered changing your artefact so that it encourages critical reflection on ableism, for instance by......
5) In line with Warm Technology, have you considered changing your artefact so that it includes several of its five elements (see below), for instance by .....

Image 16.4 Critical feedback questions students were asked to use to propose design recommendations to one other group

suggestions or solutions. The interviews conducted after the workshop indicated that students appreciated this open hands-off approach.

After the workshop, students were asked to reflect on their product and the tinkering process itself in the form of a final presentation and ensuing Q&A.

16.3 Assessing the Tinkering Exercise Through Triangulation

As stated in the introduction, two of the research questions motivating our tinkering exercise were:

RQ1: Did our collaborative tinkering exercise offer an alternative form of engineering ethics pedagogy capable of contributing to a more inclusive learning environment?

RQ2: Did our collaborative tinkering exercise stimulate moral sensitivity regarding issues of ableism, inclusivity, and accessibility in contexts of technology development?

In order to answer these questions, we opted for a triangulated method. Triangulation combines different sources of information or methods to gain a more comprehensive understanding of the subject under investigation (Jick 1979). Commonly used in mixed-methods research, the general idea behind triangulation is an epistemological one: as Moran-Ellis et al. (2006) remark, triangulation amounts to a claim about "what more can be known"(47) by combining methods in particular ways. Turner and Turner (2009) observe that triangulation is often employed "when the field of study is difficult, demanding or contentious" (171) because it can complement findings with additional data or analysis, provide supplemental corroborating evidence, and even challenge or test findings. It is frequently used in education research (Cohen et al. 2007; Altricher et al. 2005) as well as in other fields (e.g., nursing), where single methods for assessing interventions fall short. In our case, we adopt a version of triangulation to gain multiple (student, observer, and teacher) perspectives on the same activity to assess the activity based on how it achieves the educational goals referred to in our research questions. Our triangulation scheme used data collected through: (1) a participant survey (of students); (2) semi-structured interviews (with students); and (3) ethnographic observations (conducted by Marin and Franssen 2022), combined with reflections from the instructors involved in the course.[10] Two of these instructors were also involved in the research (Franssen & Van Grunsven). One was disconnected from the research project (Cristina Richie). This version of triangulation does not try to validate or confirm findings but instead puts these complementary data sources together to evaluate a pilot exercise.

Immediately after each workshop was completed, students received an email with a link to an anonymous survey which consisted of 10 questions.[11] Out of a total of 194 students taking part in the exercise, 54 of them voluntarily filled in the survey. By soliciting responses from a swath of students, the participant survey provided an important initial data source with which we gauged general student response to the exercise and learned from a large number of students what specifically they experienced as valuable.

To delve deeper behind the initial findings provided by the survey, we conducted semi-structured one-on-one interviews (30–40 min) with three participants from the workshops. During these interviews, we asked the students to reflect upon their experience during the workshop, and for suggestions on how to improve the exercise.[12] The interviewer was not involved in teaching the interviewees' course. Thus there was no possibility that the students' answers would be influenced by extraneous factors such as grading concerns.

In this project, both the participant survey and the in-depth interviews are prone to the same potential sampling bias in that the group of students reflected in both groups were self-selecting. For this reason, including further information in the form of ethnographic observations and teacher insights is essential. The

[10] The ethnographic reports can be found in the 4TU.Data repository, DOI https://doi.org/10.4121/20115983.

[11] A significant proportion of these students were from the Introduction to Responsible Innovation course. Survey answers can be found in the 4TU.Data repository, DOI https://doi.org/10.4121/20115971.

[12] Interviews are uploaded in the 4TU.Data repository and are available with a restricted license, upon request: Franssen, Trijsje (2022): Interviews about the educational exercise tinkering with technology. 4TU.ResearchData. Dataset. https://doi.org/10.4121/20020154.v1.

combination of student feedback and teacher responses/evaluation is a typical way of assessing education. We also included ethnographic observations to provide an additional source of insight into student learning and to answer our research questions. While observing, the ethnographers were not involved in teaching that class, enabling a detailed and more detached perspective on how the exercise unfolded. The ethnographic reports reflect the researcher's point of view, how they experienced the educational setting and what struck them.

Below, we will discuss how each of the methods we used interacted with each other to shed light on our research questions (Image 16.5).

16.3.1 A Triangulated Answer to RQ1

RQ1: Did our collaborative tinkering exercise offer an alternative form of engineering ethics pedagogy capable of contributing to a more inclusive learning environment?

The question on the survey that addresses RQ1 most directly is the one displayed below in Image 16.5. In this question, students were asked to rank four aspects of the workshop "from the most valuable aspect of the workshop to the least valuable" These aspects were articulated in the following four statements:

Note that options 1 and 3 either highlight or reference the *collaborative* dimension of the exercise. The other two statements emphasize the exercise's non-traditional hands-on form, with statement 4 explicitly contrasting this form with more traditional pedagogical learning formats. To be sure, many students expressed appreciation for the opportunity to collaborate with others during the workshop, both anecdotally as well as in the in-depth interviews:

Interviewee 1: if you're on your own .. doing something like this, like, having an artifact and fool around with something like that. I don't think it's going to work. ... if you have people that had these ideas and you can come up with your

#	Field	Minimum	Maximum	Mean	Std Deviation	Variance	Count
1	Tinkering with the artefact in a collaborative manner - deciding on changes to the artefact together	1.00	4.00	2.81	1.00	1.00	52
2	The fact that it encouraged a creative approach to the ethical dimensions of technological artefacts	1.00	4.00	2.35	1.14	1.30	52
3	The combination of creatively playing with the artefact and collaborating with my group	1.00	4.00	2.88	1.07	1.14	52
4	The fact that it provided an assignment that wasn't focused on reading or writing but that encouraged learning through a hands-on interactive exercise	1.00	4.00	1.96	1.00	1.00	52

Image 16.5 Survey Questions and Responses

own view on it, that's much more exciting than just being on your own. It's not the same. You don't get motivated. [1.14]

Interviewee 2: It was a good chance to have a good brainstorming session. That was a really nice thing to do with this project. ... a lot of ideas came from the brainstorming and discussion [Interviewee 2.14]

Interviewee 3: I did really enjoy the collaborative part. It was challenging but it was something that I enjoyed and that was really important [3.78]

In the same spirit, the instructors and ethnographic researchers who were present during the workshops all noted that many students "were clearly very engaged" (Marin and Franssen 2022) in their interactions; that "they were very vibrant during the tinkering" (Cristina Richie, via private email conversation). Particularly in the wake of several COVID-19 related lockdowns, students seemed eager to engage in in-person collaborative interaction.

This makes it all the more telling that students ultimately placed the collaborative dimension of the exercise at the bottom in terms of what made it valuable. Instead, the most frequently prioritized reason for valuing the exercise was that it "wasn't focused on reading or writing" but "encouraged learning through a hands-on interactive exercise." The second most valued aspect of the course was the exercise's creative dimension. Combined, this leads us to wager that there is a need among students for alternative, non-textual engineering ethics exercises; exercises capable of accommodating learning styles not frequently accounted for in traditional forms of pedagogy. The survey offers a first indication that this exercise was able to meet this need, with a significant majority of survey respondents seeing the tinkering workshop as "the most memorable part of the course." This was reiterated in the interviews:

Interviewee 1: ... "I think it's a really great concept of teaching. [...] Some teachers know a lot about their own subject and they can talk about it for hours and hours. But sometimes that doesn't really land to the students. Things like this really help [especially when you are introduced to a subject. It really helps] to gain interest and to find your own perspective on it. In my opinion it's better than just listening to someone." [1.64]

Interviewee 2: "It was a really good project to see how a tool is developed and how it can be further improved based on all the things you have learned." [2.60] "The project was really useful. I think it was the most fun part of the whole course." [2.84]

Interviewee 3: most useful was ... just the idea that you're doing something practical and creative with the group, it's something that we just don't have a lot of opportunities to do so it was. ... this mix between creativity and practical [3.75; 3.78; 3.80]

One might wonder whether students valued the exercise's creative, hands-on, non-traditional form because it is simply more entertaining. Perhaps it provided a reprieve from more 'genuine' educational activities, where the tinkering workshop offered time for play but lacked pedagogical value. Establishing if the exercise can appeal to different learning styles while having genuine merit as a form of engineering ethics pedagogy is key to RQ1. At first glance, the survey suggests that students were broadly divided over the following statement: "The workshop was fun but it wasn't of any added educational worth (I would have engaged with the course's concepts and theories in just the same way if the workshop would not have been embedded in the course)." When asked about this in the follow-up interviews, the interviewees described the activity as simultaneously fun, challenging, and interesting. Interviewee 1, for instance, describes the exercise as "really confronting and it's fun to do" [1.66]. Interviewee 2 recounts how the activity "was fun but challenging, from what I saw from other teams as well. [...] what was challenging .... was coming up with the ideas" [2.64]. They specified that "The interesting part was the whole process that we went through to develop the ideas and implement them and associate it with the values. And the fun part was the practical part where we got to use the tools, and try to mix them up, and develop the product, and the video's and pictures. [2.68]. Interviewee 3 described the workshop as "quite challenging … there is this reflection process of what values come out of this modification and is that really what we want and achieve with that?" [3.66].

Interestingly, most students – including some of those who attributed no additional educational value to the activity – agreed that the actual tinkering activity was integral to their grasp of the link between ethical values and technological artifacts: 85% of students who took the survey either agreed or strongly agreed that "New ideas about how our artifact should be altered emerged through the tinkering process," and 67% of students agreed or strongly agreed that "Engaging with the artifact in a hands-on way during the workshop (touching it, moving around it, altering it, looking at it from different angles) brought out new ethical considerations that I or my team hadn't reflected on prior to the workshop." The interviews underscored our hunch that an embodied interactive material exercise could provide a non-traditional format for engaging with engineering ethics issues. Firstly, the interviewees explicated how their ways of *seeing* the artifacts and of making choices about how to improve the artifact were co-determined by their bodily comportment and engagement with the materials available to them:

Interviewee 1: "what makes the workshop special in that kind of way is actually moving it around, and using it. … Feeling something and using makes it more confronting. So, you have a more specific way of looking at a certain artifact instead of just imagining it." [1.62]

Interviewee 2: "when we actually saw the tool by itself and what other tools we could use, I think the ideas just popped up way easier. Which was also a very nice thing of the project." … [Interviewee 2.20; 2.26–28]

Interviewee 3: "Just moving around with it really impacted how we made our decisions. Even seeing a group member walk around with it. I mean, you see like, 'Oh it actually doesn't make sense for it not to fold because if you sit down then where you are going to put the walking stick.' So, these kinds of considerations were really helped by moving with it around." [Interviewee 3.100]

These considerations emphasized the ethics domain and its relationship to technology. One of the questions asked during the interview was: "Do you believe that working with the artifact as you did right now in the workshop has somehow changed – what I could call – your moral sensitivity? Has it made you more sensitive to, well, the moral values that are embedded already in the artifact, and what you could change, and so on? The interviewees responded as follows:

Interviewee 1: *"Yeah, I think so. … I don't normally think about the ethical aspects of technology. You just use it and that's fine. But if you really look at it and you think about it, all the ethical issues come along. And you realize that your viewpoint is not the only viewpoint that there is. … That's when you realize that technology is not always necessarily always a solution, it can also be a problem. And that was something that was really interesting." [1.28]*

Interviewee 2: *I would say definitely yes. … we don't want to just see how a tool can help people, but we want to also see how the tools can be embedded inside the life of people to make their life easier. So, for example I didn't actually think about that before the project. I was just thinking that tools like this just to help us, but its more than that [2.60–62] … We were trying … to make the tool as ethically correct as possible. … We were trying to find a value that was missing and try to place it through the tools that we had [2.22; 2.24]*

Interviewee 3 stressed the important stage-setting work that the theories (Crip Technoscience and Warm Technology) had done: "I think learning about the theoretical background that we received before the workshop had already increased the moral sensitivity, especially when looking at an object and trying to see what kind of values are embedded into it [3.66] That said, they proceeded to add: "As valuable as the theory was, sometimes it's really hard to visualize it in practice if you don't do it yourself. So, I think there's obvious value to the exercise in doing something like this." [3.70]. A similar point about the combination of theory and practice was expressed by one of the instructors involved in the course: "the most meaningful was the workshop itself and the least "meaningful" was the literature review, although it was absolutely essential to the academic nature of the assignment" (Richie, via private email conversation)).

These are noteworthy results for engineering ethics educators, who often grapple with the challenge of getting students to engage with engineering's ethical dimensions.[13] We take these results as indicative of the tinkering exercise's value at the formal level, offering a non-traditional pedagogical format capable of (1) getting students to engage with ethical values and issues related to the use of these artifacts and (2) contributing to a more diverse learning environment that accommodates different learning styles, including those marked by visual–spatial and creative divergent thinking. We wager that embedding neurodiversity-acknowledging pedagogy

[13] We observed that this challenge was significant in a focus group conducted with engineering ethics educators from across the globe in Marin et al. (2022).

in engineering ethics education allows, at once, for pedagogy that is more socially just and effective.

16.3.2 A Triangulated Answer to RQ2

In order to arrive at an answer to RQ2, we traced how students actively appealed to and reflected upon the meaning of inclusivity, accessibility, and ableism during the activities surrounding the workshop (preparing for it, participating in it, and presenting on it). That students acquired new sensitivity towards the ethical issues at stake in this workshop was, as we just saw, agreed upon by a majority of students responding to the survey. But how and to what level of depth did this sensitivity manifest itself? Based on the interviews and ethnographic reports, the following things stood out:

Echoing some of Sue Austin's ideas, several students and groups linked the value of inclusivity to desirability, attempting to remove the stigma around a disability by making the artifact an object of desire. One such example was that of an aestheticized foldable walking stick. Crucially, and in line with Crip Technoscience and Warm Technology, the group working on the stick appealed to the desires and needs of actual users in linking the artifact's aesthetic look to the value of inclusivity. As Interviewee 3, who belonged to this group, explains:

> *"My cousin uses it ... she's quite young and she really didn't want to use it because it's associated with old age. So, that was kind of one of the main issues that I wanted to bring into the group discussion." [3.6] ...[With] the creative design ... we hoped to kind of increase this value of ... – I'm not sure – ideas of identity and ...allow them to express themselves through the walking stick. [3.14]*

In addition to consulting their cousin, interviewee 3 and the rest of their group also read user-testimonial blogs and talked to aging stakeholders they knew personally. It was in doing so that they discovered that:

> *one of the main problems why people refuse to use walking sticks is that they don't want to be considered old [...]. It's not congruent to their self-identity, they don't want to be seen as old, so the decoration part was kind of targeting that". [3.56]*

Strikingly, interviewee 3 is retrospectively critical of some of their group's design choices. Specifically, they describe how, after a suggestion made in the critical feedback stage of the workshop, the group was tempted to add a voice-controlled GPS tracker to the walking stick:

> *looking back on it, it is quite easy to see – maybe we were working with some biases about older individuals. And also, ... warm technology ... amplifies this idea that the technology must be really easy to use. And if we were targeting older individuals then maybe voice control would not have been the best way to approach a GPS function. [3.26]*

We tentatively take the interviewee's retrospective remark about biases informing their group's choices as an indication of the workshop's potential to stimulate sustained critical reflection. The ethnographic observations also noted efforts to connect biases and obstacles to inclusivity with aesthetic design choices. Observing a group who tinkered with a hearing aid by decorating it with a pink ribbon, an ethnographer noted:

> *This group was the most reflective one, going into deep discussions. They didn't modify the artefact much, but they did notice how the hearing aid has much in common with their airpods. They wondered why stigma is associated with the hearing aid and a lifestyle choice with having airpods in your ears? They wondered how to make it less stigmatising to wear a hearing aid. Since hearing aids are associated with aging people, one idea was to make it fashionable for young people to wear them. They noticed the difference in look and design ("the case looks medical" for the hearing aids). ... "the real issue is not the functionality but the stigma associated with it." But they also wondered how would the disabled people feel if everyone was wearing these?*

Another group, who tinkered with the picture memory phone, operationalized inclusivity by simplifying the design and usability of the artifact. As they stated in their final presentation:

> *We tried to reach optimal inclusivity by making the Design of the phone simple and with Easy-to-Learn functions which should enable all different kinds of people with dementia to use the phone. [...] We tried to focus on the person with dementia and their needs. People with dementia tend to be easily distracted and confused and can get a sensory overload quite quickly. So we realized they needed to have a phone that is not complex. That is why we got rid of all the unnecessary buttons. (see Image 16.6)*

The group reaffirmed this design choice, which was grounded in the Warm Technology framework, by engaging a key stakeholder of the artifact, namely a primary caretaker of people with advanced Alzheimer's disease. As the ethnographic report notes:

> *One of the most motivated groups that took the assignment very seriously; one of the students' mother worked as a nurse with people with Alzheimer. During the workshop, they called her (several times, I believe) in order to make the best design choices. Her recommendation in nutshell was to get rid of anything unnecessary.*

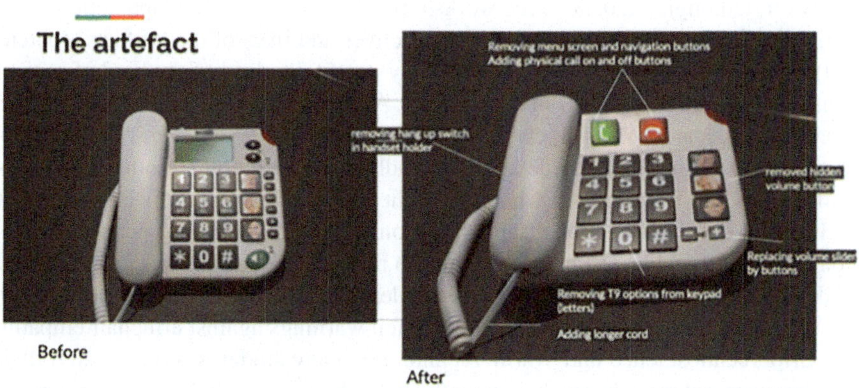

Image 16.6 Picture memory phone before and after modifications

Unfortunately, the perspective of a person with dementia was absent from the redesign process. Indeed, this was a recurring issue for many of the tinkering projects. The challenge of getting students to critically engage with the ethics of appropriate stakeholder inclusion is most apparent when we consider whether the tinkering exercise encouraged students to empathize with the potential users of their tinkering artifacts. In one sense, several interviewees noted this exercise's potential to encourage such empathy. For instance, a group who tinkered with a tricycle walker came up with a scenario in which an aging adult using the artifact would have to walk up a hill:

> *Interviewee 1: "You can have more empathy going on. You can think about it more. I guess it is more mind-opening, because you can sort of theatre what you're actually doing. I mean that was a whole kind of other way of looking at something like that than if you just wrote a report, for instance [1.6] ... "We all were imagining how that would turn out. Like how would that person go up the hill with the tricycle and what did that person need to have a more comfortable way of using it, for using the artifact. So we eventually just, we were all thinking about that and discussing what kind of scenario would that person be in, what if it was my grandma, how would she react? [1.10].*

Two interviewees suggest that this kind of empathic perspective-taking, which decenters you from your own, often taken-for-granted, point of view, could help combat biases:

Interviewee 1: *"if you're on your own for instance and you don't have the workshop, you don't have these scenarios you can think of, you don't have the way that we work together, then you stay in your own bubble and you just think that you can just do whatever is good in your view, but you don't necessarily take into account other people's view and other people's experiences." [1.12]*

Interviewee 3: "*the object there makes it really concrete what the object would be capable of or not. And you can put yourself more into the shoes of someone who would use the object" [3.62]. [...]"If you don't have an experience using these things then you also do not have the sensitivity to what actually are the necessities of the people using it." [3.100]*

However, although we believe the workshop enlivened a certain empathetic imagination for and identification with the perspectives and lives of the artifacts' (potential) users, we want to underscore that this by itself falls short of the kind of critical sensitivity we aimed to foster in our students. As discussed in Sect. 16.2.1, both Crip Technoscience and Warm Technology offer emancipatory ableism-resisting perspectives on technology and design. They adhere to the dictum "nothing about us without us," and warn against what one might call *armchair empathy* in contexts of technology development for disabled, (chronically) ill, and aging people. In this context, it is arguably just as problematic to rely upon a (well-intended) imagined understanding of the needs of one's stakeholders as it is to disregard them altogether.

Yet, despite providing students with explicit warnings against armchair empathy via Crip Technoscience and Warm Technology, many students did not catch their own engagement in such armchair empathy but only reflected on this when prompted by an instructor. Furthermore, several student groups failed to engage the end-users of the artifacts as experience experts, despite repeated reminders by the instructors of the course about the importance of doing so (and despite the fact that this was included in the grading rubric for the activity, to which students had access in advance). With respect to RQ2, we must thus critically ask to what degree the tinkering activity encouraged students to appreciate the ethical requirements and intricacies of promoting inclusivity and accessibility through technology. As Crip Technoscience warns, disability technoscience is a pervasive posture. How can the potential of the exercise de-center future engineers from their own taken-for-granted perspectives? And how can empathy with relevant stakeholders be stimulated in a manner that is critical of ableist disability technoscience tendencies? In the next and final section, we offer a few suggestions for mitigating this concern and several other possibilities for improving the exercise.

16.4 Areas for Improvement

In this final section, we discuss several areas for improving the exercise itself, as well as limitations to the research conducted about the exercise.

16.4.1 Improving the Tinkering Exercise

In light of the abovementioned worry, the primary area that needs improving concerns avoiding armchair empathy. Two key steps towards this are: (1) ensuring genuine engagement with primary stakeholders and (2) facilitating critical reflection on students' own biases and the degree to which they align with disability technoscience.

One option is to explicitly build in stakeholder engagement prior to the workshop. In the pilot, students were required to consult testimonial material (blogs, videos, personally conducted interviews). Curiously, as mentioned, many students seemed unaware of this requirement. Interviewee 1, for instance, suggested that: "you can also, for instance, make [the workshop] even more empathetic, for example. … [Adding] some way of interviewing, or whatever" [1.32]. This might be because, as interviewee 2 acknowledges, "students before the workshop don't do as much a preparation as needed, I believe" [2.56]. To ensure students engage actively in the pre-workshop requirements, such as interviewing and other forms of stakeholder engagement, we believe a separate class should be dedicated to shared reflection on the gathered testimonial material. As interviewee 3 rightfully pointed out:

> *there was maybe not enough structured time to think about these biases, and these values, and what exactly we could do with it. [3.30] Just in terms of the span of the activity. It was all in one day and we did have instructions to prepare for it, but it wasn't structured time. And I think it would've been more useful to have this initial brainstorming within a structured setting. [3.32]*

These reflections should be presented in class to deepen the student's engagement with the importance as well as the challenges of genuinely incorporating the perspectives of stakeholders. In doing so, students should be encouraged to reflect on how these testimonials challenged their own biases, mapping their findings onto disability vs. crip technoscientific and Warm Technology outlooks. These reflective presentations should be graded on the quality of the acquired material and the critical depth of their reflections.

Another route towards avoiding armchair empathy is to work solely with artifacts derived from students' own lives. Two of the interviewees suggested ideas along these lines:

Interviewee 1: *"I would personally just give the students two weeks or something to figure out what kind of artifact they want to use and just give them complete freedom of it. ... just have them to say like, "What kind of instrument/artifact do you think is lacking stuff and how can it be more useful or ethical or responsible." And I personally believe that if you give them that freedom then a lot of creativity can exist." [1.70]*

Interviewee 2: *"if we had to come up with an artifact ourselves, then I think that would actually make us investigate more. So, I think that's a good idea to also make us do our own research before coming to the workshop. [2.56] ... Perhaps, if we let students just bring their experiences on the table and try to develop an artifact through that, instead of doing it other way around, I think more interesting ideas will come to the project." [2.80]*

Although selecting their own artifact was given as an option to the students, it is possible that students opted for the purchased artifacts because it can seem like a safer choice, especially at the beginning of a course when student groups are just getting to know each other. By removing the choice altogether and building the assignment around an artifact of their own, students are less tempted by armchair empathy because the artifact either belongs to themselves or one of their direct group members. We recommend that, in much the same way as the previous option, this approach should still build in an additional class prior to the workshop, in which students use the theoretical concepts of disability vs. crip technoscience and Warm Technology to reflect on how the artifact they have selected might reflect ableist biases and support or undermine accessibility and inclusivity.

Additional areas of attention, noted in the ethnographic reports and by the interviewees, concerned the role played by the physical environment and the material artifacts themselves. In one course, the ethnographer notes that "the room was not conducive to group work – a lecture hall, they had no common table to gather the group around." Additionally, they observe that "Several artifacts seemed too small or simple to keep students busy for the entire workshop." This was echoed in the interviews, underscoring the potential to improve the exercise by asking students to select their own artifacts. One could argue that the critical epistemic, pedagogical role played by interactive, embodied, spatial engagements with the artifacts was revealed in a negative sense, when conditions for learning through the tinkering were sub-optimal.

16.4.2 Limitations of the Research

By using a triangulation method, we were able to develop a robust examination of our pilot exercise. In virtue of triangulating between sources, we were, for instance, able to see that *despite* finding the embodied practice of the tinkering exercise valuable, and despite seeing the exercise as encouraging empathic engagement with stakeholders, many students still failed to appreciate the central implications of Crip technoscience and disability studies that center disabled bodies and perspectives in design. The combination of information from student feedback from surveys and interviews, with instructor and observer analysis, yields this complex finding, enabling us to identify important steps for improving the exercise.

That said, our research method also had its limitations, two of which we will touch on. Firstly, we did not work with a control group of students who were taught the same content but did not participate in the exercise. This could be an effective additional way of confirming the exercise's pedagogical value. Secondly, a more in-depth answer to RQ1 would require that we interview students who identify as neuro-divergent. When researching the next iteration of the exercise, we intend to incorporate these areas of improvement.

16.5 Conclusion

In this chapter, we discussed a pilot engineering ethics education exercise. The exercise aimed to promote inclusivity at the levels of form and content. At the *content* level, students were invited to critically engage with inclusivity-undermining ableist assumptions in technology development. This took shape, at the *formal* level, through a hands-on 'material tinkering' workshop in which students collaboratively and creatively altered (or 'hacked') artifacts used in contexts of disability and healthcare, so as to operationalize values of inclusivity and accessibility. Our hunch was that this hands-on tinkering workshop would simultaneously encourage meaningful engagement with these ethical issues and values while also enacting a more inclusive learning environment, enriching the range of pedagogical activities and learning formats available to our students. As we showed in this chapter, this hunch largely panned out, particularly regarding RQ1. There are clear areas for improvement pertaining to RQ2, particularly concerning the worry of students engaging in mere armchair empathy. Two recommendations for mitigating this worry and two areas for improving our research on the exercise were identified. This will inform the next iteration of what we see as a promising new exercise in inclusive experiential engineering ethics education.

References

Altricher, H., P. Posch, and B. Somekh. 2005. *Teachers investigate their work: An introduction to the methods of action research*, E-book edition. London: Routledge.

Armstrong, T. 2009. *Multiple intelligences in the classroom*. Alexandria: ASCD.

Austin, S. 2012. Deep sea diving … in a wheel chair. *TED talk*. https://www.ted.com/talks/sue_austin_deep_sea_diving_in_a_wheelchair?language=en

Chapman, R. 2020. Defining neurodiversity for research and practice. In *Neurodiversity studies – A new critical paradigm*, ed. Hanna Bertilsdotter Rosqvist, Nick Chown, and Anna Stenning, 218–220. London: Routledge.

Cohen, L., L. Manion, and K. Morrison. 2007. *Research methods in education*. 6th ed. Routledge: London.

Davis, R.D. 1997. *The gift of dyslexia*. London: London Souvenir Press.

Doorn, N., and J.O. Kroesen. 2013. Using and developing role plays in teaching aimed at preparing for social responsibility. *Science and Engineering Ethics* 19 (4): 1513–1527. https://doi.org/10.1007/s11948-011-9335-6.

Franssen, T. 2022. *Interviews about the educational exercise tinkering with technology*. 4TU. Research Data. Dataset. https://doi.org/10.4121/20020154.v1.

Gardner, H.E. 2000. *Intelligence reframed: Multiple intelligences for the 21st century*. Hachette UK.

Grandin, T. 2009. How does visual thinking work in the mind of a person with autism? A personal account. *Philosophical Transactions of the Royal Society B: Biological Sciences* 364 (1522): 1437–1442.

———. 2023. Society is failing visual thinkers, and that. Hurts us all. In *New York Times, opinion, guest essay*, January 9th 2023. Accessed 9 Jan 2023.

Hamraie, A., and K. Fritsch. 2019. Crip technoscience manifesto. *Catalyst: Feminism, Theory, Technoscience* 5 (1): 1–33.

Hendren, S. 2020. *What can a body do? How we meet the built world*. New York: Riverhead Books.

IJsselsteijn, W., A. Tummers-Heemels, and R. Brankaert. 2020. Warm technology: A novel perspective on design for and with people living with dementia. In: *HCI and design in the context of dementia*, Human–computer interaction series, ed. R. Brankaert, G. Kenning. Cham: Springer. https://doi.org/10.1007/978-3-030-32835-13

Jick, T.D. 1979. Mixing qualitative and quantitative methods: Triangulation in action. *Administrative Science Quarterly* 24 (4): 602–611.

Marin, L., and T. Franssen. 2022. *Ethnographic reports of the tinkering exercise*. 4TU. Research Data. Dataset. https://doi.org/10.4121/20115983.v1.

Marin, Lavinia, Stone, Taylor, Van Grunsven, Janna 2022. *Transcripts of focus group with educators on the topic of teaching ethics to engineers*. 4TU. Research Data. Dataset. https://doi.org/10.4121/19657161.v1.

Moran-Ellis, J., V.D. Alexander, A. Cronin, M. Dickinson, J. Fielding, J. Sleney, and H. Thomas. 2006. Triangulation and integration: Processes, claims and implications. *Qualitative Research* 6 (1): L45–L59. https://doi.org/10.1177/1468794106058870.

Saunders-Smits, G.N., and M.E.D. van den Bogaard. 2019. Inclusiveness of neurodiverse students in engineering education curriculum design and delivery. In *SEFI 47th annual conference: Varietas Delectat... Complexity is the new normality, proceedings*, SEFI 47th annual conference: Varietas Delectat... Complexity is the New Normality, Proceedings, ed. B.V. Nagy, M. Murphy, H.-M. Jarvinen, and A. Kalmanpp, 2084–2095. European Society for Engineering Education (SEFI). https://doi.org/10.13140/RG.2.2.26738.45768.

Shew, A. 2020. Ableism, technoableism, and future AI. *IEEE Technology and Society Magazine* 39 (1): 40–85.

Shmulsky, S., G. Ken, and V. Steven. 2021a. Culturally relevant pedagogy for neurodiversity. *Community College Journal of Research and Practice* 46 (9): 681–685. https://doi.org/10.1080/10668926.2021.1972362.

Shmulsky, Solvegi, Ken Gobbo, Andy Donahue, and Frank Klucken. 2021b. Do Neurodivergent college students forge a disability identity? A snapshot and implications. *Journal of Postsecondary Education and Disability* 34 (1): 53–63.

The National Association for Multicultural Education. 2021. https://www.nameorg.org/definitions_of_multicultural_e.php.

Turner, P., and S. Turner. 2009. Triangulation in practice. *Virtual Reality* 13: 171–181. https://doi.org/10.1007/s10055-009-0117-2.

Van Grunsven, J.B. 2020. Perceiving 'other' minds: Autism, 4E cognition, and the idea of neurodiversity. *Journal of Consciousness Studies* 27 (7–8): 115–143.

Van Grunsven, J.B., and W.A. IJsselstein. 2023. Confronting ableism in a post-COVID world: Designing for world-familiarity through acts of defamiliarization. In *Values for a post-pandemic future*, ed. M.J. Dennis, G. Ishmaev, S. Umbrello, and J. van den Hoven. Springer. https://doi.org/10.1007/978-3-031-08424-9.

Van Grunsven, J.B., and S. Roeser. 2022. AAC technology, autism, and the empathic turn. *Social Epistemology* 36 (1): 95–110. https://doi.org/10.1080/02691728.2021.1897189.

van Grunsven, J., and W. IJsselsteijn. 2022. Confronting ableism in a post-COVID world: Designing for world-familiarity through acts of defamiliarization. In *Values for a post-pandemic future*, 185–200. Cham: Springer.

Van Grunsven, J.B., L. Marin, T.W. Stone, S. Roeser, and N. Doorn. 2021. How to teach engineering ethics?: A retrospective and prospective sketch of TU Delft's approach to engineering ethics education. *Advances in Engineering Education* 9 (4) https://advances.asee.org/how-to-teach-engineering-ethics-a-retrospective-and-prospective-sketch-of-tu-delfts-approach-to-engineering-ethics-education/.

White, H.A., and P. Shah. 2006. Uninhibited imaginations: Creativity in adults with attention-deficit/hyperactivity disorder. *Personality and Individual Differences* 40 (6): 1121–1131. https://doi.org/10.1016/j.paid.2005.11.007.

Chapter 17
At the Verge of 'Is' and 'Could Be': Storytelling as Medium to Develop Critical Ethical Skills

Marietjie Botes and Arianna Rossi

Abstract Stories about the socio-technical impact of the outcomes of science, technology, engineering, and mathematics (STEM) on the world may create a concrete sense of the future consequences they may cause. Such stories may offer students and early career researchers who (will) develop, test, and deploy such technologies insights into their effects on individuals, other living beings, the environment, and society. In this paper, we argue that this method, known as storytelling in academic literature, can be effectively used to embed ethics in higher learning curricula through its power to appeal to students' hearts and minds. Through storytelling, ethics lessons that emerged throughout history can be taught in a memorable way and raise empathetic skills, while fictional stories may allow students to identify with the characters and contemplate possible ethical outcomes of a decision in a safe manner. We provide an overview of various fictional and nonfictional forms that can be used to improve students' motivation and equip them with some innovative tools that can support ethical decision-making, including science fiction, digital storytelling, Virtual Reality, and videogames. In addition, we recommend that ethics training via storytelling should be embedded as early as possible as integrated nodes into existing STEM curricula to stimulate empathy and critical thinking skills in students and to supplement existing rule-based education.

Keywords Storytelling · Fiction for specific purposes · Ethical reflection · STEM pedagogy · Narratives · Science fiction · Scientific training

---

M. Botes (✉)
SnT Interdisciplinary Center for Security, Reliability and Trust, University of Luxembourg, Esch-sur-Alzette, Luxembourg

College of Law, University of KwaZulu Natal, Durban, South Africa
e-mail: maria.botes@uni.lu

A. Rossi
SnT Interdisciplinary Center for Security, Reliability and Trust, University of Luxembourg, Esch-sur-Alzette, Luxembourg
e-mail: arianna.rossi@uni.lu

E. Hildt et al. (eds.), *Building Inclusive Ethical Cultures in STEM*, The International Library of Ethics, Law and Technology 42,
https://doi.org/10.1007/978-3-031-51560-6_17

17.1 Introduction

> *Stories appeal to us not because they are untrue, but rather because they might be true, and therefore provide useful information or ideas in our navigation through eventual, possible, realities. If the present could be thought of as a stationary line crossing an intricate mesh of vectors shooting across time – some ancient, some new, some yet to be – then stories are a form of reality attached, as it were, to one of these passing vectors: possible pasts, possible presents, possible futures – all arrows in the epic battle for reality […]. Fiction (but also design and architecture) creates provisional realities for others to occupy. Durfee (2015, p. 11)*

17.2 Background and Purpose

Historically, three waves of ethics revival emerged after World War II that sparked the need for better ethics training for future medical practitioners, which eventually also included Science, Technology, Engineering, and Math (STEM) scientists.

Firstly, the most significant ethical concerns centered around the use of humans and non-human animals for scientific experiments. Ethical concerns about scientific studies employing non-human animals date back to the nineteenth century (Briggle and Mitcham 2012) and continued well into the twentieth century with ethical concerns about breeding sick, mutated, or transgenic animals and the creation of chimeras (Macer 1991). It was only after the trial of the Nazi physicians at Nuremberg in 1947 that experiments using human subjects were scrutinized. Therefore, the Nuremberg Code codified ten ethical principles that cover the necessity of informed consent and protective measures, and the avoidance of unnecessary suffering and risks to health and life. Many of these principles have since been integrated into international ethical guidelines, such as the World Medical Association's Declaration of Helsinki (BMJ 1996). Regardless of these guidelines, the Belmont Report had to be authorized to further address the protection and rights of human research subjects after details of the unethical longitudinal Tuskegee Syphilis Study were disclosed in 1972 (Adashi et al. 2018).

The second wave of ethical revival was marked by several unethical experiments conducted during the late nineteenth century across different scientific fields involving both humans and non-humans. Many of these experiments targeted and exploited vulnerable people, such as patients committed to mental health institutions, prisons, or children in orphanages, and were conducted without the consent of those parties. Thus, they raised questions about researchers' general awareness and approach to research ethics (Broad and Wade 1982). To do more to protect the rights of study participants, various governments established and formalized biomedical research ethics committees and transformed ethics into a discipline that mandates overseeing and approval (Hedgecoe 2009). The third ethics revival wave seemed to be driven by rapid scientific advancements and their unintended or unconsidered social consequences, such as the atomic bomb, genetic engineering, and space travel in the late 1900s (Goldstein 2017). Ethical concerns in respect of genomic research escalated to the point where the Human Genome Project (HGP 1990) launched a formal ethical, legal, and social implications (ELSI) research program which also covered areas such as synthetic biology, geoengineering, and nanotechnology (Langfelder and Juengst 1993).

This growing emphasis on ethics led the U.S. National Institutes of Health (NIH) to insist on formal ethics training for all graduate students and researchers of institutions that apply for NIH funding (NIH 1989). Other countries echoed similar needs for STEM ethics training, as discussed in the German report Proposals for Safeguarding Good Scientific Practice (DFG 1998 updated in 2013). However, as much as the need is acknowledged, formal methods of providing ethics education to STEM scientists or incorporating it in higher education curricula have not been addressed (Steneck and Bulger 2007), and the efforts to develop a standard approach in this respect remain insufficient. Consequently, a patchwork of different pedagogical tools has been used, varying between traditional forms of content delivery to case-based discussions, including pedagogical games, seminars, workshops, and one-on-one interactions (Swazey and Bird 1997).

However, if all these ethics education efforts are in place, why are ethical principles in scientific practice still being disregarded on a regular basis? Firstly, ethical principles are guidelines that still require the sound human judgment of a scientist to decide how these principles are to be interpreted and applied to particular circumstances. Sound judgment or intellectual-moral skills are not easily taught but are critical reasoning skills that must be developed and practiced daily: the goal is to "teach how, not what, to think" (Burton et al. 2018, p. 58). These capacities are rather dynamic: they must evolve in sync with the advancements of science and social values, as the ethical conundrums of today will be different from those of tomorrow, sometimes in unpredictable manners (Burton et al. 2018).

In essence, STEM scientists not only need intellect, moral virtues, and ethical knowledge to do the right thing, they also need to be equipped with ethical reasoning skills to go past a blind application of fixed sets of beliefs and mature a sense of critical inquiry (Callahan 1980).

In this article, we argue that storytelling can serve as an applied teaching approach to actively engage students and future scientists with the broader impacts of scientific practices on human and non-human beings, society, and the environment and to train them to apply ethical principles.

17.3 The Storytelling Approach

The mere imparting of knowledge, rules, principles, and methods are insufficient to prepare future scientists and practitioners with practical skills to successfully navigate uncertain situations. A 2014 report by the National Research Council and the National Academy of Engineering in the US identified good judgment as one of the most important "mechanisms" for ethical decision-making (NRC 2014). For some scientists, a return to virtue ethics, with an emphasis on practical moral reasoning, may ensure the required intellectual skills to determine the correct course of action in an uncertain situation and the moral skills to proceed with the necessary courage to act on such determinations, regardless of external pressures (Annas 2002). In this regard, the telling of stories to "illuminate fault lines, highlight oddities, and paint a picture of the past, present, and future that is both compelling and easily understandable" follows the same moral developmental methodology as virtue ethics (Suzuki et al. 2018).

Research is rife with stories. Researchers tell stories when asking organizations to fund their research projects, during the development of their research proposals, and in the articles, they write to disseminate their research results. Researchers also listen to the stories of others when they conduct interviews with research participants or study the collected information to find the stories that such data 'tell'. The story of Henrietta Lacks, as told by Rebecca Skloot, serves as an example of how storytelling can allow anyone to empathize with this patient whose aggressive cancer cells were harvested - without her consent - and developed into the immortal HeLa cell line that is still widely used in laboratories across the world today (Skloot 2010). This story showed scientists the emotional toll that unethical sample collection may have on a family and how to deal with similar circumstances. Additionally, remembering Henrietta Lacks in this way also enabled her family to come to terms with the existence of her cell line, which they perceived as a prolonging of a portion of her life after she died in 1951 (Skloot 2010).

From a neurological perspective, Hansen (2021) reminds us that reading stories strengthens brain connectivity and can therefore enhance creativity. It can also increase awareness of others' perspectives and, thereby, arguably increase empathy. Studies demonstrate how the responses in the listeners' brains correlate with the responses in the narrator's brain, and the stronger this neural alignment is, the better the communication between the listeners and the narrator (Stephens et al. 2010). Reason and Hawkins explain this connectedness as the ability of a story to move the responses felt by individual members of a group, after deeply listening to the story of others, to being part of a collective (Reason and Hawkins 1988). According to Griffin, this connection with others is based on the story becoming part of oneself, as opposed to hearing the story simply for the purposes of transmitting information (Griffin 1993).

17.4 Storytelling to Develop Empathy and Understanding

One of the powers of stories is to act as a catalyst to embody STEM ethics via its ability to create empathy, defined as the projection of one's own personality into the personality of another to understand them better (Webster 1979). The most important change brought about by stories is the paradigm shift experienced by the listener when he or she gains an understanding of the lives of others, providing an opportunity to imagine alternatives to a given situation (Neile 2009). This empathetic way of looking at another's situation is made possible by the storyteller who shares his or her subjective experiences and thinking to create a strong bond between the listener and teller based on a shared vision of what was told.

In contemporary society, where internet connectivity determines most people's social interactions and communications, the development of the capacity to empathize plays an increasingly important role. Our hyper connectedness ironically does not seem to be proportionate to an increase in global empathy. Instead, the lack of empathy is visible in widespread online crime and socially unacceptable behavior such as hate speech, cyberbullying, cyber harassment, and discrimination, the targeted spreading of false information during the recent Covid-19 pandemic, and the short-sighted behavior of the so-called "me generation" (Manney 2008). The erosion of empathy contradicts the historical relationship between empathy and technology sparked by the development of other information technologies. For instance, written language allowed the recording of individual thoughts, thereby offering insight into the reflections and worldviews of others. It was the invention of the printing press that popularized vernacular literature as a mass-medium of communication which allowed the dissemination of powerful counter-cultural and liberalizing ideas, including novels that served as the first mass entertainment medium (Davis 2004). This technological marvel of the day enabled people to gain insights into the lives and experiences of others and opened their eyes to different possibilities. Futurist Paul Saffo already voiced concerns about the destructive nature that social media had on people's ability to develop and show empathy as far back as 2005 (Saffo 2005). His concerns were founded on the habits of people to saturate their information spaces exclusively with information that reinforced their existing world views, shutting out everything that may contradict or challenge people's own perceptions. He states that shared knowledge and information are "the stuff that caused people to change their opinions and to empathize with others" (Saffo 2005). Most current STEM students have only known a world dominated by social media in which they encounter more of themselves via selfies or posts about themselves or are following or liking people with similar points of view. Social interactions of this nature do not increase empathy because they do not open them up to different belief systems and lived experiences. Therefore, we need stories - to force people into the discomfort of strange new worlds to teach them to empathize with "the other".

But why do stories harbor the ability to connect people and foster empathy when near-constant electronic connectedness fails to do so? The essence is seated in the imagination of the readers or viewers of stories which allows them to translate the story into an imagined world, as seen and experienced through the storyteller's eyes, thoughts, and feelings. This imaginary experience is given a real-life feel through sharing certain basic human needs such as recognition, respect, and freedom, which often form the focus of the conflict experienced by the character in the presented story. Stories provide readers or viewers with a portal to experience events that they would never experience otherwise (Hansen 2021), even harmful ones such as exploitative medical experimentation or cultural victimization, and from a safe distance. In her Pulitzer-prize-winning book, *Thirteen Ways of Looking at the Novel*, the author explains how a novel fosters the creation of empathetic responses in a reader through entering a relationship with the story's characters. When this relationship formation is repeated through extensive reading of many different stories, a generally empathetic personality is created in the reader (Smiley 2006). The idea is that if you regularly put yourself in the shoes of different characters, their experiences will evoke in the reader or viewer the ethical issues that these characters struggle with and, over time, will lead to the (further) development of empathy for these characters and people in similar situations.

Empathy is thus not a mere abstract construct of the mind but evolves by increasing direct contact with people and their, as well as our own, worldly experiences. For example, when reading the story of Oliver Twist, the reader's feelings of sadness are not actually based on wanting to come to the aid of a purely fictional character; instead, these feelings are stimulated by ancient instinctive and empathetic impulses fostered culturally and genetically through many generations, to care for and protect members of your tribe, your genetic offspring, or fellow human beings (Manney 2008).

Empathy is thus of critical ethical relevance. Without empathy, STEM scientists may conduct experiments and design technological systems and products without understanding the needs, wants, and fears of the people directly impacted by their research results and developments, thereby risking future ethical disasters or harms such as those seen historically and referred to in the introduction. In addition, if scientists are unable to empathize, they will also be unable to protect the interests of those who may be solely reliant on the ethical judgments of scientists at that moment, such as people suffering from adverse effects in pharmacological trails and those unable to make informed decisions. Finally, empathizing with colleagues who have different views may avert unnecessary conflict relating to ethical decisions and other disagreements, which may improve cooperation and ultimately lead to better decision-making.

17.5 Truth or Fiction

Should the stories used for ethical training be true, or would imagined scenarios presented in fictional stories be valuable?

True stories, such as those reported in the media, are previous sources to identify issues of ethical interest, debate conflicting ideas, and challenge what is deemed the most appropriate decision in those circumstances. In education, these stories are called case studies and are accepted teaching tools that favor the link between theory and practice in various disciplines: case studies can initiate discussions, questioning, and arguments. Even the disagreements that follow from the comparison and analysis of students' reasons and values should be welcomed as being critical for the formation of their ethical thinking skills.

However, one of the constraints of using true stories to teach ethics is the need to protect the privacy of the characters of these stories – real patients for example. In this regard, fictional stories do not intrude on any person's privacy, whilst offering perfect opportunities for honing one's moral and intellectual muscles, grounded on the values and ethical decisions in real life. Some scholars, such as Currie, are doubtful about 'the extent to which fictions are ever reliable sources' of cognitive improvement or them serving as thought experiments, because of the complexity of some literature that contain too many potentially significant details for readers who may be less able to generate convergent responses from it (Currie 2020). In this context the literary style of a work is important due to its potential emotional power which may unduly influence readers and its complex stimulus which may make it hard for readers to fully understand with what ethical issues they are. Or should be engaging with. For these reasons, it is critical that institutions who wants to use fiction to teach STEM ethics, to pay careful attention to the style and complexity of the literature they use, and to facilitate discussion and engagement with such stories in a highly focused and structure way to ensure that the ethical issues touched upon in the story is highlighted in a substantive way. Although, Nussbaum is of the opinion that the literature style 'expresses its own sense of what matters', which may way to emotional manipulation, distortion, and the confusing dispersal of focus (Nussbaum 1990), Currie does acknowledge the reflective potential of fiction which, together with proper discussion and engagement facilitation as discussed above, is exactly what fiction should be used for in STEM ethics education and debate.

It thus follows that fictional stories are an ideal medium for STEM students and scientists to gain insight into the lives of an even wider variety of characters to learn empathy, develop their critical thinking skills, and imagine the possible worlds that these characters may need to live in if they make certain decisions, without causing real-life harm to anyone. In other words, fictional narratives can act as a sandbox where students can "test various scenarios without risking too much" (Vermeule 2011). Literature-based approaches to ethics also offer the advantage of facilitating the overcoming of a mere survey of rules and best practices (Burton et al. 2018) and embracing a more profound, open-ended education in critical reasoning. This method of engaging scientists in ethical debates is increasingly being employed. The Global Bioethics Initiative screened films such as *Three Identical Strangers* and *The Wisdom of Trauma* during their annual summer schools to center debates about genetic origin, privacy, and trauma (GBI 2021). Jodi Picoult's best-selling novel *My Sister's Keeper* (Picoult 2008) has also been used as a case study by Santa Clara University's Markkula Center for Applied Ethics to explore the ethical dilemmas relating to genetically designed babies, savior siblings, and parent versus child autonomy (Markkula Center for Applied Ethics 2009).

Another approach to using fictional stories as tools to teach STEM ethics is the act of writing, in the first person, as the individual whose story is being told, where the scenes follow a decision that the writer of the story would make. In addition, STEM students could also write different endings to existing novels or films that result from different decisions taken by the fictional characters, which decision are decided by the STEM students in their written works. Writing out the consequences of a decision this way not only focuses the mind but also forces the writer to practice empathy. Engaging with the characters of a story in this manner allows one to experience the fear, anxiety, anger, or confusion these characters may have and gain better insight and a much richer understanding of the professional and ethical situations that one may possibly face in future, as opposed to clinically discussing the facts of each case.

17.6 Different Forms of Storytelling

There exist various forms of storytelling that can be used as mediums to promote STEM students' engagement with stories that can motivate them to make better ethically informed decisions.

17.6.1 Science Fiction

When discussing the potential of storytelling to analyze and solve the possible frictions between science and ethics, it is impossible to abstain from mentioning the genre that epitomizes such tensions: science fiction, which is a kind of writing that creates "prototypes of other worlds, other experiences, other contexts" (Bleecker 2022). Maynard (2018) dedicates an entire book to how sci-fi movies allow us to discuss the morality of technology across domains, backgrounds, and expertise, often in unexpected and innovative ways. It is not a surprise, then, that science fiction movies, novels, and television shows figure in the ethics pedagogy of computer science curricula (Burton et al. 2018; Fiesler et al. 2020) and are acknowledged as anticipation tools in the medical profession (Pomidor and Pomidor 2006). The science fiction genre can contribute to enhancing people's ability to recognize, evaluate and assess ethical issues (Fiesler et al. 2020), foresee unanticipated consequences of existing and emerging technologies that present both high potential and high risks (Fiesler 2021), and ward off acritical authority-based views of scientific truth (Burton et al. 2018). Hansen (2021) saliently argues that, rather than negative examples, ethical education should provide models to encourage the emulation of positive behaviors and step away from the perception that unethicality is the norm. It is maintained that fictional worlds based on "what ifs …", hypothetical scenarios, and thought experiments can gain a proper place next to traditional means of knowledge creation based on the past, the present, and purely "justified true beliefs" (Thornley et al. 2021). After all, science starts with fictional hypotheses (Byrne and Schouweiler 2015).

Moreover, sci-fi storytelling has a more accessible and compelling nature with respect to other kinds of knowledge production and discussion (Thornley et al. 2021). Therefore, it carries the potential of broadening the scientific debate beyond the closed doors of research laboratories, involve a wider non-specialized audience and spark more inclusive societal debates (Delgado et al. 2012) on the values embedded in new technologies and practices while countering prevailing, often unchallenged, Western techno positivist and techno solutionist perspectives (Ryan

2019). Together with other design fiction (Sterling 2009) tools, participatory narratives are increasingly being experimented, as they leverage imagination and critical thinking to envisage alternatives visions of the future and of the present (Tharp and Tharp 2019) and bolster the sense of agency of non-scientists over them. An excellent example is presented by the initiative Narratopia,[1] that gathers and promotes multilingual creative practices, based on arts and fiction such as stories, fragments, design fiction, and games, to propose new narratives and thereby enable systemic transformations towards alternative futures.

The future-oriented scenarios evoked in sci-fi literature and cinema, as well as their "plausible" otherness, can act as "intellectual protheses" (Tharp and Tharp 2019) to strengthen and deepen scientific and argumentative reasoning by bolstering creativity, foreshadowing alternative scenarios, and expanding imagination through the generation of hypotheticals and conceptual metaphors (Fiesler 2021). Narrative speculation is not a mere intellectual, abstract act. On the contrary: it can exert a tangible influence on technological advancements. Through the insertion of diegetic prototypes into a fictional work (namely depictions of future technologies), the public can see how people interact with them in a social context and may be induced to desire their actual creation, thereby supporting it (Byrne 2019). A classical diegetic prototype is the gestural interface technology appearing in the 2002 movie *Minority Report*, which has brought about financial investments and has heavenly driven the widespread development of such a technology. It is true that "today's technology is yesterday's science fiction" (Fiesler 2021), and this can happen so fast that its creators must embrace ethical foresight and oversight since regulations tend to be reactive and their enforcement slow-paced.

Sci-fi storytelling is essentially based on "imagined futures," and the world-building activity that characterizes it often offers diverse scenarios to speculate about and critically present ethical and societal questionings from the perspectives of various characters. Such imagined worlds may act as positive, inspirational drivers of human imagination and behaviors toward "desirable futures", as well as present cautionary tales of undesirable futures (Dunne and Raby 2013). In both cases, they can function as material catalyzers of the actions that should or should not be performed for that future to become a reality. Without such tangible imagination, risks may appear too far away, abstract, and complex to be sufficiently and readily mitigated. An example that lends itself well for didactic purposes is the television series "The One" by Howard Overman (2021), which enables the spectators to explore elaborate subjects like genetic determinism and its implications, as well as examine more mundane ethical topics concerning scientific oversight and research integrity, including sexism, data safety, and (the lack of) informed consent to research participation. The common factor with other sci-fi sources that are relevant to ethical discussions concerns the dual use of new or existing technologies, where the depicted socio-technical ramifications are in the cone of plausible futures (Auger 2013) – or even in the cone of the present.

[1] https://www.plurality-university.org/projects/narratopias

Sci-fi literature enables us to reflect, for example, on our society's growing tendency to body enhancement and the blurred border between human beings and AI. Therefore, it raises deep, serious questions about the very concept of humanity (Delgado et al. 2012) and the nature of personhood (Thornley et al. 2021). In the popular anthology series Black Mirror by Charlie Brooker (2011-present), multiple, often dystopian scenarios allow for the exploration of controversial themes like the mayhem deriving from social scoring and hyperconnected lives, as well as the amplification of hate, violence, suffering, and human fragility through technology. Some of the stories told in the anthology series are strangely familiar as they constitute distorted mirrors of our lives: their plausibility is more troubling than pure fiction. Such twisted stories evoke a sense of discomfort, that is also typical of the tangible fictional artefacts created by design fiction to explore legal, ethical, and societal implications of near-future technologies, devise alternative scenarios, challenge narrow assumptions, preconceptions, and givens (Dunne and Raby 2013). In this way, such discourses exit laboratories and research centers to engage other relevant stakeholders, including those that oversee and make appropriate future-proof policies and anticipatorily assess the risks of emerging technologies to guide their development in a proactive manner (Rossi et al. 2022).

Although the literature cited in this section provides practical examples of how to leverage science fiction in STEM training, it may prove impossible to create a standard program of sci-fi resources because each educational curriculum will have its own needs: inviting data science students to engage with the opportunities and perils of AI is likely more meaningful than asking them to confront biomedical ethics issues, even though the development of moral judgments as well as research ethics principles are transversal to domains and applications. It may also be difficult to assess the usefulness of a certain selection of sources rather than a different one on the learning outcomes: it is the moral question embedded in the narration as well as the speculative exercise per se that can help students develop original moral arguments, engage with the alternative endings of certain decisions, and foresee the entailed risks.

17.6.2 Digital Storytelling

Digital storytelling offers an immersive experience through a mixed media approach that allows an author to tell their story by creating a short video that may include photographs, images, music, and personal narrative to communicate an experience that is of ethical significance (Lambert 2013). This method has been used by researchers to support a variety of purposes and outcomes, including its application as a pedagogical tool (Sandars and Murray 2009); a method of translating knowledge into practice (Scott et al. 2013); and to foster ongoing community engagement (Gubrium 2009) by sharing stories and experiences in local, regional, and national settings (de Jager et al. 2017). The value of this form of storytelling is based on the participatory and visual nature of this method in which the storyteller controls the camera and decides about the content and the visuals that are included in the video (Chalfen 2012). This can be used by people who either experienced unethical incidents, or STEM students who wish to place themselves in the circumstances of people who may be vulnerable to ethical exploitations. In digital storytelling, the storyteller can describe the meaning behind their chosen visuals, or in collaboration with the relevant scientist, may offer some insight into that person's lived experience, which in turn offers valuable insight into the ethical issues and perspective of a specific incident the storyteller experienced (Foster-Fishman et al. 2005). The process through which these short videos are crafted refers to Freire's theory of critical consciousness which is created through dialogue, reflection, and action (Mahmoudi et al. 2014) and that is an important skill to develop for STEM students to embed ethics in their daily practices.

Considering modern-day accessibility to and the ease of filming with mobile phones, so-called "cellphilms" are increasingly branded as a tool "that can combat the assumption that marginalized individuals need an intermediary to tell their stories" (MacEntee et al. 2016). This means that everyone can voice, show, and tell their story, even those who may have been previously considered to be marginalized, unseen, or unheard. In addition, the positive political act of "attention and attunement to personal narratives" is committed by attending to someone's "cellphilm" through active viewing and listening to the views, perspectives, and story of such a person (Matthews and Sunderland 2017). Because the act of storytelling originates from and is shaped by social, cultural, and historical contexts that allow both the listener (via interpretation) as well as the storyteller to make sense of the experience, the meaning, and, more specifically, the ethical meaning of the story is transformed through active listening. This requires people who occupy positions of power, such as scientists, to actively listen to the experiences of those on the margins, such as research participants or patients.

17.6.3 Virtual Reality

Midway between truth and fiction, virtual reality (VR) has transformed the stories from telling them into the art of showing them. On this basis, VR, which has previously been used to desensitize people suffering from phobias (in clinical settings) and violence (used by the military), is now being repurposed to create insight and empathy by showing the stories of people suffering from illness and physical or psychological trauma (Maples-Keller et al. 2017). Former psychotherapist and artist Rita Addison was written off by medical professionals as untreatable due to brain damage she suffered during an accident in 1992, which left her neurologically and visually impaired and unable to work because they could not quantify or understand her trauma. However, in 1994 she teamed up with MIT (David Zeltzer) and the University of Illinois/Chicago (Marcus Thiebaux) to create the VR CAVE installation, Detour: Brain Deconstruction Ahead which allows viewers to experience the "story" of her accident through her artwork, and most importantly its effects on her world and how she viewed and experienced this world as depicted in her visual artwork (Addison 1995). This tool allowed Addison to show the medical professionals what they could not understand or quantify. The medical professionals were now able to treat Addison with dignity and respect and to provide her with medical treatment that would benefit her – in other words treating her ethically.

These days VR illness simulators provide professionals with a peek into the world of patients who suffer from heart disease (AstraZeneca's Heart FX Pod), macular degeneration (Virtual Reality in Medicine 8 Lab, University of Illinois/ Chicago), and stroke (Addison and Umea University) (Aldous 2006). These simulators enable caregivers to understand and appreciate their patients' experiences, which leads to an increased quality of care. On the same basis, STEM students who have access to these simulators and patient experiences may be able to make better ethical decisions about their research protocols or ethical challenges they are faced with during their research.

Although this method is too technical and costly at date for purposes of embedding ethics in STEM training, it does illustrate how technological advancements in storytelling can be developed in the future to improve ethics training in a more experiential manner.

17.6.4 Video Games

Where VR allows one a glimpse into the experience of another, video games enable one to participate in such experiences actively, not only to gain empathy through storytelling but also to fictitiously participate in the decision-making process of the person faced with complex moral questions. A game like *Darfur is Dying*, where players must escape the Janjaweed and find supplies to survive or save their village, has a huge effect on how players experience this world, which is often very foreign to their own experiences, and how it shapes their attitudes towards people living through such events (Fairweather 2006). For example, in the game *PeaceMaker,* players are placed either in the role of the Israeli Prime Minister, or the Palestinian President. However, when the creator of the game had real Israelis and Palestinians switch roles and play the game, "they developed a more nuanced sense of why the other side acted as it did" and "they kind of understood more the pressures the Israeli Prime Minister has" (Thompson 2006). This type of interactive storytelling is defined as the development of media in which the narrative and its evolution can be influenced, in real-time, by the user (Cavazza et al. 2007).

This differs from conventional narratives in which the author maintains exclusive control over the story's events and characters. For purposes of creating empathy, this deviation from the conventional narrative is critical for allowing a player to identify with characters, their experiences, and decisions making processes, and because players are so directly involved in the story, they can easily attribute events, or the implications of their decisions, to themselves, directly contributing to their insights and empathy into the experiences of others.

Video games can thus be designed to immerse STEM students into situations that simulate the pressures and other aspects of the decision contexts regularly faced by scientists, in which they are forced to practice active listening, courage, and moral decision judgements. The potential of video games to allow students to take risks, without causing any real-life harm, and its ability to mine deeper levels of cognition, emotion, and reflection have been offered as motivation for the further development of so-called applied ethics games (McDaniel and Fiore 2010).

17.7 Embedding Storytelling into STEM Ethics Training

Various STEM fields have evaluated the benefits of embedding ethics throughout the entire curricula to encourage students to develop critical reasoning skills that can naturally emerge in professional situations (Grosz et al. 2019), as opposed to limiting such teaching to stand-alone courses (Fiesler et al. 2020). We are in favor of the development of stories that can be used as intellectual prostheses to discuss ethical principles and values as and when needed in the curricula of the specific STEM course. This approach allows us to immediately focus on, debate, and embed these principles in parallel and integration with technical and scientific learnings. It should also encourage the notion among STEM students that ethics is a constant and not something that needs to be addressed at the end of a project as a mere afterthought – like a privacy-by-design approach.

However, the proposal of widening the "scientific dreaming" (Durfee 2015) through various forms of storytelling presents its own challenges. For instance, if ethical training should be embedded into other teachings, teachers should have the necessary awareness, skills, knowledge, resources, and motivation to select and discuss ethics principles and raise appropriate questions to lead the ethical learning endeavors of their students effectively. For example, how might we expect STEM teachers to appreciate sci-fi literature, create videos, or acquire video games and VR tools? Currently, there are limited narrative learning resources available. Fortunately, significant efforts have been made by the European Values for Ethics in Digital Technology project, which developed educational units called bricks, that contain downloadable resources, including case studies, that may be used to teach digital ethics to STEM students (Ethics4EU 2022), while some authors map out ethical questions in existing literature and movies (Fiesler 2021; Hansen 2021; Thornley et al. 2021) and provide related training material. However, since the questions raised should be topical or even use case-specific, the learning materials should also be developed ad hoc over time and kept constantly updated to keep pace with STEM developments and applications, which may only add to the time constraints and other pressures already experienced by teachers in general.

Another question worth raising concerns the evaluation criteria that should be used to assess students' learning. If embedding ethics into STEM education has the dual goal of enhancing students' capacity to reflect on morality (Burton et al. 2018) and ask the right questions while bringing about positive, forward-looking change in the development of new technologies, it follows that both a short-term individual-oriented and a long term socially orientated evaluation is needed. In the first case, the student's individual capacity to critically examine and respond to the relevant ethical principles applicable to specific situations must be evaluated, while in the second case, the socio-technical implications that this method of education may produce over time in the research culture and in society should be measured. Although methods of evaluation in these regards may vary, it is important that students effectively prove their capabilities to identify ethical and social implications when presented with various use cases and offer sound, thoroughly considered arguments for the answers or solutions they reach – with due consideration of the caveat that any effective application of such knowledge outside the classroom may be limited (Fiesler et al. 2020) and that the learning outcomes will be highly subjective. With regards to the socio-technical implications of the implementation of any STEM products or services, it is recommended that longitudinal studies be conducted to investigate whether the number of ethical issues addressed in published STEM articles have substantially increased, whether there are more ethics-related jobs in public and private organizations, whether the development of new technologies included more ethical considerations and adjustments in accordance with such considerations, to name a few.

17.8 Conclusion

Telling stories is not only a good way to attract the attention of students, but because of the emotive foundation on which stories are built, they can also communicate important ethical principles in a more engaging and impactful manner. Stories can show how principles that are already known to students can be applied in life-like scenarios, and the possible consequences of failing to apply ethical principles, thereby helping students to bridge the gap between theory and practice while developing their own critical thinking skills (Yang and Wu 2012). Because our moral understanding is limited to the experiences we have had, which extrapolate into the lives we eventually lead and professions we practice, storytelling provides a way of interacting with different kinds of situations, whether real or fictitious, to allow us to develop our moral beliefs and exercise the application of ethical principles without causing real harm when dealing with such situations in the real world (Fairbairn 2002). For these stories to have a real impact, we must explore them in as many forms of media as possible in the hope that even though we may not be alike, we may gain understanding after having listened to another's story. During the 2006 Academy Awards broadcast, Paul Haggis, the writer/producer/director of Crash (a movie about the need for social empathy) quoted Bertolt Brecht: "[a]rt is not a mirror held up to reality, but a hammer with which to shape it". Accordingly, let's use the ancient form of storytelling to shape the application of ethics in our futures.

References

Adashi, Eli Y., LeRoy B. Walters, and Jerry A. Menikoff. 2018. The Belmont report at 40: Reckoning with time. *American Journal of Public Health.* 108 (100): 1345–1348. https://doi.org/10.2105/AJPH.2018.304580.

Addison, R. 1995. Detour: Brain deconstruction ahead. *IEEE Computer Graphics and Applications.* 15 (2): 14–17. https://doi.org/10.1109/38.364998.

Aldous, Peter. 2006. Illness simulators boost medical empathy. *New Scientist.* 2546. https://www.newscientist.com/article/mg19025466-400-illness-simulators-boost-medical-empathy/. Accessed 16 Jan 2023.

Annas, Julia. 2002. Moral knowledge as practical knowledge. *Social Philosophy and Policy* 18 (2): 236–256. https://doi.org/10.1017/S0265052500002971.

Auger, James. 2013. Speculative design: Crafting the speculation. *Digital Creativity* 24 (1): 11–35. https://doi.org/10.1080/14626268.2013.767276.

Bleecker, Julian. 2022. Design fiction. In *Machine learning and the city*, 561–578. Hoboken: Wiley. https://doi.org/10.1002/9781119815075.ch47.

BMJ. 1996. The Nuremberg code (1947) and declaration of Helsinki (1964). *British Medical Journal.* 313 (7070): 1448. https://doi.org/10.1136/bmj.313.7070.1448.

Briggle, Adam, and Carl Mitcham. 2012. *Ethics and science: An introduction.* Cambridge: Cambridge University Press.

Broad, William J., and Nicholas Wade. 1982. *Betrayers of the truth: Fraud and deceit in the halls of science.* New York: Simon and Schuster.

Burton, Emanuelle, J. Goldsmith, and Nicholas Mattei. 2018. How to teach computer ethics through science fiction. *Communications of the ACM.* 61 (8): 54–64.

Byrne, Emmet. 2019. Too much truth: A conversation with David Kirby. In *Designs for different futures*, 160–164. Yale University Press.

Byrne, Emmet, and Susannah Schouweiler. 2015. The future never gets old. On Julian Bleecker. In *Made up: Design's fictions*, 28–33. Actar Art Center Graduate Press.

Callahan, Daniel. 1980. Goals in the teaching of ethics. In *Ethics teaching in higher education*, 61–80. Boston: Springer.

Cavazza, Marc, Jean-Luc Lugrin, David Pizzi, and Fred Charles. 2007. Madame Bovary on the Holodeck: Immersive interactive storytelling. In *ACM Multimedia*, 651–660. https://doi.org/10.1145/1291233.1291387.

Chalfen, R. 2012. Differentiating practices of participatory visual media production. In *Margolis: The Sage handbook of visual research methods*, 186–200. https://doi.org/10.4135/9781446268278.N10.

Currie, Gregory. 2020. *Imagining and knowing: The shape of fiction.* Vol. 81, 142–147. Oxford University Press. https://doi.org/10.1093/mind/fzaa081.

Davis, Erik. 2004. *TechGnosis.* New York: Harmony Books.

de Jager, Adele, Andrea Fogarty, Anna Tewson, Caroline Lenette, Katherine Boydell, and M. 2017. Digital storytelling in research: A systematic review. *The Qualitative Report* 22 (10): 2548–2582. http://nsuworks.nova.edu/tqr/vol22/iss10/3?utm_source=nsuworks.nova.edu%2Ftqr%2Fvol22%2Fiss10%2F3&utm_medium=PDF&utm_campaign=PDFCoverPages. Accessed 16 Jan 2023.

Delgado, Ana, Kjetil Rommetveit, Miquel Barceló, and Louis Lemkow. 2012. Imagining high-tech bodies: Science fiction and the ethics of enhancement. *Science Communication* 34 (2): 200–240. https://doi.org/10.1177/1075547011408928.

Dunne, Anthony, and Fiona Raby. 2013. *Speculative everything: Design, fiction, and social dreaming.* MIT press.

Durfee, Tim. 2015. Aspirants to reality. Possible essays on the made up. In *Made up: Design's fictions*, 10–15. Los Angeles: Actar Art Center Graduate Press. https://www.timdurfee.com/Aspirants-to-Reality-Possible-Essays-on-the-Made-Up. Accessed 16 Jan 2023.

Ethics4EU. 2022. *Welcome to our open educational resources platform!* http://ethics4eu.eu/. Accessed 16 Jan 2023.

Fairbairn, Gavin. 2002. Ethics, empathy, and storytelling in professional development. *Learning in Health and Social Care.* 1 (1): 22–32. https://doi.org/10.1046/j.1473-6861.2002.00004.x.

Fairweather, Jack. 2006. 'Ethical' computer games take on shoot'em-up classics. *The Daily Telegraph.*. https://www.telegraph.co.uk/news/worldnews/northamerica/usa/1522642/Ethical-computer-games-take-on-shoot-em-up-classics.html. Accessed 16 Jan 2023.

Fiesler, Casey. 2021. Innovating like an optimist, preparing like a pessimist: Ethical speculation and the legal imagination. *Colorado Technology Law Journal* 19 (1): 1–18. https://ssrn.com/abstract=3779036. Accessed 16 Jan 2023.

Fiesler, Casey, Natalie Garrett, and Nathan Beard. 2020. What do we teach when we teach tech ethics? A syllabi analysis. In *Proceedings of the 51st ACM technical symposium on computer science education*, 289–295. https://doi.org/10.1145/3328778.3366825.

Foster-Fishman, Pennie, Branda Nowell, Zermarie Deacon, M. Angela Nievar, and Peggy McCann. 2005. Using methods that matter: The impact of reflection, dialogue, and voice. *American Journal of Community Psychology* 36 (3–4): 275–291. https://doi.org/10.1007/s10464-005-8626-y.

Global Bioethics Initiative. 2021. *Summer school 2021.* https://globalbioethics.org/?s=summer+school. Accessed 29 July 2021.

Goldstein, Melissa M. 2017. Revising the common rule: Ethics, scientific advancement, and public policy in conflict. *The Journal of Law, Medicine & Ethics.* 45: 452–459. https://doi.org/10.1177/1073110517737546.

Griffin, Susan. 1993. *A chorus of stones: The private life of war.* New York: Anchor.

Grosz, Barbara J., David Gray Grant, Kate Vredenburgh, Jeff Behrends, Lily Hu, Alison Simmons, and Jim Waldo. 2019. Embedded EthiCS: Integrating ethics across CS education. *Communications of the ACM.* 62 (8): 54–61. https://cacm.acm.org/magazines/2019/8/238345-embedded-ethics/abstract. Accessed 16 Jan 2023.

Gubrium, Aline. 2009. Digital storytelling: An emergent method for health promotion research and practice. *Health Promotion Practice.* 10 (2): 186–191. https://doi.org/10.1177/1524839909332600.

Hansen, Kathryn Strong. 2021. Optimistic fiction as a tool for ethical reflection in STEM. *Journal of Academic Ethics.* 19 (3): 425–439. https://doi.org/10.1007/s10805-021-09405-5.

Hedgecoe, Adam. 2009. A 'form of practical machinery': The origins of research ethics committees in the UK, 1967–1972. *Medical History* 53 (3): 331–350. https://doi.org/10.1017/2Fs0025727300000211.

Human Genome Project. 1990. National Human Genome Research Institute. https://www.genome.gov/human-genome-project. Accessed 22 July 2022.

International Collaboration for Participatory Health Research. 2013. *Position paper 1: What is participatory health research?* http://www.icphr.org/uploads/2/0/3/9/20399575/ichpr_position_paper_1_defintion_-_version_may_2013.pdf. Accessed 23 Aug 2021.

Lambert, Joe. 2013. *Digital storytelling: Capturing lives, creating community.* New York: Routledge. https://doi.org/10.4324/9780203102329.

Langfelder, Elinor J., and Eric T. Juengst. 1993. Ethical, Legal, and Social Implications (ELSI) program, National center for human genome research, National Institutes of Health. *Politics and the Life Sciences* 12 (2): 273–275. https://doi.org/10.1017/s0730938400024217.

MacEntee, Katie, Casey Burkholder, and Joshua Schwab-Cartas. 2016. *What's a Cellphilm?* Integrating mobile phone technology into participatory visual research and activism: Sense Publishers. https://doi.org/10.1007/978-94-6300-573-9.

Macer, Darryl R. 1991. New creations? *The Hastings Center Report* 21 (1): 32–35. https://doi.org/10.2307/3563344.

Mahmoudi, Ayoub, Ali Khushnood, and Abdolrazagh Babaei. 2014. Paulo Freire critical pedagogy and its implications in curriculum planning. *Journal of Education and Practice.* 5 (14):

86–92. https://www.academia.edu/87360600/Paulo_Freire_Critical_Pedagogy_and_its_Implications_in_Curriculum_Planning. Accessed 16 Jan 2023.
Manney, P.J. 2008. Empathy in the time of technology: How storytelling is the key to empathy. *Journal of Evolution and Technology* 19 (1): 51–61. https://jetpress.org/v19/manney.htm. Accessed 16 Jan. 2023.
Maples-Keller, Jessica, Brian Bunnell, and E., Kim, Sae-Jin., Rothbaum, Barbara. 2017. The use of virtual reality technology in the treatment of anxiety and other psychiatric disorders. *Harvard review of psychiatry.* 25 (3): 103–113. https://doi.org/10.1097/hrp.0000000000000138.
Markkula Center for Applied Ethics. 2009. *Making an ethical decision.* https://www.scu.edu/media/ethics-center/resources/making.pdf. Accessed 28 July 2021.
Matthews, Nicole, and Naomi Sunderland. 2017. *Digital storytelling in health and social policy.* Routledge.
Maynard, Andrew. 2018. *Films from the future: The technology and morality of Sci-Fi movies.* Coral Gables: Mango Media Inc.
McDaniel, Rudy, and Stephen M. Fiore. 2010. *Applied ethics game design: Some practical guidelines in ethics and game design: Teaching values through play*, Information Science Reference. IGI Global. https://doi.org/10.4018/978-1-61520-845-6.ch015.
National Institutes of Health. 1989. *NIH guide for grants and contracts.* https://grants.nih.gov/grants/guide/historical/1989_12_22_Vol_18_No_45.pdf. Accessed 22 July 2021.
National Research Council and National Academy of Engineering. 2014. *Emerging and readily available technologies and national security a framework for addressing ethical, legal, and societal issues.* Washington, DC: The National Academies Press.
Neile, Caren Schnur. 2009. Storytelling and social change: Introduction to the special issue. *Storytelling, Self, Society.* 5 (2): 69–71. https://doi.org/10.1080/15505340902828068.
Nussbaum, Martha. 1990. *Love's knowledge*, 3. New York: Oxford University Press.
Picoult, Jody. 2008. *My sister's keeper.* Paw Prints.
Pomidor, Bill, and Alice K. Pomidor. 2006. Essay: With great power... the relevance of science fiction to the practice and progress of medicine. *The Lancet.* 368: S13–S14. https://doi.org/10.1016/S0140-6736(06)69908-X.
Reason, Peter. 1988. Storytelling as inquiry. In *Human inquiry in action.* London: Sage.
Rossi, Arianna, Régis Chatellier, Stefano Leucci, Rossana Ducato, and Estelle Hary. 2022. What if data protection embraced foresight and speculative design? *DRS2022: Bilbao.* https://doi.org/10.21606/drs.2022.681.
Ryan, Zoë. 2019. The design imagination. In *Designs for different futures*, 41–47. New Haven: Yale University Press.
Saffo, Paul. 2005. *Farewell information, it's the media age.* http://www.saffo.com/essays/essay_farewellinfo.pdf. Accessed 23 Aug 2021.
Sandars, John, and Christopher Murray. 2009. Digital storytelling for reflection in undergraduate medical education: A pilot study. *Education for Primary Care.* 20: 441–444. https://doi.org/10.1080/14739879.2009.11493832.
Scott, Shannon, Pamela Brett-MacLean, Mandy Archibald, and Lisa Hartling. 2013. Protocol for a systematic review of the use of narrative storytelling and visual-arts-based approaches as knowledge translation tools in healthcare. *Systematic Reviews.* 2: 19–25. https://doi.org/10.1186/2046-4053-2-19.
Skloot, Rebecca. 2010. *The immortal life of Henrietta lacks.* Crown: US.
Smiley, Jane. 2006. *Thirteen ways of looking at the novel.* Germany: Anchor.
Steneck, Nicholas, and Ruth Bulger. 2007. The history, purpose, and future of instruction in the responsible conduct of research. *Academic Medicine.* 82 (9): 829–834. https://doi.org/10.1097/acm.0b013e31812f7d4d.
Stephens, Greg J., Lauren J. Silbert, and Uri Hasson. 2010. Speaker–listener neural coupling underlies successful communication. *PNAS* 107 (32): 14425–14430. https://doi.org/10.1073/pnas.1008662107.

Sterling, Bruce. 2009. Cover story design fiction. *Interactions* 16 (3): 20–24. https://dl.acm.org/doi/pdf/10.1145/1516016. Accessed 16 Jan 2023.

Suzuki, W.A., M.I. Feliú-Mójer, U. Hasson, R. Yehuda, and J.M. Zarate. 2018. Dialogues: The science and power of storytelling. *The Journal of Neuroscience.* 38 (44): 9468–9470. https://doi.org/10.1523/jneurosci.1942-18.2018.

Swazey, Judith P., and Stephanie Bird. 1997. *Teaching and learning research ethics*, Research ethics: A reader. Dartmouth: University Press of New England.

Tharp, Bruce M., and Stephanie M. Tharp. 2019. *Discursive design: Critical, speculative, and alternative things*. MIT Press.

Thompson, Clive. 2006. Saving the world, one video game at a time. *New York Times*. https://www.nytimes.com/2006/07/23/arts/23thom.html. Accessed 24 Aug 2021.

Thornley, Clare, Shane McLoughlin, and Sinead Murnane. 2021. 'At the round earth's imagined corners': The power of Science Fiction to enrich ethical knowledge creation for responsible innovation. In *Proceedings of 22nd European conference on knowledge management, ECKM 2021*. https://doi.org/10.34190/EKM.21.059.

Vermeule, Blakey. 2011. *Why do we care about literary characters?* Johns Hopkins University Press.

Webster's New Universal Unabridged Dictionary. 1979. *Empathy*. 2nd ed.

Yang, Ya-Ting C., and Wan-Chi I. Wu. 2012. Digital storytelling for enhancing student academic achievement, critical thinking, and learning motivation: A year-long experimental study. *Computers and Education*. 59 (2): 339–352. https://doi.org/10.1016/j.compedu.2011.12.012.

Chapter 18
Philosophy in the Rainforest: Reflections on Integrating Philosophy and Fieldwork

Clair Morrissey

Abstract Embedding research ethics education into apprenticeship-model undergraduate research experiences can contribute to creating, and maintaining, ethical and inclusive research cultures. Occidental College's Biology and Philosophy Departments collaborated to develop a model for undergraduate ecological field research ethics education focused on promoting students' understanding of ethics as embedded within scientific research practices. The model has two primary components: (a) a philosophical reading, reflective journaling, and discussion group for both philosophy and ecology undergraduate researchers about ecological research ethics; and (b) philosophy faculty and undergraduate researchers embedded within and assisting with ecological fieldwork, while also pursuing their philosophical fieldwork projects. This project highlights a range of ways of embedding ethics in research experiences that can be adapted to other contexts, including sustained and structured reflective journaling focused one's scientific practice and regular philosophical discussions involving the entire research group.

Keywords Research ethics education · Undergraduate research · Fieldwork · Ecology · Field philosophy

18.1 Introduction

Given their importance to the development of students' understanding of the nature of scientific practice and themselves as scientists, robust and effective research ethics education embedded into apprenticeship-model undergraduate research experiences is a promising site for interventions aimed at developing ethical and inclusive research cultures. This chapter describes a long-term collaboration undertaken by

C. Morrissey (✉)
Occidental College, Los Angeles, CA, USA
e-mail: cmorrissey@oxy.edu

E. Hildt et al. (eds.), *Building Inclusive Ethical Cultures in STEM*, The International Library of Ethics, Law and Technology 42,
https://doi.org/10.1007/978-3-031-51560-6_18

Occidental College's Biology and Philosophy Departments to develop a model for undergraduate ecological field research ethics education.

In the first section, I describe the context of apprenticeship-model undergraduate research experiences, concerns about current approaches to undergraduate research ethics training, as well as the specific challenges of teaching ecological research ethics. I suggest that philosophy, when practiced as field philosophy, that is, as itself a kind of fieldwork, can be a helpful partner in cultivating the kinds of discussions and habits of reflection that ecologists identify as critical to ethical and inclusive research practices. To illustrate one way that field philosophy can do so, in the second section I describe the collaboration and provide several samples of the kinds of philosophical reflections and discussions conducted in the field.

18.2 Teaching Undergraduate Ecological Field Research Ethics

18.2.1 The Importance of Undergraduate Research Experiences & Limitations of Traditional Responsible Conduct of Research Training

Nearly 25 years ago, the Boyer Commission Report called on universities to make research-based learning central to a college education (Boyer 1998). Since then, institutions of higher education and funding agencies have devoted significant resources to developing undergraduate research programs and curricular experiences. Summer research programs have become a pillar of science, technology, engineering, and mathematics (STEM) higher education. In a study of apprenticeship-model summer undergraduate research programs, Hunter et al. (2007) found that both students and faculty reported significant gains in students' professional socialization into the sciences; to their "becoming scientists." Undergraduate research programs are a site of significant student learning about what it means to be a member of the professional STEM community. That these experiences encourage students, especially those from marginalized communities, to pursue STEM careers confirms their importance for the pipeline to the profession (Hernandez et al. 2018; Lopatto 2017; Carpi et al. 2016; Eagan Jr. et al. 2013; Russell et al. 2007).

Research ethics is often neglected in undergraduate STEM education (Hendrickson 2015; Olimpo et al. 2017; Marocco 2000). This is especially problematic considering the importance of the "hidden curriculum" to what students learn (Fryer-Edwards 2002). Students learn from what we say and do and what we do not say. A notable barrier, then, to the cultivation of a robust research culture that values ethics and social responsibility is the perception of some scientists that ethics are arbitrary and distinct from science (Wolpe 2006). McCormick et al. (2012) report that among the barriers scientists identified to thinking about the ethical and social dimensions of their research is that they see the reflection on ethical issues and social responsibility as at odds with "the daily practice of science." Reasons for holding this view include the extra burden of the obligation to do so and the characterization of ethicists as "outsiders" who slow down or interfere with scientific work against researchers' professional best interests.

Although conceptions of appropriate outcomes and strategies for effective Responsible Conduct of Research (RCR) or research ethics education vary (Kalichman and Plemmons 2007; Steneck and Bulger 2007), common approaches to RCR education may contribute to the perception of ethics as external to scientific practice, as it often focuses on promulgating legal or quasi-legal rules governing minimal requirements for acceptable behavior (Pennock and O'Rourke 2017). Shortcomings attributed to traditional approaches include that they encourage researchers to understand ethics merely bureaucratically and as external to, rather than as embedded in, the production of scientific research and analysis (Pennock 2015; Schienke et al. 2011; Devereaux 2014). Furthermore, studies suggest that while there is some evidence that participants in RCR courses acquire knowledge about regulations and compliance, participants demonstrate little improvement in ethical decision-making skills, attitudes, and behaviors (Plemmons et al. 2006; Powell et al. 2007). Moreover, RCR training is typically not developmental. In a study of RCR plans for National Science Foundation compliance from a sample of 108 U.S. institutions designated as Carnegie "very high research activity," only 31% had content and requirements that differed by career stage (Phillips et al. 2018). Worse yet, Heitman et al. (2007) found that previous undergraduate hands-on research or undergraduate research mentorship experiences had no significant effect on new graduate students' responses to test questions about core RCR concepts and standards.

18.2.2 Ecological Field Research Ethics & Field Philosophy

Ecological field research is often described as falling into an "ethics gap." Insofar as field research primarily involves and investigates systems, groups, communities, or interactions, the work is not overseen by either Institutional Review Boards (IRBs), whose purview is individual human research participants, or Institutional Animal Care and Use Committees (IACUCs), whose purview is particular non-human research subjects, and methodologies that involve manipulation of laboratory animals, primarily vertebrates. These institutions are not equipped for or designed to provide oversight for "subjects" that are whole ecosystems or species or to evaluate methodologies, such as collection or observation, that do not involve manipulation (Costello et al. 2016). This gap has, in turn, led to a set of loosely connected, often incomplete or "patchwork" guidelines from different bodies within the ecological research community (e.g., particular regulatory agencies, journals, and professional societies), and regular appeals by those in the field for the development of a comprehensive ethics strategy or framework (Costello et al. 2016; Crozier and Schulte-Hostedde 2015a; Minteer and Collins 2008) Recently, there has also been considerable attention to the harms of colonial or parachute science, the practice of developed-world scientists using the resources of the developing world in a merely extractive way or with only minimal or short-term relationships with the communities in which fieldwork is conducted; this approach to scientific practice was dominant during the colonial period of the nineteenth Century, and the balance of power, resources, and knowledge, persists (Johnson et al. 2022; Trios et al. 2021; Asase et al. 2022).

Just as the calls for reform in ecological field research ethics have been robust and perennial, proposed solutions have also been robust. Some have argued that:

- IACUCs should broaden their purview to include environmental effects (Costello et al. 2016; Marsh and Kenchington 2004);
- scientific societies should develop specialist ethical codes (Costello et al. 2016; Marsh and Kenchington 2004);
- journals should require authors to conform with particular ethical guidelines to publish their work (Costello et al. 2016; Marsh and Kenchington 2004), including potentially requiring that ethical criteria used to make decisions about methodologies, field locations, etc. be discussed in the methods sections of research publications (Crozier and Schulte-Hostedde 2015a);
- agencies that issue field permits should create their committees to review the ethics of research proposals (Costello et al. 2016; Marsh and Kenchington 2004);
- ecologists should increase ethics training that prioritizes discussion and active engagement (Crozier and Schulte-Hostedde 2015b), especially for graduate students and junior scientists (Curzer et al. 2011).
- ecologists should actively decolonize their practices (Trios et al. 2021), work toward global science (Asase et al. 2022), and build scientific capacity in the developing world (Johnson et al. 2022).

At the heart of many, if not all, of these recommendations, is the need for an ongoing process of ethical reflection led by and for members of the ecological research community (Farnsworth and Rosovsky 1993; Minteer and Collins 2005a, b), in collaboration with members of other disciplines (Costello et al. 2016; Farnsworth and Rosovsky 1993), and with key members and stakeholders from the broader community.[1]

If this is right, regardless of what more specific policies, practices, or proposals are developed, it will be essential to train young ecologists in a way that makes such ethical reflection and discussion part and parcel of their conception of ecological research. Insofar as taking the perennial calls to develop ecological ethics seriously requires that ecologists develop language and conceptual resources for discussing ethical practices alongside a robust habit of incorporating ethical reflection into

[1] Some recommend that this reflection be put in service of the creation of a comprehensive ethical framework and the creation of an extensive case database for use in reflection and ethics education (Minteer and Collins 2005a, b), while others argue for the development of more concrete core principles modeled on Beauchamp and Childress' *Principles of Biomedical Ethics* and IACUC's "3R's" (Crozier and Schulte-Hostedde 2015a; Curzer et al. 2013).

their research practices, it requires taking seriously the challenge of educating young ecologists to understand ethical reflection as intrinsic to their scientific practice. Given this, interdisciplinary collaboration is a promising direction to pursue, especially with philosophers.

However, as it is often practiced, philosophy may lead some students to perceive more, rather than less, distance between ethics and the everyday concerns of practicing scientists. This is especially true in ecology, as environmental ethics was traditionally primarily concerned with metaphysical questions about value that are quite removed from everyday decision-making (Light 2002), and more recent work on environmental justice, including climate justice, while urgent and relevant to policy, may not be immediately or obviously relevant to many of the ethical concerns about how ecological science is practiced on the ground. Moreover, given the siloing of disciplines and ever-increasing sub-specialization across the academy, philosophers often have little experience with the daily practice of science. They have to contend with the pressure of writing for and talking exclusively to other philosophers (Briggle and Frodeman 2016). It is this feature of the traditional practice of philosophy that Frodeman has in mind when he claims that the goals articulated by those who call for increased ecological ethics are, themselves, strengthened by a concomitant critique of "the institution of academic philosophy" (Frodeman 2008, p. 600). The kind of collaboration that is needed requires that philosophers engage in new (to some) practices[2] and increasingly to learn how to do philosophy "with" rather than philosophy "of" (Hansson 2008, p. 479). That is, to abandon the position of mere observer and characterizer of institutions and practices and to embrace actively employing philosophical reflection and argument as part of those practices and institutions.

The term "field philosophy" has been adopted by many to describe collaborations that center and prioritize doing philosophy directly with non-philosophers (Brister 2020, 1). More specifically:

> Field philosophy, or philosophical fieldwork, is philosophical scholarship that makes 'timely and effective contributions to contemporary discussions,' carried out through 'actual presence in the field, engaged in an ongoing, day-to-day or week-to-week fashion with non-philosophers.'" (Briggle and Frodeman 2016, 24).

In this way, field philosophy is contrasted with "applied philosophy," which involves philosophers commenting on or providing guidance concerning "real world" problems through independently developed philosophical frameworks. What philosophers bring to the table is *not* a theory that can be used to solve research ethics problems, but, instead, "our understanding and techniques for others *to use*" (Brister and Frodeman 2020).

To the project of developing young ecologists' capacity for ethical reflection and the broader project of strengthening an ethical and diverse STEM research culture, the unique contribution of philosophers is not only a kind of subject matter expertise

[2] For examples of other projects that feature this kind of collaboration, see: Frodeman 2008; Rozzi et al. 2008, 2012.

(i.e., familiarity with the language and conceptual resources for making sense of complex and varied ethical commitments) but also a practice of adopting a critical stance concerning one's habits and practices. In this way, the unique (and often parodied) philosophical habit of asking questions can be practiced in service of the goal of cultivating the reflection of young ecologists and facilitating communication between research mentors and their students. Discussion and reflection, facilitated by a philosopher positioned as a kind of "outsider within" can help a research group autonomously and authentically to articulate previously unexpressed values and commitments, as well as buried assumptions, to reveal gaps in individual or collective appreciation of or understanding of something; and, to expose where members agree and where they disagree, or where they have consistent and inconsistent goals and expectations.

18.3 Occidental's Field Ecology-Philosophy Collaboration

From 2017 to 2019, I collaborated with Dr. Elizabeth Braker, Dr. Shana Goffredi, and Dr. Gretchen North from Occidental College's Biology Department to add a new element to their long-standing and successful summer undergraduate field research program. At the heart of our collaboration is the belief that research ethics and social responsibility education is more likely to be effective if it focuses on promoting students' understanding of ethics as embedded within scientific research practices. The project had two primary components: (a) a philosophical reading, reflective journaling, and discussion group for both philosophy and ecology undergraduate researchers about ecological research ethics; and (b) philosophy faculty and undergraduate researchers embedded within and assisting with ecological fieldwork, while also pursuing their philosophical fieldwork projects.

The field location was La Selva Biological Research Station in Puerto Viejo de Sarapiquí, Costa Rica, run and managed by the Organization for Tropical Studies. It is located in the Caribbean lowlands in the northern part of the country. The station is nearly 4000 acres, with 73% of the area being primary tropical rainforest. It is one of the most productive field stations in the world for tropical forest research; every year, it hosts, on average, 300 scientists, from 100 institutions.

For each program iteration, the research team was on station for 2–3 weeks. The faculty and students then returned to Occidental to complete their research projects as part of the College's more extensive residential 10-week Summer Research Program. For each iteration of the collaboration, participants included 3–4 faculty members (1 philosopher) and 8–12 students (2 philosophy students). Student participants were recruited directly by the individual faculty members from ongoing research labs and coursework.

On the station, the research experience for the research team was genuinely immersive. Members of the research group lived together, ate every meal together, and formed a true community organized around common goals. A typical daily schedule would be to wake up for breakfast served from 6–7 am, head into the field quickly thereafter, and return for lunch from 12–1 pm. In the afternoons, the team would either head back into the field or complete lab work in the station's facilities. Dinner was served from 6–7 pm, and our group discussions, rotated with downtime, were held from 7–9 pm. Meals and laboratory spaces were shared with all researchers at the station.

Ecology students engaged in field research directly with their faculty mentors. Students worked on a range of projects, from studies on heat tolerance of leaves of different tree species, to investigate the gut microbiomes of insects that eat latex-producing plants to contributing to a multi-year longitudinal study of oil bean tree (*Pentaclethra macroloba*) saplings found in different locations and conditions across the station. The philosophy students assisted their colleagues with their fieldwork: measuring sapling heights, taking images of canopies to measure the light that reached individual saplings, or hunting for latex-producing plants and collecting the insects found feeding on them. They also conducted philosophical fieldwork projects of their own. They interviewed research scientists about ethics in science (2017 cohort), collaborated with an art student to write the text for a bilingual children's coloring book about the scientific work at the station (2018 cohort), and interviewed Costa Rican scientists, naturalists, and station staff about their work, the station, and the relationship between it and local and national communities to write a memo to the Organization for Tropical Studies leadership sharing their findings (2019 cohort).

The virtue ethics tradition and the work done by social scientists and philosophers to explore and establish the importance of character traits, habits, and dispositions to how scientists create and maintain robust, successful, trusting and trustworthy working communities informed our pedagogical approach (Reilly and Stapleford 2018; Bellon 2014; Daston 1995; Hicks and Stapleford 2016; Keller 1984; Bezuidenhout 2017; Resnik 2012; Pennock 2018). With respect to research ethics education, this tradition focuses on cultivating students' individual good judgment, sensitivity to various moral values, and the ability to discern ethical considerations within the context of practicing scientific research. The approach fits especially well with our starting commitment to embedding ethics within scientific research practices, as the habits and character traits identified as those of "good scientists" are grounded squarely in the experiences and knowledge of those who work in the field.

The philosophy discussions were organized around a set of reflective journal assignments integrated into (i) the students' preparation for the program, (ii) their time at the station, and (iii) their completion of their fieldwork and return to Occidental. Short readings, practice-based activities, and group discussions were integrated throughout. This approach is supported by a wealth of literature that suggests the following strategies cultivate the kind of practical judgment central to ethical decision-making: active-learning (Clements et al. 2013); engagement with exemplars (Schnall and Roper 2012); mentorship and role modeling (Bird 2001; Vianello et al. 2010); semi-structured dialogues (Hall et al. 2018); self-reflection and open dialogue with colleagues (Deming et al. 2007); contemplative writing (Gregory 2014); focus on membership and participation in the professional community (Mitcham 2003); focus on social responsibility (Zandvoort et al. 2013); and the study of philosophical issues relevant to the specific scientific disciplines (Saltz et al. 2019).

With respect to (i) preparation for the program, students were asked to articulate their educational goals clearly and were explicitly encouraged to think about the breadth of skills cultivated through scientific research. They were also asked to learn about the community in which the field station is located. Not only were they required to learn about the nation-state of Costa Rica, but they were encouraged to engage with current events and news to learn about what was going on in the community while they were on station. See Table 18.1.

With respect to (ii) their time on station, students engaged in more targeted and focused philosophical reading and discussion. This was sometimes a challenge. Reading philosophy can be difficult when you are hot, wet, and tired from being in the field all day! The specific topics and content changed each year, and were tailored to the interests and strengths of the cohort of student participants. For

Table 18.1 Sample Pre-departure journal reflections

Taking stock Take stock of your aims and goals. Write a journal entry reflecting on the following questions: (a) Thinking in terms of skills rather than specific content knowledge: What do you hope to learn or gain from your time at La Selva? How do you anticipate this experience will be different than what you have done before? How does this experience relate to and extend coursework or research you have already done? (b) What are your strengths as a researcher? (think broadly about not only your technical skills, but also what people refer to - inappropriately – As “soft” skills related to meaningful, collaborative work.) what do you bring to the research group? (c) What do you find challenging about research? (again, think broadly about not only technical skills, but also so-called soft skills.) how can you use this experience to develop or make progress in these areas?
What’s going on in Costa Rica? On at least two different days, search for news about or from Costa Rica and read at least 2–3 articles. (you can start with google news or the *Tico times*.) write a journal entry for each day you read costa Rican news on the following questions: (a) What did you read about? Where did the articles come from? (b) What did you learn? (c) What did you find interesting, or how is this connected to things you care about or know about, either personally or academically?
The station and OTS Learn about the Organization for Tropical Studies,[a] and La Selva biological Research Station.[b] write a journal entry reflecting on the following questions: (a) What did you learn about OTS and/or La Selva? Is there anything that stuck out to you, or that you did not expect, that you found out more about, or wanted to know more about? (b) What do the organizations value? How are those values related to your work and how might you approach your work with those values in mind? (c) Do you see connections between what you learned and the background research on Costa Rica you did or the news from Costa Rica you have been reading?

[a]https://tropicalstudies.org/
[b]https://tropicalstudies.org/portfolio/la-selva-research-station/

examples, see Table 18.2. Overall, students were most excited to discuss questions related to values in science and those related to the norms of science communication.

Noticing especially strong student interest in cultivating science communication skills, we supplemented the discussions with activities that allowed them to practice explaining scientific concepts and their own research to different audiences, see Table 18.3.

Table 18.2 Sample journal reflections for on-station discussion

Scientific virtue
The short "Character traits: Scientific virtue" Career Brief from *Nature* (2016) describes preliminary results from a study of 400 members of "elite scientific societies." The study was conducted by a philosopher at Michigan State University, Robert Pennock, who studies "scientific virtue." "Scientific virtue" refers to traits that make individuals exemplary scientific researchers. "Traits" refers to character traits – <u>dispositions</u> to <u>act</u> and <u>feel</u> in particular ways that are <u>habitually developed</u> over the course of someone's life or career. "Exemplary" picks out those traits that make one <u>excellent</u> as a researcher.
Take a moment to read the (very) short article.
Questions for discussion:
Which of the 10 virtues (most) stood out to you? How do you understand what those virtues mean in the context of scientific practice? What does it look like when a researcher has those virtues? Are these virtues unique to research scientists, or are there unique ways that they are manifested by scientists?
How do these virtues relate to your experiences this summer and your goals for participating in the summer research program? Are any virtues missing that you believe are important for being an excellent scientist?
What questions did this reading and reflection raise for you (either about the study, or about virtue, or something else)? What hypotheses do you have about how results from the future studies described in the article would differ from these results?
Democratic science
Philip Kitcher, a philosopher at Columbia University, is a leading expert in the philosophy of science, especially the philosophy of biology. He is well known for his work on the role of science in democratic societies (Kitcher 2004).
In his work, he suggests that scientists maintain a number of important roles, including acting as experts in their fields as well as acting in the best interest of the public informed by their expertise *("artisans working for the public good").*
Does this fit with your understanding of science/scientists? What does it look like for scientists to be "artisans" and to aim at the public good?
What is valuable about this dual role? What kind of responsibility does this place on a scientist?
Kitcher also argues that the aim of scientific research is not mere "truth," but "significant truth." he defines the latter as always relative to a community of inquirers at a particular time, and as "those truths that provide relief from the kinds of ignorance that are properly of concern at that time."
What strikes you about this definition of the aim of science? What do you make of this shift to truths being meaningful? Meaningful to whom?
Notice: Kitcher draws attention to the role of ethics in determining not just *how* we do our research, but about *what* we do our research about in the first place.
Reflecting on your own work this summer, what significant truth does it aim at?

(continued)

Table 18.2 (continued)

Reverence for nature
Evelyn fox Keller earned her PhD in physics and is now emeritus professor of the history and philosophy of science at MIT. She is a leader in feminist philosophy of science. The book from which this reading is an excerpt, is her biography of the life and work of Barbara McClintock, a cytogeneticist and winner of the 1983 Nobel prize in physiology or medicine for the discovery of genetic transposition.[a]
Read the chapter "A Feeling for the Organism" of Keller (1984) (the final chapter in the book that summarizes the philosophical insights from the biography), and reflect on and write about the following questions:
(a) According to McClintock what do "good scientists" do? How do they do this/what is needed to enable a "good scientist" to do their work well?
(b) What virtues or intellectual skills can you identify within McClintock's account of her work as a scientist? Are these virtues or skills consistent with ideas you have reflected on in previous reflections? Do these virtues or skills resonate with how you approach your work? How so?
(c) How can "a deep reverence for nature" coexist with traits like rationality and objectivity that are conventionally associated with "good" scientific work (p. 201)? Do they coexist or are they necessary for one another? What do you think is the relationship between these traits?
(d) How do you understand the role of traditional "non-scientific" forms of knowledge within scientific discovery? How have other ways of knowing influenced your approach to scientific work? Does such an approach contribute to your conception of a good "scientist?" how so?

This focus on *A Feeling for the Organism* was developed with Gretchen North

Table 18.3 Sample science communication mini-projects in the field

Lightning talks[a]
You and your partner(s) will give 3-minute presentations on a plant-arthropod pair to your colleagues. Your goal is to convey scientific information about the two species, as well as their interaction with each other.
You should present in a clear and concise manner, and be ready to show off the species involved. You need to own the Latin names, we don't care how you pronounce them but they should roll out of your mouth like words to your favorite song.
After your presentation, your peers should be able to:
Find and identify the organisms on their own--so they will need to have key characteristics pointed out;
Know what larger groups (plant family, arthropod order) the species belong to;
Explain how the species interact.
Instagram
Create an IG post that teaches the public in the US (adults, non-science professionals) something about scientific research at La Selva. Reflect on our discussion about science communication (the deficit model v. the continuity model) and the aims and purposes of such communication.
"Draft" the post to share with others and workshop. When we are happy with them, we will post them to the lab account.[b]

This assignment was developed by Elizabeth Braker
The Instagram account is Shana Goffredi's research lab, http://sites.oxy.edu/sgoffredi/Symbiosis_Lab/Home.html, @symbioxys

Table 18.4 Sample end of fieldwork reflections

Meta-reflection Return to and reread your reflections on practices and experiences and those on the philosophy readings. Write a journal entry addressing the following questions: (a) Do you notice any patterns to what you wrote about? Any connections you did not see before? A trajectory you were on that you were not aware of? (b) How would you characterize the experiences you found noteworthy? What is important or meaningful about your experiences? How are they related to your goals for your time in La Selva? How do they relate to the philosophy discussions
Progress toward goals Return to and reread your first journal entry about your aims, goals, strengths, and areas for development. Write a journal entry addressing the following questions: (a) What progress did you make toward your goals for your time in La Selva? (b) How would you characterize your strengths and challenges as a researcher *now*? What strategies have you developed for enhancing your strengths, and addressing what you find challenging? What can you bring from this experience to your future research, education, and life beyond biology?

Finally, with respect to (iii) completing the program and returning to Occidental, students were asked to evaluate what they learned from the experience, especially on the progress they made on the goals they articulated at the start of the program. See Table 18.4.

All discussions and reflections were rooted in the faculty collaborators' commitment to treating students as full research team members and as developing scientists. Each member of the research team was expected to contribute meaningfully to every discussion. Everyone's views were taken seriously and were recognized as significant contributions to our shared intellectual community. Reinforcing this was the fact that the discussions were done *with* their faculty research mentors. Although students were expected to take the lead and primary responsibility for the discussion and reflection, they were also encouraged to see their mentors not as the final word on a matter or the source of the "answer" but, instead, as experienced, thoughtful, mentors who are learning and thinking along with them.

18.4 Conclusion

Scientific research is a community of practice, and "learning takes place through the *engagement* in that practice" (Gherardi et al. 1998, emphasis added). The meaning of the ethical norms that govern the practice are shaped and passed down by the community of practitioners (Hicks and Stapleford 2016). As such, cultivating the attitudes and skills necessary to create and maintain an ethical and inclusive research culture may be best done through a bottom-up process grounded in the experiences and knowledge of those who work in the field. This project illustrates one way of working toward this goal: seeding the importance of reflection in the practice of young scientists through embedding philosophy into apprenticeship-model undergraduate research experiences. The interdisciplinary exchange can help to frame and make progress on difficult questions where conceptual resources, distinctions, and patterns of reasoning are useful, and also to help everyone communicate productively about them. In this way, the "outsider within," who has not been enculturated into the relevant norms of the research community, can help make those norms, expectations, aims, and concerns clearer, by asking about them, drawing attention to them, and helping the research team have a collaborative conversation about them.

Fully developing a habit of ethical reflection, both for an individual and within a research group, cannot be done in 2 weeks or even 2 years. It is a practice that must be perpetually engaged in and taught to new members of a research team or community. However, the intensity of an immersive fieldwork experience can be a productive context to introduce and reinforce the importance of ethical reflection for young researchers. Moreover, this project highlights a range of ways of embedding ethics in research experiences that can be adapted to other undergraduate educational contexts, including lab-based research groups and classrooms. Namely, sustained and structured reflective journaling focused on one's scientific practice; regular (rather than one-off) philosophical discussions involving the entire research group; and involving philosophers, and other social scientists and humanists, directly in one's research group.

Acknowledgments This project was supported by the Graves Award in the Humanities and an Occidental College Faculty Enrichment Grant. Support for the student researchers was provided by Occidental's Undergraduate Research Center and the Endeavor Foundation.

References

Asase, Alex, Tiwonge I. Mzumara-Gawa, Jesse O. Owino, Andrew T. Peterson, and Erin Saupe. 2022. Replacing "parachute science" with "global science" in ecology and conservation biology. *Conservation Science and Practice* 4 (5): e517. https://doi.org/10.1111/csp2.517.

Bellon, Richard. 2014. *A sincere and teachable heart: Self-denying virtue in British intellectual life, 1736–1859*. Leiden: Brill. https://doi.org/10.1163/978900426335.

Bezuidenhout, Louise. 2017. The relational responsibilities of scientists: (re)considering science as a practice. *Research Ethics* 13 (2): 65–83. https://doi.org/10.1177/1747016117695368.

Bird, Stephanie J. 2001. Mentors, advisors, and supervisors: Their role in teaching responsible research conduct. *Science and Engineering Ethics* 7 (4): 455–468. https://doi.org/10.1007/s11948-001-0002-1.

Boyer, Ernest. 1998. Reinventing undergraduate education: A blueprint for America's research universities. In *Boyer Commission on Educating Undergraduates in the Research University, State University of New York at Stony Brook for the Carnegie Foundation for the Advancement of Teaching*. Stony Brook, NY.

Briggle, Adam, and Robert Frodeman. 2016. The institution of philosophy: Escaping disciplinary capture. *Metaphilosophy* 47 (1): 26–38. https://doi.org/10.1111/meta.12167.

Brister, Evelyn. 2020. Field philosophy and social justice. *Social Epistemology* May: 1–12. https://doi.org/10.1080/02691728.2020.1757176.

Brister, Evelyn, and Robert Frodeman. 2020. *A guide to field philosophy: Case studies and practical strategies*. New York: Routledge.

Carpi, Anthony, Darcy M. Ronan, Heather M. Falconer, and Nathan H. Lents. 2016. Cultivating minority scientists: Undergraduate research increases self-efficacy and career ambitions for underrepresented students in STEM. *Journal of Research in Science Teaching* 54 (2): 169–194. https://doi.org/10.1002/tea.21341.

Character traits: Scientific virtue. 2016. *Nature*: 532.139. https://doi.org/10.1038/nj7597-139a.

Clements, John D., Nancy D. Connell, Clarissa Dirks, Mohamed El-Faham, Alastair Hay, Elizabeth Heitman, James H. Stith, Enriqueta C. Bond, Rita R. Colwell, Lida Annestidou, Jo L. Husbands, and Jay B. Labov. 2013. Engaging actively with issues in the responsible conduct of science: Lessons from international efforts are relevant for undergraduate education in the United States. *CBE Life Sciences Education* 12 (4): 596–603. https://doi.org/10.1187/2Fcbe.13-09-0184.

Costello, Mark J., Karen H. Beard, Richard T. Corlett, Graeme S. Cumming, Vincent Devictor, Rafael Loyola, Bea Maas, Abraham J. Miller-Rushing, Robin Pakeman, and Richard B. Primack. 2016. Field work ethics in biological research. *Biological Conservation* 203 (November): 268–271. https://doi.org/10.1016/j.biocon.2016.10.008.

Crozier, G.K.D., and Albrecht I. Schulte-Hostedde. 2015a. Towards improving the ethics of ecological research. *Science and Engineering Ethics* 21 (3): 577–594. https://doi.org/10.1007/s11948-014-9558-4.

———. 2015b. Why training in ecological research must incorporate ethics education. *Hastings Center Report* 45 (5): 14–19. https://doi.org/10.1002/hast.486.

Curzer, Howard J., Mark Wallace, Gad Perry, Peter Muhlberger, and Dan Perry. 2011. Teaching wildlife research ethics: A progress report. *Teaching Ethics* 12 (1): 95–112.

Curzer, Howard J., Mark Wallace, Gad Perry, and Center for Environmental Philosophy, The University of North Texas. 2013. Environmental. *Research Ethics: Environmental Ethics* 35 (1): 95–114. https://doi.org/10.5840/enviroethics20133518.

Daston, Lorraine. 1995. The moral economy of science. *Osiris* 10: 2–24.

Deming, Nicole, Kelly Fryer-Edwards, Denise Dudzinski, Helene Starks, Julie Culver, Elizabeth Hopley, Lynne Robins, and Wylie Burke. 2007. Incorporating principles and practical wisdom in research ethics education: A preliminary study. *Academic Medicine* 82 (1): 18–23. https://doi.org/10.1097/01.ACM.0000250028.51329.6b.

Devereaux, Mary L. 2014. Rethinking the meaning of ethics in RCR education. *Journal of Microbiology & Biology Education* 15 (2): 165–168. https://doi.org/10.1128/2Fjmbe.v15i2.857.

Eagan, M., Jr., Sylvia Hurtado Kevin, Mitchell J. Chang, Gina A. Garcia, Felisha A. Herrera, and Juan C. Garibay. 2013. Making a difference in science education: The impact of undergraduate research programs. *American Educational Research Journal* 50 (4): 683–713. https://doi.org/10.3102/0002831213482038.

Farnsworth, Elizabeth J., and Judy Rosovsky. 1993. The ethics of ecological field experimentation. *Conservation Biology* 7 (3): 463–472. https://doi.org/10.1046/j.1523-1739.1993.07030463.x.

Frodeman, Robert. 2008. Redefining ecological ethics: Science, policy, and philosophy at Cape Horn. *Science and Engineering Ethics* 14 (4): 597–610. https://doi.org/10.1007/s11948-008-9100-7.

Fryer-Edwards, Kelly. 2002. Addressing the hidden curriculum in scientific research. *The American Journal of Bioethics* 2 (4): 58–59. https://doi.org/10.1162/152651602320957619.

Gherardi, Silvia, Davide Nicolini, and Francesca Odella. 1998. Toward a social understanding of how people learn in organizations: The notion of situated curriculum. *Management Learning* 29 (3): 273–297. https://doi-org.oxy.idm.oclc.org/10.1177/1350507698293002.

Gregory, M. 2014. Ethics education as philosophical practice: The case from Socratic, critical and contemplative pedagogies. *Teaching Ethics* 15:1. https://doi.org/10.5840/tej201410173.

Hall, Eric E., Helen Walkington, Jenny Olin Shanahan, Elizabeth Ackley and Kearsley A. Stewart. 2018. Mentor perspectives on the place of undergraduate research mentoring in academic identity and career development: An analysis of award winning mentors. *International Journal for Academic Development* 23 (1): 15–27. https://doi.org/10.1080/1360144X.2017.1412972.

Hansson, Sven Ove. 2008. Philosophy and other disciplines. *Metaphilosophy* 39 (4–5): 472–483. https://doi.org/10.1111/j.1467-9973.2008.00567.x.

Heitman, Elizabeth, Cara H. Olsen, Lida Anestidou, and Ruth Ellen Bulger. 2007. New graduate students' baseline knowledge of the responsible conduct of research. *Academic Medicine* 82 (9): 838–845. https://doi.org/10.1097/ACM.0b013e31812f7956.

Hendrickson, Tamara L. 2015. Integrating responsible conduct of research education into undergraduate biochemistry and molecular biology laboratory curricula. *Biochemistry & molecular biology education* 43 (2): 68–75. https://doi.org/10.1002/bmb.20857.

Hernandez, Paul R., Anna Woodcock, Mica Estrada, and P. Wesley Schultz. 2018. Undergraduate research experiences broaden diversity in the scientific workforce. *Bioscience* 68 (3): 204–211. https://doi.org/10.1093/biosci/bix163.

Hicks, Daniel J., and Thomas A. Stapleford. 2016. The virtues of scientific practice: MacIntyre, virtue ethics, and the historiography of science. *Isis* 107 (3): 449–472. https://doi.org/10.1086/688346.

Hunter, Anne-Barrie, Sandra L. Laursen, and Elaine Seymour. 2007. Becoming a scientist: The role of undergraduate research in students' cognitive, personal, and professional development. *Science Education* 91 (1): 36–74. https://doi.org/10.1002/sce.20173.

Johnson, Arlyne, Santi Saypanya, Troy Hansel, and Madhu Rao. 2022. More than academic expertise: Structuring international partnerships to build research and professional capacity for conservation impact. *Conservation Science and Practice* 4 (5): e539. https://doi.org/10.1111/csp2.539.

Kalichman, Michael W., and Dena K. Plemmons. 2007. Reported goals for responsible conduct of research courses. *Academic Medicine* 82 (9): 846–852. https://doi.org/10.1525/jer.2013.8.2.95.

Keller, Evelyn Fox. 1984. *A feeling for the organism*. Times Books.

Kitcher, Philip. 2004. Responsible biology. *Bioscience* 54 (4): 331–336. https://doi.org/10.1641/0006-3568(2004)054[0331:RB]2.0.CO;2.

Light, Andrew. 2002. Contemporary environmental ethics from metaethics to public philosophy. *Metaphilosophy* 33 (4): 426–449. https://doi.org/10.1111/1467-9973.00238.

Lopatto, David. 2017. Undergraduate research experiences support science career decisions and active learning. *CBE-Life Science Education* 6 (4): 297–306. https://doi.org/10.1187/cbe.07-06-0039.

Marocco, Dominick A. 2000. Biology for the 21st century: The search for a core. *The American Biology Teacher* 62 (8): 565–569.

Marsh, Helene, and Richard Kenchington. 2004. The role of ethics in experimental marine biology and ecology. *Journal of Experimental Marine Biology and Ecology* 300 (1–2): 5–14. https://doi.org/10.1016/j.jembe.2003.11.024.

McCormick, Jennifer Blair, Angie M. Boyce, Jennifer M. Ladd, and Mildred K. Cho. 2012. Barriers to considering ethical and societal implications of research: Perceptions of life scientists. *AJOB Primary Research* 3 (3): 40–50. https://doi.org/10.1080/21507716.2012.680651.

Minteer, Ben A., and James P. Collins. 2005a. Why we need an 'ecological ethics.'. *Frontiers in Ecology and the Environment* 3 (6): 332–337. https://doi.org/10.1890/1540-9295(2005)003.

———. 2005b. Ecological ethics: Building a new tool kit for ecologists and biodiversity managers. *Conservation Biology* 19 (6): 1803–1812. https://doi.org/10.1111/j.1523-1739.2005.00281.x.

———. 2008. From environmental to ecological ethics: Toward a practical ethics for ecologists and conservationists. *Science and Engineering Ethics* 14 (4): 483–501.

Mitcham, Carl. 2003. Co-responsibility for research integrity. *Science and Engineering Ethics* 9 (2): 273–290. https://doi.org/10.1007/s11948-003-0014-0.

Olimpo, Jeffrey T., Laura A. Diaz-Martinez, Jay M. Bhatt, and Christina E. D'Arcy. 2017. Integration of RCR and ethics education into course-based undergraduate research experiences in the biological sciences: A needed discussion. *Journal of Microbiology & Biology Education* 18 (2): 1–8. https://doi.org/10.1128/jmbe.v18i2.1344.

Pennock, Robert T. 2015. Fostering a culture of scientific integrity: Legalistic vs. scientific virtue-based approaches. *Professional Ethics Report* 28 (2): 1–3. https://doi.org/10.1126/srhrl.acr8257.

———. 2018. Beyond research ethics: How scientific virtue theory reframes and extends responsible conduct of research. In *Cultivating moral character and virtue in professional practice*, ed. David Carr, 166–177. London: Routledge. https://doi.org/10.4324/9781315182414.

Pennock, Robert T., and Michael O'Rourke. 2017. Developing a scientific virtue-based approach to science ethics training. *Science and Engineering Ethics* 23 (1): 243–262. https://doi.org/10.1007/s11948-016-9757-2.

Phillips, Trisha, Franchesca Nestor, Gillian Beach, and Elizabeth Heitman. 2018. America COMPETES at 5 years: An analysis of research-intensive universities' RCR training plans. *Science and Engineering Ethics* 24 (1): 227–249. https://doi.org/10.1007/s11948-017-9883-5.

Plemmons, Dena K., Suzanne A. Brody, and Michael W. Kalichman. 2006. Student perceptions of the effectiveness of education in the responsible conduct of research. *Science and Engineering Ethics* 12 (3): 571–582. https://doi.org/10.1007/s11948-006-0055-2.

Powell, Sean T., Matthew A. Allison, and Michael W. Kalichman. 2007. Effectiveness of a responsible conduct of research course: A preliminary study. *Science and Engineering Ethics* 13 (2): 249–264. https://doi.org/10.1007/s11948-007-9012-y.

Reilly, Timothy S., and Thomas A. Stapleford. 2018. Science, virtue, and moral formation. *Journal of Moral Education* 47 (3): 267–271. https://doi.org/10.1080/03057240.2018.1484583.

Resnik, David. 2012. Ethical virtues in scientific research. *Accountability in Research* 19 (6): 329–343. https://doi.org/10.1080/2F08989621.2012.728908.

Rozzi, Ricardo, Ximena Arango, Francisca Massardo, Christopher Anderson, Kurt Heidinger, and Kelli Moses. 2008. Field environmental philosophy and biocultural conservation: The Omora Ethnobotanical Park educational program. *Environmental Ethics* 30: 12–336.

Rozzi, Ricardo, Juan J. Armesto, Julio R. Gutiérrez, Francisca Massardo, Gene E. Likens, Christopher B. Anderson, Alexandria Poole, et al. 2012. Integrating ecology and environmental ethics: Earth stewardship in the southern end of the Americas. *Bioscience* 62 (3): 226–236. https://doi.org/10.1525/bio.2012.62.3.4.

Russell, Susan H., Mary P. Hancock, and James McCullough. 2007. Benefits of undergraduate research experiences. *Science* 316: 548–549. https://doi.org/10.1126/science.1140384.

Saltz, David, James Justus, and Buddy Huffaker. 2019. The crucial but underrepresented role of philosophy in conservation science curricula. *Conservation Biology* 33 (1): 217–220. https://doi.org/10.1111/cobi.13162.

Schienke, Erich W., Seth D. Baum, Nancy Tuana, Kenneth J. Davis, and Klaus Keller. 2011. Intrinsic ethics regarding integrated assessment models for climate management. *Science and Engineering Ethics* 17 (3): 503–523. https://doi.org/10.1007/s11948-010-9209-3.

Schnall, Simone, and Jean Roper. 2012. Elevation puts moral values into action. *Social Psychological and Personality Science* 3 (3): 373–378. https://doi.org/10.1177/1948550611423595.

Steneck, Nicholas H., and Ruth Ellen Bulger. 2007. The history, purpose, and future of instruction in the responsible conduct of research. *Academic Medicine* 82 (9): 829–834. https://doi.org/10.1097/ACM.0b013e31812f7d4d.

Trios, Christopher H., Jess Auerbach, and Madhusudan Katti. 2021. Decoloniality and anti-oppressive practices for a more moral ecology. *Nature Ecology & Evolution* 5: 1205–1212. https://doi.org/10.1038/s41559-021-01460-w.

Vianello, Michelangelo, Elisa Maria Galliani, and Jonathan Haidt. 2010. Elevation at work: The effects of leaders' moral excellence. *The Journal of Positive Psychology* 5 (5): 390–411. https://doi.org/10.1080/17439760.2010.516764.

Wolpe, Paul Root. 2006. Reasons scientists avoid thinking about ethics. *Cell* 125 (6): 1023–1025. https://doi.org/10.1016/j.cell.2006.06.001.

Zandvoort, Henk, Tom Børsen, Michael Deneke, and Stephanie J. Bird. 2013. Editors' overview perspectives on teaching social responsibility to students in science and engineering. *Science and Engineering Ethics* 19 (4): 1413–1438. https://doi.org/10.1007/s11948-013-9495-7.

Chapter 19
Building Inclusive Cultures Through Community Research

Jennifer F. Nyland, Timothy Stock, and Michele M. Schlehofer

Abstract The science, technology, engineering, and mathematics (STEM) classroom is an ideal site for implementing community-based ethics resources. Doing so fulfills programmatic requirements in the social reality of science and demonstrates increased applicability of science concepts to issues of immediate community concern. This chapter elaborates on the Re-envisioning Ethics Access and Community Humanities (REACH) initiative at Salisbury University, its community research methodology, and the implementation of community-sourced ethics cases in the biology classroom. Preliminary student and instructor feedback is summarized. As opposed to using more traditional ethics resources, such as cases sourced from the national media or case-based ethics readers, the use of community-based cases demonstrates promise in increasing student engagement in discussion, the applicability of course content, and a sense of the social importance of science in addressing contemporary ethics issues. We end with future implications of this pedagogical approach for revising ethics instruction in the STEM classroom.

Keywords STEM · Biology · Ethics · Community-based research · CASE studies · Pedagogy

J. F. Nyland (✉)
Department of Biological Sciences, Salisbury University, Salisbury, MD, USA
e-mail: jfnyland@salisbury.edu

T. Stock
Department of Philosophy, Salisbury University, Salisbury, MD, USA
e-mail: testock@salisbury.edu

M. M. Schlehofer
Department of Psychology, Salisbury University, Salisbury, MD, USA
e-mail: mmschlehofer@salisbury.edu

E. Hildt et al. (eds.), *Building Inclusive Ethical Cultures in STEM*, The International Library of Ethics, Law and Technology 42,
https://doi.org/10.1007/978-3-031-51560-6_19

19.1 Building Inclusive Cultures Through Community Research

Teaching ethics as part of undergraduate science, technology, engineering, and mathematics (STEM) education is one way of demonstrating science as an enterprise that confers substantial benefits to society and delineates the ways in which scientific practice can increase its benefit and reduce its harm. The ethical justification of science itself is intrinsic to science education (Gauch 2012). At the same time, the scientific method prescribes no specific ethical framework beyond a commitment to truth and reason, and so introducing ethics into the science classroom entails a wide range of problems and pitfalls, not the least of which is a broadening of science education away from core scientific content and into philosophical territory, moral theory, or broader social and political realities. Even when such broader philosophical concerns are raised, the science educator runs into a well-documented problem of framing ethics in a way that respects the divergence and controversial nature of ethics while at the same time communicating the ways in which science is directly implicated in these controversies.

One could say that there are problems on the philosophical side of ethics as well, such as debates over naturalism (the idea that morality and science describe the "same world") that have occupied several hundred years of debate. This problem inspired Appiah's *Experiments in Ethics* (Appiah 2008a), which renewed the call for philosophers to work more closely with psychology and embrace experimental methodologies (Appiah 2008b, 2010). Appiah's distinction between empirical and philosophical methods is instructive: "the perspectives in question are not a philosophical perspective and a scientific one, but one in which we look at the world as agents and one in which we look at it as objects" (Appiah 2010, 239). Appiah's articulation on the importance of integrating philosophical and scientific questions was a turning point for philosophers and scientists who seek to move beyond the dogmatic use of sharp distinctions between the natural and human world; the impact of this work has produced a wide range of new experimental philosophical projects, graduate programs, and public philosophy initiatives.

The purpose of introducing ethics work by way of relevant case discussion is not to encroach on intellectual terrain so much as to introduce the possibility of viewing science students, instructors, and the institutions of science as moral agents, rather than merely third-party, neutral observers or describers of an objective world. Our focus on descriptive ethics research in the community immediately surrounding the Salisbury University classroom provides a natural limit to the scope and nature of concerns raised, give focus, context, and a picture of the living normative landscape that our students find themselves in. These cases add a community-sourced check on our intuitions, enhancing what is called a "naturalism of plentitude" (Appiah 2010; Weinberg and Wang 2010), where the intuitions or associations between science and ethics concepts are mapped out in advance by people with a stake in what is decided (again, studying "agents" rather than "objects"). In this way, our project makes available real viewpoints, ideas, and realities that students can productively test their own learning against.

Beyond questions of the relationship between the intellectual domains of ethics and science, our approach to the STEM classroom is consistent with that offered by Gauch (2012): a communitarian approach following the work of philosophers McIntyre and Sandel. This communitarian approach "locates persons within a community or tradition providing resources for developing ethics," and emphasizes that ethics in science must include (per Sandel) the possibility of civic, moral, and political engagement (Gauch 2012, 222). Yet even if, as Gauch holds, ethics is placed within the scientific method itself, the conceptual resources available typically derive from governing norms of experimental or laboratory research, as opposed to civic engagement or community research. Likewise, in many scientific professions, ethics is typically taught from a starting point of professional codes or legal requirements rather than via the communities in which students will ultimately find themselves. For example, ethics instruction surrounding vaccination development might focus on required processes for conducting human clinical trials; less common are discussions and instruction surrounding community-level assumptions of the value of vaccinations or trust in science and how those nuances influence the process or outcomes of human clinical trials. This conceptualization is broadly consistent with communitarian arguments made by Sandel and others, wherein questions of rights and responsibilities cannot be articulated without reference to a specific community's conception of what is "good" or needed, broadly construed (Sandel 1998, 186).

We implemented a community-sourced ethics case, "When to vaccinate, when to educate?", prompted by an active community concern about the role of non-profit health workers and public health authorities as they played an ethically challenging role in mitigating vaccine hesitancy during the COVID-19 pandemic. In this way, we sought to articulate a community research project that brings our particular community's view of fundamental concepts around vaccination, medical science, and ethical decision-making into the classroom. Our project integrates community-based ethics research into the science classroom as a way of demonstrating how such engagement is possible, even as the ethical dimension of science (such as the ethical decisions made during the COVID-19 vaccine roll-out) unfolds in real-time.

If the communitarian framing of ethics requires us to acknowledge plural conceptions of the good that are distinct in each local community, community-based research is also required to bring concrete and particular understandings of ethical concepts into the classroom. A generic conception of "ethics", or even one determined by a particular professional or research context, does not fully address how communities (especially those affected by science, as opposed to the communities that members of a classroom represent) and their divergent or convergent understanding of fundamental ethical concepts are to be heard, understood, and included. Too often within STEM classrooms, ethical deliberation is constrained to abstract scenarios, limited to ethics compliance procedures in scientific research, and is framed as a concern tertiary to content and methods (Bisbee 1994; Benya 2013). There is also a bias towards engagement primarily at the professional level, e.g., it is assumed that it is primarily or even exclusively the responsibility of front-line professionals (such as doctors and nurses within health ethics or IT professionals for data ethics) to engage ethically with their communities.

We argue that the opposite should be the case: science students and scholars should be engaging directly with community issues and demonstrating the value of their training and methods directly to practitioners, organizational leaders, and activists. From our perspective, ethics teaching materials should be grounded in real-world, local community experiences, and student learning in the classroom should be more deeply connected to experiential learning, such as internships, community projects, and student and faculty research and scholarship, which is returned to the community. Almost universally, student and faculty research in ethics is isolated from community-situated practice and the realities of ethical decision-making. In the worst case, the isolation of science from its immediate community introduces a new level of precarity to the STEM fields insofar as science as an institution is taken to be in opposition to specific and strongly held ethical views and fails to prioritize a community's ethical concerns. In response, science fields must contemplate community-based evidence as to how it is that science enters into broader normative debates. Ethical justifications for the scientific method that rest solely on presumptive claims to "truth," "reason," and "common sense" are themselves contestable on normative grounds. Hence, scientific practice runs the risk of making an unwarranted and untenable assumption of the very moral neutrality or beneficence that ethics education in science ought to address critically.

Failure to consider broader normative understandings of science among the public can also have significant negative real-world implications. An example of this played out in real-time during the COVID-19 pandemic, as appeals to the public to engage in behaviors to reduce the spread of COVID-19 (social distancing, wearing masks, receiving the vaccine once it became available) were met by some communities with resistance. Some of this resistance was justified on the grounds that community health leaders had larger agendas of social or political control. In contrast, others speculated that safety protocols in vaccine development may not have been in use during the development of the COVID-19 vaccine and the historical, scientific exploitation of Black communities. Failure to understand the normative grounds on which public health actions were reliant led to a slower vaccine rollout and a less effective pandemic response. Producing real-time case studies that reflect these debates as they arise via concrete ethical decision-making by stakeholders in our community allowed for this unfolding debate to be directly addressed in STEM classes on Immunology and other related fields.

19.2 The Importance of Undergraduate Ethics Education

We argue that the undergraduate classroom is the natural setting to engage in broader ethical discussions and to introduce and address areas of community concern, as well as the broader social impact of science. The learning goals at the undergraduate level are to confer a healthy respect for moral and ethical divergence, an increase in respect for the range of normative issues and positions implicated by scientific practice, and an increased sense of the applicability of science content to matters of pressing concern. Whereas legal and professional environments will inherently constrain the discussion of ethics within graduate-level training in specific fields of study, the undergraduate classroom provides space for the exploration of ethics of science and society more broadly. While individuals who enter graduate programs, and eventually professional settings, will need to develop awareness about what impacts specific communities or clients, their own blind spots, etc., the undergraduate classroom is a place where fundamental questions about the role of science in various social landscapes can be contemplated, contested goods that science raises can be explored, and spaces of moral disagreement can be made available and pedagogically productive.

Such an approach is consistent with the undergraduate educational goals set forth by the American Association for the Advancement of Science (AAAS) and supported by the National Science Foundation (AAAS 2018). Further, this approach addresses the deficiencies identified in the most comprehensive evaluation of STEM teaching practices in North American universities to date (Stains et al. 2018). Specifically, while best practices involve a multitude of student-centered teaching modalities, with targeted efforts to develop students holistically rather than satisfying content requirements, to our knowledge, the majority of STEM classrooms rely heavily on lectures and are content-driven. The Vision & Change in Undergraduate Biology Education framework (AAAS 2011) identified outcome-oriented, inquiry-driven, active teaching practices, organized in modified student learning outcomes, and centered on core competencies (skills) and concepts as critical to success in biology-based careers. Interestingly, two of the core competencies identified in this framework are the "ability to communicate and collaborate with other disciplines" and the "ability to understand the relationship between science and society." This is one example of a STEM field guidance framework recognizing science's pluralistic and intersectional nature and supports the integration of materials and methodologies from other fields beyond STEM in STEM education.

But a major barrier arises: in our experience, supported by preliminary data we collected (Schlehofer and Stock 2021), the standard delineation of topics within applied ethics (e.g., data ethics, effective altruism, bioethics) can seem like abstractions, or even distractions, from what the greater community considers a priority ethical concern. In contrast, our work with community stakeholders has identified a need for developing ethical leaders who are competent in translating abstract ethical concepts into real-world practice. Ethical leadership requires both *ethics literacy* (a form of social/emotional learning conferring the ability to identify points of ethical decision-making; see Burroughs and Barkauskas 2017) and *ethical agency* (the ability to engage in actions, projects, or policies that could realize ethical change). At an extreme, it is possible that those with the highest degree of ethical literacy would be the ones furthest from agency or authority over a given concern, and those with the maximal agency, even if well-intentioned, may be unable to recognize when or why ethical change is necessary. This dilemma is articulated by the fact that any given community's conception of ethics must achieve some degree of objective validity while seeking change and transformation around the issues being described (Hardina 2004). We believe ethics should not be approached as merely a subcomponent of various STEM disciplines at the undergraduate level but rather as an undergraduate educational outcome that is intertwined with other skills. Ethical problems are not mere technical problems but rather reflect the integration of professional disciplines into a broader social world (Gregory 2014).

19.3 Community-Based Ethics: An Approach to Undergraduate STEM Education

To address these concerns and improve ethics education in the science curriculum, we have created a public-facing, community-based, and community-inclusive process for generating curricular content centered on ethics. Our approach begins with the recognition that our faculty, as possessors and disseminators of ethics expertise (including theoretical literacy, research, dialogue, and decision-making tools), lack the ethical agency for sustained community-based action. Correspondingly, our decision-makers and stakeholders in the broader community often suffer from a lack of ethical literacy (or have strictly delimited senses of ethics as only applying in certain cases, according to certain professional norms, or within certain institutional roles). Opportunities for broader dialogue about social change or institutional transformation are, at best, ignored and, at worst, left unnoticed. Our work, conducted under the auspices of Salisbury University's *Re-envisioning Ethics Access and Community Humanities (REACH) Initiative*, bridges this gap and brings stakeholders new resources while at the same time incorporating real-world, local, community-based ethical issues into our teaching across the STEM curriculum. Our process utilizes community psychology, philosophy, and other social science methodologies to develop case studies for use in the STEM classroom. We provide students the opportunity to apply concepts from these other fields in discussions around ethical dilemmas. Thus, integrating case studies or coursework that centers ethical conversations in community-sourced materials strengthens student engagement with the course material and integrates content across multiple disciplines. The method could be recontextualized as a means of making the importance of the humanities to ethical education central to STEM curriculum development, to integrate STEM and liberal arts schools within a university, and to bridge the gap between university campuses and their surrounding communities.

19.4 Why Community-Based Learning?

Real community needs and conversations provide clarity and focus to a curriculum (Chaffee 1980). Evidence shows that community-based learning can be exceptionally effective in conveying the importance of ethics for professions (Castleberry 2007). Indeed, many professional codes of ethics include features such as empathy and respect that must be enculturated by direct experience (Quinn et al. 2001).

Students benefit from discussing the ethical beliefs of the local community simply because they are real, and doing so provides the opportunity to articulate and address those beliefs via the knowledge of their discipline.

Communities, in turn, benefit from the participation of faculty and students reflecting on these beliefs and developing "collective hermeneutical resources" (Fricker 2007, 6), that is, conceptual resources that allow for a range of valid interpretations of normative concerns, and these resources can be returned to the community to help address the concerns that the community itself has articulated. In this way, we develop an approach to ethics curriculum that responds directly to epistemic injustice as articulated by Miranda Fricker, defined as a form of harm done to individuals *as* knowers that occurs when people's first-hand accounts of a social or community concern are dismissed (Fricker 2007). For instance, authorities repeatedly disregarded reports from the residents of Flint, Michigan, that their tap water was unsuitable for drinking, cooking, or bathing, leading to a dangerously high accumulation of lead levels resulting in significant negative health outcomes among people in the community (Bosman 2016). One way of understanding this would be a failure of oversight, but this would ignore the epistemic harm of residents being discounted as unreliable knowers of their own water quality. This real-world situation underscores that ethical discussion and decision-making are too often seen as requiring professional or other qualifications, resulting in significant gaps in our shared tools of interpretation.

Our implementation expresses a commitment to epistemic justice insofar as we seek to facilitate "critical social awareness that is distinctively reflexive" (Congdon 2017, 246). Our community-inclusive approach institutes epistemically just practices (Fricker 2007) by empowering community members as fully constituent knowers of ethics, facilitating *public participation* with ethics resources. Ultimately, this approach has the potential to build procedural justice, in which people have equal and fair involvement in the decision-making process itself (Lind and Tyler 1988).

19.5 The REACH Process

REACH is a transdisciplinary initiative formed between faculty in philosophy, social sciences, and STEM sciences. The initiative is designed to address the gap between ethics literacy and ethics agency, ultimately improving ethics education in the sciences via community-engaged learning. We accomplish this via a cyclical four-step process by which we engage community stakeholders in generating input on what types of ethical concerns exist locally, which is then used to inform classroom-based practices.

First, we start with *community listening sessions* conducted by the REACH team. During the sessions, representatives from community-based organizations (CBOs) engage in guided, confidential discussions around ethics, ethical dilemmas, and community solutions. Listening sessions are open to participation by any and all in our broader community, and community representatives can participate in as many listening sessions as they like. Each session is two hours long and is held at a convenient time for attendees. As we formed REACH during the COVID-19 pandemic, sessions were held over Zoom, with an anticipated later transition to in-person sessions. The sessions are recorded, with consent. Discussion topics and questions are intentionally written to be broad and open-ended in order to elicit dynamic conversation among participants. For instance, questions might ask participants to discuss "Do you feel your organization is capable of making transformative change for the good? Why or why not?" "How does your organization differentiate between a difference of opinion and a conflict? When do differences of opinion become conflicts?" and "Are community ethics standards open to change, or are they fixed?" The sessions assist REACH faculty in identifying priority issues, dilemmas, and situations facing the community, guidance on curriculum focus, and content for community-sourced teaching materials.

Our next step is to *connect listening session content to the curriculum.* At our institution, we formed a Faculty Learning Community (FLC) focused on Ethics Across the Curriculum to assist with these efforts. Members of the FLC were strategically recruited from across the university to ensure multi-disciplinary representation, including representation of business school faculty and faculty teaching in the humanities and social sciences, in addition to STEM fields. Members of the FLC engage in discussion over ethics pedagogy in the classroom. The FLC identifies courses across the university which have a substantial ethical component, provides input on which discussions in the listening sessions might make for exceptionally strong ethics lessons in their classrooms and their disciplines, and provides feedback on teaching tools developed from the listening sessions.

In Step three, we create *course-based assignments* based on FLC input and using content generated from the listening sessions as a guide. The content of the listening sessions is then mapped onto central concepts in philosophical ethics (e.g., trust, accountability, justice), and sets of facts are selected from the listening sessions to create cases that weigh competing interests in philosophical terms. We write ethics cases according to these facts, focusing on the normative tensions our community partners describe while at the same time altering incidental details. The mapping follows a community of inquiry teaching methods commonly found in classes on philosophical ethics. Community of inquiry is facilitated, but not directed, by a trained philosopher and has the goal of "producing criticism," such as the stimulation of differences in students' views, developing ideas that do not align with any student's directly held opinions, and framing opposing views as being in dialogue with each other (Kienstra et al. 2014). The emphasis is on clarifying moral differences and identifying knowledge gaps and presuppositions rather than discovering the "right answer". In our project, the case discussion is facilitated by the instructor and an outside faculty member from the REACH team or the FLC.

Finally, our process includes mechanisms of both *campus and community return.* Campus return practices include presentations to the campus community on the REACH process and resultant case studies and faculty training on integrating case-based ethics curricula into the classroom. As part of our commitment to community accountability, we also provide ethics workshops and training to representatives of CBOs in our local community, quarterly newsletters updating community members about our activities, and conduct ethics analyses for CBOs upon request.

19.6 Sample Case Study: *When to Vaccinate, When to Educate?*

As one example of how the REACH process looks in practice, our work in the FLC has led to an interdisciplinary partnership with faculty in biology. Assignments were developed for two synchronous online undergraduate courses: Introductory Biology for Majors (BIOL 210: Biology Concepts and Methods) and an upper-level biology course (BIOL 432: Immunology) and consisted of a community case study regarding the adoption of a mandatory vaccination policy at a large-sized employer with a high-risk, marginalized employee population. The case, "When to Vaccinate, When to Educate?" (see Appendix to this chapter), was derived from the listening sessions. However, facts were generalized to preserve confidentiality. Classroom activities associated with the case included philosophical discussion, including the identification of salient moral concepts (e.g., care, justice, health, and authority) and having the students articulate what they see as the central moral question of the case (e.g., "Should non-profit funders have a say in health policy?" and "Does a health authority have the right to influence how a religious community discusses vaccination?") and then justifying why their questions are important. A subsequent reflection assignment consisted of having students identify key areas of community and campus engagement around topics generated from the case study.

19.7 Student and Instructor Feedback

To assess the effectiveness of the REACH approach, we collected evaluative feedback from 17 students enrolled in BIOL 201: Introduction to Biology: Molecular and Cellular Biology. Students completed post-discussion rating forms coupled with short, written reflections. The majority were first-year (n = 8; 47.1%) or second-year (n = 6; 35.3%) students. Students were diverse in gender: nine (52.9%) were female, seven (41.2%) were male, and one (5.9%) was nonbinary. Almost all students identified as White (n = 16; 94.1%); one (5.9%) identified as Brazilian. Four (23.5%) self-identified as first-generation students.

Student feedback on quantitative items is summarized in Table 19.1. Students reported that the case study deepened their understanding of the connections between the sciences and the humanities, how the course is related to other disciplines, and that the case study helped them think more critically about the content they are learning in their major. Students also perceived the classroom environment as a safe space in which to freely discuss ethics.

Table 19.1 Average ratings from the student evaluation form

Item	Mean (SD)
The case study helped me understand the content taught in this course.	3.12 (1.36)
The case study helped me see the connections between material taught in the course and issues in our community.	3.41 (1.18)
The case study deepened my understanding of how this course is related to other disciplines.	3.71 (.99)
The case study helped me understand the connections between the sciences and the humanities.	4.12 (.86)
The case study helped me understand the connections between my future career goals or profession and the humanities.	3.41 (1.12)
The case study increased my interest in this course.	2.65 (1.17)
The case study increased my interest in taking ethics courses or participating in co-curricular ethics activities.	2.76 (1.39)
The case study encouraged me to think more critically about the course content.	3.18 (1.19)
The case study encouraged me to think more critically about the content I am learning in my major.	3.53 (1.01)
I enjoyed learning via the case study.	3.24 (.97)
I felt the ethics discussion was productive.	3.41 (1.06)
I felt the classroom environment was a safe space in which to freely discuss ethics.	3.88 (.96)

Student responses to open-ended items provided more profound insight into their classroom experience. They suggested that the REACH approach allowed students to discuss a biology-related topic in a context relevant to their everyday lives while also exploring their own ethical frameworks. Students could better see the connections between the science they were learning as a STEM major and philosophical concepts pertaining to ethics. Students could clearly articulate how ethical concepts such as humanity and providing care related to public health. For instance, in describing their choice to prioritize providing care, a student said, "providing care should be the primary goal here, as the people this board are trying to serve might not be able to on their receive the care that they need." Another echoed this perspective in their response: "…it is important to make sure you are actually caring for people and are not trying to harm them."

Student feedback also suggested that the case study helped reinforce the need to have a firm understanding of science when communicating to the public. Students better understood that, as STEM majors, it would be their role and responsibility to communicate scientific information to the public in a thorough and accurate way, whose message was aligned to specific communities and in which the importance of adopting public health measures was communicated. For instance, in response to being asked "If you had to take responsibility for the issues present in this case tomorrow, would you feel you have the right tools or skills? What additional tools might you need?" a student articulated the importance of communication as follows: "I feel like I would need more of an idea on exactly what types of education we would be trying to provide to people…I feel like I would need more information on exactly why certain political parties and religions oppose the vaccine." Similarly, another student articulated that they needed "more information about the cultural issues behind the hesitancy." Finally, students also explicitly mentioned the benefit of establishing ethical cultures and discussions in the STEM classroom and how an understanding of ethics improves science practices. For instance, one student said that the content learned in the course "can help us address issues with stuff like neutrality, funding, bias, access, religion, the right type of questions to ask, education, etc." Another said that "the content I am learning…allows me to know how to bring science to communities that may not be educated on health." In general, student feedback was positive and demonstrated a high level of engagement. The issues presented were not seen as abstract or disconnected from their learning in the course, and similar ethics activities would be welcomed in the future.

Instructor feedback, collected informally via conversations about the in-class discussion of the case, was also very positive, noting that ethics discussions in the STEM classroom tend to focus on research ethics in isolation from immediate community concerns. They held that biologists typically address ethical decision-making with respect to the application of genetic modifications to organisms or the utilization of animal models to answer scientific questions. These discussions were noted to be less nuanced, with black-and-white framings, "dos" and "don'ts," and tend to frame ethics in absolute terms (e.g., "good" vs. "evil"). Typically, students are not afforded the opportunity to explore their own ethical culture and belief systems or their own agency in matters of moral concern. Instead, the conversations center around the justifications for the scientific experiment or decision and whether these justifications are valid in light of the students' understanding of the biology content. Using REACH cases created a sharp and well-received contrast; community-sourced case discussion allowed students to apply their scientific knowledge while also exploring their own ethical frameworks, developing a reflective and critical awareness of the justification of the same. From the feedback collected, it appears that students listened more, shared more, and ultimately engaged more with the scientific content due to their participation in the case discussion. We are currently in the process of empirically validating these outcomes in a more significant number of classes and comparing the impact of community-situated ethics cases with traditional case study approaches.

19.8 Comparison with Traditional Approaches to Ethics Case Studies

The benefits of the REACH approach become apparent when comparing the experiences in Introductory Biology with that in a Psychology course in which a traditional case study approach was used to engage students in discussion over ethics in research. The Psychology course used a national case, that of the Cambridge

Analytica scandal (Meredith 2018), in which a political analytics firm used the personal data of Facebook users without user consent for targeted political campaigns as the basis for discussion. The case touched on multiple ethical concerns, including the dominance of commercial interests in the decision to collect Facebook user data. The case has particular relevance for Psychology students, given that the data sourced from Facebook was developed and analyzed by a psychology graduate student for the purposes of completing his dissertation. The case touches on issues of informed consent, the social responsibility of psychological scientists, and whether psychological research should strive for political neutrality, all of which are directly relevant to the content taught in the Research Methods course.

Students were upper-division psychology majors enrolled in a synchronous online Research Methods course, and the case was introduced in conjunction with course coverage of research ethics. The case study was written in a style similar to the community-sourced cases, and students were led through a guided discussion of the case by the same faculty member who conducted the Biology drop-in sessions, using the same discussion plan (identification of concepts, articulation, and justification of moral questions). Unlike the students using the community-sourced cases, however, students in the Research Methods course struggled to connect ethical concepts used in class, such as procedures for informed consent and ethical use of personal data, to the Cambridge Analytica case. Student conversation lagged, and students appeared disinterested in the case and the accompanying discussion. Students repeatedly returned to meta-ethical questions about why privacy should be protected, whether discussing ethics should matter, or whether anything can be done to enforce ethical norms. Perhaps more concerning, the general sentiment among students after discussion of the case was that psychologists are detached from and not responsible for any social implications of their work.

This comparison also indicates the hazard of discussing ethics cases that students cannot identify with, feel disassociated from, or cannot immediately understand as relevant. In this case, ethics or normative concerns, in general, are perceived as something that scientists should insulate themselves from and that science should limit the ambit of its concern to "just the facts" or just "doing your job" or "following the rules" as a researcher. In this way, the purported neutrality of science appears not as a positive value but as a defensive and reactive mechanism for eschewing responsibility for broader social issues. We posit that when sufficiently contextualized within broader social and community concerns, ethics interventions create a clear connection between the responsibilities of scientists and the pursuit of scientifically-derived knowledge and the outcomes and consequences of scientific endeavors. However, when not sufficiently contextualized or taken as serious (or taken as something extraneous to other course content), then ethics education can actually have the opposite effect than intended, leading to students situating science and scientists outside of broader social and community concerns, and even eroding confidence in the principles of beneficence from which science education is justified.

19.9 Next Steps and Future Directions

The initial feedback we have received demonstrates the promise and potential of community-based ethics education to transform undergraduate education in the sciences, helping to meet student learning goals for undergraduate science majors and better preparing them to grapple with real-world ethical issues in their local communities. We have several next steps for our work, including the development of a more formalized assessment of our approach, additional assignments, and the creation of an online repository in which pedagogical materials can be housed and accessed by faculty.

Our top priority is to develop a formal, curriculum-embedded assessment of our pedagogical approach at both the course and university level. Our assessment plans include capturing the impact of using community-based ethics content on multiple student outcomes, including but not limited to students' sense of belonging, their level of engagement with the content of their course or major, development of critical thinking skills, and ability to perspective-take, and retention in the major.

We are also looking ahead to creating additional pedagogical resources from our community-sourced ethics content. While we believe case studies are an important pedagogical tool, we recognize that creating a multitude of tools will only increase faculty engagement with community-based ethics tools. Thus, in addition to creating additional case studies, we are working to develop a set of "ethics bites" which are similar in size and scope to "data bites" that are currently provided to biology students as "food for thought" content in study guides and homework assignments. In these activities, students are provided a snapshot of data or an interesting finding relevant to the content of the day's classroom topic. Typically, these "food for thought" activities stretch the students' understanding of the material, delve into common misconceptions, and push students to apply information in unexpected scenarios. "Ethics bite" activities provide an additional opportunity to incorporate community-based ethics tools in the undergraduate classroom, helping supplement other community-based activities.

In future iterations, these cases, discussions, and "ethics bites" could form the basis of durable student work. Non-disposable assignments (NDAs, sometimes referred to as "renewable assignments" or "open pedagogy") are a priority for development, as these assignments have real-world value and impact outside of the classroom (Seraphin et al. 2019). Examples of NDAs are products such as infographics, scientific op-eds, Wikipedia entries and blogs, community action-research projects, podcasts, the development of lesson plans for elementary school teachers, and more (Wiley and Hilton 2018). As opposed to disposable assignments such as term papers, which are thrown away after the conclusion of a course, NDAs are designed for use by future students, the instructor, and local community members after the course ends. As such, NDAs provide a process by which students and faculty can create usable products to address the exact issues that community organizations are facing locally. The NDAs could also serve as a primary platform for campus and community return, as the products are available after the end of the course, are revisable and updatable by future students, and can be tested for continued utility to a variety of stakeholders outside of the classroom. NDAs additionally have the benefit of serving as a sample product that students can add to their resume or professional portfolio and can discuss when interviewing for jobs or applying for graduate school.

We are also working to expand our work's reach via creating institutional mechanisms to support the use of community-based ethics tools across the curriculum. This includes a distributed ethics curriculum where faculty can be trained on ethics case writing and form their own framings of community issues, a faculty learning community fostering Ethics Across the Curriculum professional development, an 8-week workshop series for faculty to adapt assignments to include ethics-focused assignments and activities, as well as a case repository so that a history of ethics cases can be searchable and accessible campus-wide.

19.10 Conclusion

The undergraduate classroom is the appropriate place for broad discussions about ethics in science, despite cultural, intellectual, and professional practices that broadly separate ethics and normative concerns from teaching science. However, community-based ethics cases achieve more directly the goals that ethics education in science ought to foster: critical awareness of science's impact on society, the agency of science students in accomplishing social good, and the applicability of scientific content to pressing ethical concerns. Community-based research has the potential to integrate campus activities with its surrounding community, articulate clearer assessment standards for ethics, and foster deeper partnerships. Ultimately community-based ethics addresses the social role of science, as well as philosophical concerns about the need for more diverse conceptual tools and greater hermeneutical resources to address both the perception that philosophical ethics are abstract and not applicable to real concerns as well as the purported neutrality of science and epistemic harms that can flow from a lack of community engagement and accountability. As Appiah notes, "there is no clear dividing line between nature and culture … human nature is cultural twice over. First, it is the historical result of earlier social practices … Second, human behavioral possibilities are, in part, the result of what concepts are available to people" (Appiah 2008b, 125). Community-based ethics education situates students in a continuity of such a history of local practices. It seeks to make new moral concepts available to explore new possibilities for an ethical culture in science. This promising approach to undergraduate ethics education has the potential to not only better prepare STEM undergraduate majors for graduate training or future STEM careers but to facilitate interdisciplinary solutions to pressing ethical issues facing our communities and society as a whole.

Appendixes

Salisbury UNIVERSITY
REACH
Re-Envisioning Ethics Access and Community Humanities

Community-Based Ethics Resources – Terms of Use The following case represents ethical issues identified by the Community Ethics Network of the Salisbury University REACH initiative. The facts of the case are generalized and stipulated in such a way that the case represents a real-world ethics conflict at the same time that it does not disclose confidential information. The purpose of this case is for use in coursework and community discussions around vaccination and public health.

Case 1: When to Vaccinate, When to Educate?

A health non-profit organization serving vulnerable populations during the COVID-19 epidemic defines its mission as providing access to healthcare and education about safety and public health during the pandemic. The communities they serve have significant numbers of homeless individuals, lower-income workers without access to healthcare, and migrant populations. Because the organization addresses health care and health literacy, they serve these populations simultaneously with access to treatment and health education.

As the leadership team and board of directors considers plans for assisting in the rollout of COVID-19 vaccines, they conduct a meeting to construct a comprehensive health plan in line with their organizational mission. From the perspective of community health, this organization sees its role as exceptionally important, as members of the population they serve could either substantially accelerate or inhibit community-level spread or "herd immunity" and the prospects of the community enduring returned lockdowns and other negative consequences.

For this reason, a plan emerges to publish educational materials for distribution along with the vaccine, in particular, information on COVID-19 rates in the local area, guidelines on mask-wearing even after complete vaccination, and recommendations against unnecessary travel. Additionally, the materials contain basic information on the mechanics of the vaccine and its efficacy and the need for clients to encourage members of their families, workplaces, and churches to schedule a vaccine appointment.

A board member raises two concerns. The first is that major grant funding for the initiative depends on all information regarding the virus to "avoid political, religious and other forms of speech that could introduce bias", making the case that encouraging people to wear masks and restrict travel could be seen as political advocacy, rather than advancing legitimate health education. The second concern is that advocating for clients to advertise the vaccine in their churches, some of which may have religious objections to vaccination, would be tantamount to establishing a religiously biased program.

References

American Association for the Advancement of Science. 2011. *Vision and change in undergraduate biology education: A call to action*. Washington, D.C. https://visionandchange.org/wp-content/uploads/2013/11/aaas-VISchange-web1113.pdf

———. 2018. *Vision and change in undergraduate biology education: Unpacking a movement and sharing lessons learned.* Washington, D.C. https://visionandchange.org/wp-content/uploads/2018/09/VandC-2018-finrr.pdf

Appiah, A. 2008a. *Experiments in ethics*. Cambridge, MA: Harvard University Press.

Appiah, K.A. 2008b. Experimental philosophy. *Proceedings and Addresses of the American Philosophical Association* 82 (2): 7–22.

———. 2010. More experiments in ethics. *Neurotics* 3 (3): 233–242. https://doi.org/10.1007/s12152-010-9062-8.

Benya, F. F. (Ed). 2013. *Practical guidance on science and engineering ethics education for instructors and administrators: Papers and summary from a workshop.* December 12, 2012. National Academies Press.

Bisbee, L.A. 1994. Ethics in the science classroom: An ethics-across-the-curriculum program in Colorado that gets an A+ for results. *Journal of College Science Teaching* 24 (2): 132–134. https://www.jstor.org/stable/42989585.

Bosman, J. 2016, March 22. Flint water crisis inquiry finds state ignored warning signs. *New York Times*. https://www.nytimes.com/2016/03/24/us/flint-water-crisis.html

Burroughs, M.D., and N.J. Barkauskas. 2017. Educating the whole child: Social-emotional learning and ethics education. *Ethics & Education* 12 (2): 218–232. https://doi.org/10.1080/17449642.2017.1287388.

Castleberry, S. 2007. Prison field trips: Can white-collar criminals positively affect the ethical and legal behavior of marketing and MBA students? *Journal of Marketing Education* 29 (1): 5–17.

Chaffee, J. 1980. The role of ethics in education: A practical example. *ERIC: Institute for Education, Sciences* [ED205249]. https://eric.ed.gov/?id=ED205249

Congdon, M. 2017. What's wrong with epistemic injustice? Harm, vice, objectification, misrecognition. In *The Routledge handbook of epistemic injustice*, ed. B.J. Kidd, J. Medina, and G. Pohlhaus. London: Routledge.

Fricker, M. 2007. *Epistemic injustice*. Oxford: Oxford University Press.

Gauch, H.G. 2012. *Scientific method in brief.* Cambridge: Cambridge University Press.

Gregory, M. 2014. Ethics education as philosophical practice: The case from Socratic, critical, and contemplative pedagogies. *Teaching Ethics* 15 (1).

Hardina, D. 2004. Guidelines for ethical practice in community organization. *Social Work* 49 (4): 595–604.

Kienstra, N., M. Karskens, and J. Imants. 2014. Three approaches to doing philosophy: A proposal for grouping philosophical exercises in classroom teaching. *Metaphilosophy* 45 (2): 288–318. https://doi.org/10.1111/meta.12085.

Lind, E.A., and T.R. Tyler. 1988. *The social psychology of procedural justice*. New York: Plenum Press.

Meredith, S. 2018, March 21. Here's everything you need to know about the Cambridge Analytica scandal. *CNBC News*. https://www.cnbc.com/2018/03/21/facebook-cambridge-analytica-scandal-everything-you-need-to-know.html

Quinn, S., D. Gamble, and A. Denham. 2001. Ethics and community-based education: Balancing respect for the community with professional preparation. *Family & Community Health* 23 (4): 9–23.

Sandel, M. 1998. *Liberalism and the limits of justice*. Cambridge: Cambridge University Press.

Schlehofer, M. M., and T. Stock, 2021, June 22–26. *Community ethics: A multidimensional framework*. Poster presented to the Society for Community Research and Action Biennial Conference.

Seraphin, S.B., J.A. Grizzell, A. Kerr-German, M.A. Perkins, P.R. Grzanka, and E.E. Hardin. 2019. A conceptual framework for non-disposable assignments: Inspiring innovation, implementation, and research. *Psychology Teaching & Learning* 18: 84–97. https://doi.org/10.1177/1475725718811711.

Stains, M., J. Harshman, M.K. Barker, S.V. Chasteen, R. Cole, S.E. DeChenne-Peters, and M. Levis-Fitzgerald. 2018. Anatomy of STEM teaching in North American universities. *Science* 359 (6383): 1468–1470.

Weinberg, J.M., and E. Wang. 2010. Naturalism's perils, naturalism's promises: A comment on Appiah's experiments in ethics. *Neuroethics* 3 (3): 215–222. https://doi.org/10.1007/s12152-010-9065-5.

Wiley, J., and J.L. Hilton. 2018. Defining OER-enabled pedagogy. *The International Review of Research in Open and Distributed Learning* 19 (4) http://www.irrodl.org/index.php/irrodl/article/view/3601.

The manufacturer's authorised representative in the EU is Springer Nature Customer Service Centre GmbH, Europaplatz 3, 69115 Heidelberg, Germany. If you have any concerns regarding our products, please contact ProductSafety@springernature.com

Printed and bound by CPI Group (UK) Ltd, Croydon, CR0 4YY
28/09/2026
02224065-0002